T0336059

METHODS IN MOLECULAR BIOLOGY

Series Editor
John M. Walker
School of Life and Medical Sciences
University of Hertfordshire
Hatfield, Hertfordshire, AL10 9AB, UK

For further volumes:
http://www.springer.com/series/7651

RNA-Protein Complexes and Interactions

Methods and Protocols

Edited by

Ren-Jang Lin

*Irell & Manella Graduate School of Biological Sciences of the City of Hope, Duarte, CA, USA;
Department of Molecular and Cellular Biology, Beckman Research Institute of the City of Hope,
Duarte, CA, USA*

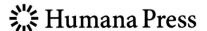 Humana Press

Editor
Ren-Jang Lin
Irell & Manella Graduate School of Biological
    Sciences of the City of Hope
Duarte, CA, USA

Department of Molecular and Cellular Biology
Beckman Research Institute of the City of Hope
Duarte, CA, USA

ISSN 1064-3745   ISSN 1940-6029   (electronic)
Methods in Molecular Biology
ISBN 978-1-4939-3589-5   ISBN 978-1-4939-3591-8   (eBook)
DOI 10.1007/978-1-4939-3591-8

Library of Congress Control Number: 2016931597

Springer New York Heidelberg Dordrecht London
© Springer Science+Business Media New York 2016
This work is subject to copyright. All rights are reserved by the Publisher, whether the whole or part of the material is concerned, specifically the rights of translation, reprinting, reuse of illustrations, recitation, broadcasting, reproduction on microfilms or in any other physical way, and transmission or information storage and retrieval, electronic adaptation, computer software, or by similar or dissimilar methodology now known or hereafter developed.
The use of general descriptive names, registered names, trademarks, service marks, etc. in this publication does not imply, even in the absence of a specific statement, that such names are exempt from the relevant protective laws and regulations and therefore free for general use.
The publisher, the authors and the editors are safe to assume that the advice and information in this book are believed to be true and accurate at the date of publication. Neither the publisher nor the authors or the editors give a warranty, express or implied, with respect to the material contained herein or for any errors or omissions that may have been made.

Printed on acid-free paper

Cover illustration: An image illustrating the complexity of RNA-protein interactions involved in the assembly of a RNA splicing complex. Designed by Ren-Jang Lin.

Humana Press is a brand of Springer
Springer Science+Business Media LLC New York is part of Springer Science+Business Media (www.springer.com)

Dedication

To my beloved wife and children, who find my ivory-tower work interesting.

Preface

I hesitated a bit when I received an email from the Series Editor, John Walker, inviting me to consider producing an updated version of the *RNA–Protein Interaction Protocols* that I edited in 2008. I was pleased to know the volume was a success—at least in his opinion, but it was a long process and I might have gotten some authors frustrated because of my slow pace during editing. I finally decided to take on the task for I saw protocols for studying RNA–protein interaction are still being refined or developed anew, and the number of investigators engaging in RNA research continues to increase—many of them were in diverse fields and would benefit from detailed protocols.

I solicited protocols from authors who either developed or were very experienced with the method; I am very grateful for their willingness to contribute to this collection. By way of introduction, the first chapter uses two closely related RNA-binding proteins as examples to illustrate how various biochemical methods were applied to study the interactions between protein and RNA. Nineteen protocol chapters then follow, which can be categorized roughly in four sections.

The first section, from Chapters 2 to 5, describes ways to purify RNA–protein complexes assembled in cells or in isolated cellular extracts. The common strategy is to tag either the RNA with a biotin residue or the protein with the MS2 or the FLAG moiety so that the RNA–protein complexes can be readily purified by affinity.

The second section, from Chapters 6 to 9, describes methods for measuring various biochemical activities of RNA-interacting proteins or ribonucleoproteins. Protocols to assay the loading of small RNA onto argonaute protein, the role of ubiquitination in the assembly and disassembly of RNA–protein complexes, the kinetics of RNA methylation, and the reconstitution of small nucleolar RNPs are represented. The protocols contain specifics for individual experiments, yet they are written with a broad application in mind and are applicable to related studies.

The third section, from Chapters 10 to 14, features biochemical methods for measuring direct RNA–protein contact. Crosslinking protein and RNA in situ using chemicals followed by deep sequencing or proteomic detection generates data with high specificity and resolution. Together with Northwestern and polysomal profiling methods, these chapters may be viewed as the bread and butter of a typical RNA–protein methodology book.

The fourth section, from Chapters 15 to 20, collects various new or innovative methods pertinent to the subject. Cellular incorporation of a modified amino acid to a specified protein residue is adopted to pinpoint protein–protein interaction within an RNA–protein complex. A cell-based SELEX method is sophisticatedly expanded to isolate novel RNA molecules that can distinguish cell types. Elegant bioinformatic methods are developed to predict RNA–protein interactions and to design/manipulate RNA-binding proteins. Clever and effective reporters are constructed to assay spliceosome-mediated RNA splicing specificity or changes inside cells.

It would bring satisfaction if this book, in conjunction with the two previous editions of the *RNA–Protein Interaction Protocols* in the *Methods in Molecular Biology* series, provides a set of useful protocols, basic or advanced, to even a small number of researchers working with RNA and RNA-interacting proteins.

I would like to thank all the authors for their excellent contributions to this book, John Walker for his guidance, and the Springer editorial staff for their assistance. I am indebted to my students and colleagues who helped with this work and my wife for unconditional support.

Duarte, CA   *Ren-Jang Lin*
On Thanksgiving of 2015

Contents

Contributors

JAE-WOONG CHANG • *Department of Biochemistry, Molecular Biology and Biophysics, University of Minnesota Twin Cities, Minneapolis, MN, USA*

TIEN-HSIEN CHANG • *Genomics Research Center, Academia Sinica, Nankang, Taipei, Taiwan*

SUNGHEE CHO • *School of Life Science, Gwangju Institute of Science and Technology, Gwangju, South Korea*

KEITH T. GAGNON • *Department of Biochemistry and Molecular Biology, School of Medicine, Southern Illinois University, Carbondale, IL, USA; Department of Chemistry and Biochemistry, Southern Illinois University, Carbondale, IL, USA*

MYRIAM GOROSPE • *Department of Biochemistry and Molecular Biology, Medical University of South Carolina, Laboratory of Genetics and Genomics, National Institute on Aging-Intramural Research Program, National Institutes of Health, Charleston, SC, USA*

SHUAI HOU • *Institute of Cancer Stem Cell, Cancer Center, Dalian Medical University, Dalian, Liaoning, People's Republic of China*

YA-MING HOU • *Department of Biochemistry and Molecular Biology, Thomas Jefferson University, Philadelphia, PA, USA*

CHAO HUANG • *Process Science Downstream, Bristol-Myers Squibb Company, East Syracuse, NY, USA*

HA NA JANG • *School of Life Science, Gwangju Institute of Science and Technology, Gwangju, South Korea*

FEDOR V. KARGINOV • *Department of Cell Biology and Neuroscience, University of California, Riverside, CA, USA*

NAOYUKI KATAOKA • *Medical Innovation Center, Laboratory for Malignancy Control Research, Kyoto University Graduate School of Medicine, Kyoto, Japan*

SARAVANAN KOLANDAIVELU • *Department of Biochemistry, School of Medicine, West Virginia University, Morgantown, WV, USA*

MAREK KUDLA • *Department of Cell Biology and Neuroscience, University of California, Riverside, CA, USA*

HAIXIN LEI • *Institute of Cancer Stem Cell, Cancer Center, Dalian Medical University, Dalian, Liaoning, People's Republic of China*

REN-JANG LIN • *Irell & Manella Graduate School of Biological Sciences of the City of Hope, Duarte, CA, USA; Department of Molecular and Cellular Biology, Beckman Research Institute of the City of Hope, Duarte, CA, USA*

TIING JEN LOH • *School of Life Science, Gwangju Institute of Science and Technology, Gwangju, South Korea*

HUA LOU • *Department of Genetics and Genome Sciences, Case Western Reserve University, Cleveland, OH, USA*

SARAH F. MITCHELL • *Department of Chemistry and Biochemistry, Howard Hughes Medical Institute, University of Colorado Boulder, Boulder, CO, USA*

HEEGYUM MOON • *School of Life Science, Gwangju Institute of Science and Technology, Gwangju, South Korea*

DANIEL MURPHY • *Department of Biochemistry, School of Medicine, West Virginia University, Morgantown, WV, USA*

ROY PARKER • *Department of Chemistry and Biochemistry, Howard Hughes Medical Institute, University of Colorado Boulder, Boulder, CO, USA*

VISVANATHAN RAMAMURTHY • *Department of Biochemistry, School of Medicine, West Virginia University, Morgantown, WV, USA*

XAVIER ROCA • *School of Biological Sciences, Nanyang Technological University, Singapore, Singapore*

JOHN J. ROSSI • *Division of Molecular and Cellular Biology, Beckman Research Institute of City of Hope, Duarte, CA, USA; Irell and Manella Graduate School of Biological Sciences, Beckman Research Institute of City of Hope, Duarte, CA, USA*

HAIHONG SHEN • *School of Life Science, Gwangju Institute of Science and Technology, Gwangju, South Korea*

LEI SHI • *Institute of Cancer Stem Cell, Cancer Center, Dalian Medical University, Dalian, Liaoning, People's Republic of China*

LINDSEY SKRDLANT • *Irell & Manella Graduate School of Biological Sciences of the City of Hope, Duarte, CA, USA; Department of Molecular and Cellular Biology, Beckman Research Institute of the City of Hope, Duarte, CA, USA*

PETER STOILOV • *Department of Biochemistry, School of Medicine, West Virginia University, Morgantown, WV, USA*

CHERYL STORK • *Division of Biomedical Sciences, School of Medicine, University of California at Riverside, Riverside, CA, USA*

JIAZI TAN • *School of Biological Sciences, Nanyang Technological University, Singapore, Singapore*

LUH TUNG • *Genomics Research Center, Academia Sinica, Nankang, Taipei, Taiwan*

YANG WANG • *Institute of Cancer Stem Cell, Cancer Center, Dalian Medical University, Dalian, China*

ZEFENG WANG • *Department of Pharmacology, School of Medicine, University of North Carolina at Chapel Hill, Chapel Hill, NC, USA*

SEBASTIEN M. WEYN-VANHENTENRYCK • *Department of Systems Biology, Department of Biochemistry and Molecular Biophysics, Center for Motor Neuron Biology and Disease, Columbia University, New York, NY, USA*

GUOWEI WU • *Department of Cellular and Molecular Medicine, University of California, San Diego, La Jolla, CA, USA*

FU-LUNG YEH • *Genomics Research Center, Academia Sinica, Nankang, Taipei, Taiwan*

HSIN-SUNG YEH • *Department of Biochemistry, Molecular Biology and Biophysics, University of Minnesota Twin Cities, Minneapolis, MN, USA*

JEONGSIK YONG • *Department of Biochemistry, Molecular Biology and Biophysics, University of Minnesota Twin Cities, Minneapolis, MN, USA*

JE-HYUN YOON • *Department of Biochemistry and Molecular Biology, Medical University of South Carolina, Charleston, SC, USA*

YI-TAO YU • *Department of Biochemistry and Biophysics, Center for RNA Biology, University of Rochester Medical Center, Rochester, NY, USA*

SHANGBING ZANG • *Department of Molecular and Cellular Biology, Beckman Research Institute of the City of Hope, Duarte, CA, USA*

CHAOLIN ZHANG • *Department of Systems Biology, Department of Biochemistry and Molecular Biophysics, Center for Motor Neuron Biology and Disease, Columbia University, New York, NY, USA*

SIKA ZHENG • *Division of Biomedical Sciences, School of Medicine, University of California at Riverside, Riverside, CA, USA*

XUEXIU ZHENG • *School of Life Science, Gwangju Institute of Science and Technology, Gwangju, South Korea*

HUA-LIN ZHOU • *Department of Genetics and Genome Sciences, Case Western Reserve University, Cleveland, OH, USA*

JIEHUA ZHOU • *Division of Molecular and Cellular Biology, Beckman Research Institute of City of Hope, Duarte, CA, USA*

Chapter 1

Characterization of RNA–Protein Interactions: Lessons from Two RNA-Binding Proteins, SRSF1 and SRSF2

Lindsey Skrdlant and Ren-Jang Lin

Abstract

SR proteins are a class of RNA-binding proteins whose RNA-binding ability is required for both constitutive and alternative splicing. While members of the SR protein family were once thought to have redundant functions, in-depth biochemical analysis of their RNA-binding abilities has revealed distinct binding profiles for each SR protein, that often lead to either synergistic or antagonistic functions. SR protein family members SRSF1 and SRSF2 are two of the most highly studied RNA-binding proteins. Here we examine the various methods used to differentiate SRSF1 and SRSF2 RNA-binding ability. We discuss the benefits and type of information that can be determined using each method.

Key words Splicing, RNA binding protein, SRSF1, SRSF2, UV crosslinking, SELEX, Enzymatic footprinting, RNA affinity chromatography, Single cell recruitment assay, Solution NMR, CLIP-seq

1 Introduction

Pre-mRNA splicing is an important and highly regulated process for eukaryotic cells. Nearly all protein coding genes in the human genome have constitutively spliced introns, and 90–95 % of these genes also have one or more alternatively spliced introns. These alternative splicing events are what provide the vast array of protein diversity in human cells by removing or adding different coding regions. This alternative splicing is regulated throughout development and tissue definition. Although canonical 5′ and 3′ splice site sequences, branch point sites, and polypyrimidine tracts are useful for defining introns, they are not sufficient for identifying all introns, particularly those of alternatively spliced genes. Alternatively spliced sites are often determined by binding of exonic or intronic splicing enhancers or silencers by trans-acting splicing regulators.

There are two large groups of trans-acting splicing regulators, the hnRNP protein family and the SR protein family. While hnRNP proteins primarily inhibit spliceosome formation at a particular site

Ren-Jang Lin (ed.), *RNA-Protein Complexes and Interactions: Methods and Protocols*, Methods in Molecular Biology, vol. 1421, DOI 10.1007/978-1-4939-3591-8_1, © Springer Science+Business Media New York 2016

of the pre-mRNA, SR proteins primarily promote spliceosome formation. These SR proteins, particularly SRSF1 (formerly known as ASF/SF2) and SRSF2 (formerly known as SC35), are the primary focus of this chapter.

SR proteins are defined as having one or more RNA recognition motifs (RRM) followed by an arginine-serine rich domain (RS) required for protein–protein interaction [1, 2]. SR proteins are required for both constitutive and alternative splicing, specifically for the formation of complex A of the early spliceosome [3–7]. SR protein binding leads to the commitment of the pre-mRNA to the splicing pathway [8]. The commitment is facilitated by the fact that SR proteins interact with the U1 and U2 snRNPs through protein–protein interactions [4, 9, 10]. Specifically, SR proteins are required for the interactions of both the U1 snRNP with the 5′ and 3′ splice sites and the U2 snRNP with the branch point site [4, 6, 11, 12].

Initially, it was thought that SR family members, particularly SRSF1 and SRSF2, were redundant because their function was indistinguishable in complementing S100 extracts to splice a β-globin minigene covering exons 1 and 2 and their intervening intron (HβΔ6) during in vitro reactions [13–15]. Furthermore, both SRSF1 and SRSF2 favor proximal 5′ splice sites and can be counteracted by incubation of the pre-mRNA with hnRNP A1 prior to addition to the in vitro splicing reaction [13]. However, further studies showed important differences between SRSF1 and SRSF2 function.

SRSF1 has 2 RRM domains and 1 RS domain [16–19]. SRSF1 promotes the splicing of bovine growth hormone intron D through specific interaction with the FP element, a purine-rich splicing enhancer, in the terminal exon (exon 5) and can also splice the tat minigene that contains exons 2 and 3 and their intervening intron, but SRSF2 is not active in those reactions [14, 20]. Conversely, SRSF2 has 1 RRM and 1 RS domain [11]. SRSF2 is required for the splicing of the third intron of the constant region of IgM pre-mRNA and SRSF1 cannot reconstitute this reaction upon depletion of SRSF2 [14]. Furthermore, SRSF1 stimulates inclusion of exon 6A of the chicken β-tropomyosin gene, while SRSF2 antagonizes exon 6A inclusion [21]. It has been determined that the substrate specificity of SR proteins requires the RRM domain(s) of the protein [22]. Therefore, it is imperative to study the interactions of SR proteins with their substrate RNAs in order to understand alternative splicing regulation.

Here we reviewed a variety of RNA–protein binding assays that have been used, both alone and in conjunction with each other, to study the differences of RNA-binding capabilities of SRSF1 and SRSF2. The methods used include electromobility shift assay (EMSA), UV crosslinking, UV crosslinking followed by immunoprecipitation (CLIP), Systematic evolution of ligands by exponential enrichment (SELEX), RNA affinity chromatography, quantitative single cell recruitment assay, enzymatic footprinting, nuclear magnetic resonance (NMR), and CLIP followed by deep sequencing of the RNA (CLIP-seq).

2   Biochemical Methods Used to Study SRSF1 and SRSF2 Protein–RNA Interactions

2.1   UV Crosslinking

UV crosslinking was one of the very first methods used to study the protein–RNA interactions of SRSF1 and SRSF2. UV crosslinking works by covalently linking amino acid residues of the protein to the nucleotides of the RNA with which they are in direct contact. Not only does this fix the protein–RNA interaction to allow the complex to go through harsh purification steps, it also allows the researcher to identify the specific RNA regions that are directly involved in protein–RNA interaction.

Initially, ^{32}P-labeled RNA was incubated with nuclear (contains SR proteins) or S100 (devoid of SR proteins) extracts and then UV irradiated to induce crosslinking. After RNase digestion, the mixtures were resolved on a polyacrylamide gel. A ^{32}P-labeled RNA–protein complex with ~30 kDa in size was observed in nuclear extracts but not in the S100 extract. The sizes of SRSF1 and SRSF2 are both at ~30 kDa. To decipher the identity of the 30 kDa crosslinked complex, antibodies specific to SRSF1 or SRSF2 were used to observe the ability of each to interact with the ^{32}P-labeled complex by immunoprecipitation. This UV-XL-IP protocol was used to establish the first known difference in SRSF1 and SRSF2: SRSF1 is able to promote the splicing of intron D of the bovine growth hormone gene by binding a purine-rich exonic splicing enhancer (ESE) present in exon 5, while SRSF2 not only was unable to bind this ESE, it also could not promote splicing of intron D [20]. Later, SRSF1 binding was also found to synergistically increase with the addition of SRSF7 (formerly known as 9G8) [23].

This experimental workflow of UV-crosslinking followed by immunoprecipitation has been repeated for many splicing events including SRSF1 binding of c-*src* exon N1 [24], SRSF1 and SRSF2 binding to the Pu2 sequence of the adenovirus E1A pre-mRNA bidirectional splicing enhancer model [25], SRSF1 and SRSF2 binding to the exon 3 ESE of human apolipoprotein A-II [26]. Furthermore, this method has also been used to determine the molecular mechanism of splicing-related diseases, such as the mutation of an ESE within *PLP1* exon 3B that is associated with the neurodegenerative Pelizaeus–Merzbacher Disease. SRSF1, but not SRSF2, was shown to bind this ESE. The disease-related mutation reduced the SRSF1 binding and the SRSF1-dependent alternative splicing of exon 3 that produces the regulated protein isoform [27].

2.2 Systematic Evolution of Ligands by Exponential Enrichment (SELEX)

In 1989, a method was developed for determining the common binding sequences of a DNA-binding protein by using serial rounds of in vitro selection, using immunoprecipitation against purified yeast GCN4, and PCR amplification, from a pool of random DNA sequences [28]. In 1990, this method was adapted to study common RNA-binding sequences of the bacteriophage T4 DNA polymerase [29]. And in 1995, this SELEX method was used to study the most common RNA binding motifs of SRSF1 and SRSF2 [30]. Three truncated proteins lacking the RS domain were used: the portion of SRSF1 that contains RRM1 and RRM2, the portion of SRSF1 that contains only the RRM1, and the portion of SRSF2 that contains its sole RRM. This experiment confirmed that SRSF1 prefers to bind purine-rich domains, such as the octamer RGAAGAAC or decamers AGGACAGAGC or AGGACGAAGC. Binding of SRSF1 to these oligos was confirmed by gel mobility shift assays using ^{32}P-labeled RNA probes, and removal of the second RRM drastically changed the consensus binding motif to ACGCGCA. Therefore, it was determined that both of SRSF1's RRMs are required for its binding specificity. In contrast, SRSF2 binds motifs that are more balanced in their nucleotide makeup, such as AGSAGAGUA or GUUCGAGUA (Table 1) [30].

In 1998, the conventional SELEX protocol was further modified for the study of SRSF1 to select for splicing activity rather than binding alone [31]. This functional SELEX assay used a model splicing event that requires a known ESE. This known ESE was

Table 1
IUPAC nucleotide code

IUPAC nucleotide code	Base
A	Adenine
T	Thymine
C	Cytosine
G	Guanine
K	G or T
S	G or C
R	A or G
W	A or T
M	A or C
V	A, G, or C
H	A, C, or T

then removed and replaced by random RNA sequences. S100 extracts complemented with purified SR proteins were used to then examine which random sequences could function as an ESE for that particular SR protein. It was found that SRSF1 prefers ESEs resembling the motif of SRSASGA [31]. This same method was later used to determine that the preferred ESE for SRSF2 was UGCNGYY or GRYYcSYR [32, 33].

2.3 Enzymatic Footprinting: Protecting RNA from RNase by the Interacting Protein

Enzymatic digestion of RNA by RNase after incubation with an RNA-binding protein is another common way to determine the exact binding region of a protein to a pre-mRNA. Residues of the pre-mRNA that are bound by the protein will be protected from RNase digestion of the RNA, leaving a short, specific sequence that is directly contacted by the protein's amino acid residues. Enzymatic digestion can be used with or without prior UV-crosslinking. In characterizing SRSF2, ESE deletion/mutation experiments delineated that ESE sequences UGCCGUU (exon 1) and UGCUGUU (exon 2) are required for SRSF2-dependent splicing of β-globin minigene [34]. In order to determine direct interaction of SRSF2 with these sequences, pre-mRNA was labeled with ^{32}P at specific nucleotide positions. After UV-crosslinking, the RNA–protein complexes were digested with RNases and resolved by gel electrophoresis. Only ^{32}P-labeled regions that were protected by the binding of and crosslinked to SRSF2 would yield a ^{32}P-labeled RNA–protein complex [34].

Enzymatic digestion has also been used in the absence of crosslinking to study the binding of SRSF2 protein to HIV-1 RNA [35]. In this case, RNA–protein complexes were subjected to limited digestions using RNases T1 and T2 that cleave unprotected single-stranded RNA or RNase V1 that cleaves unprotected double-stranded RNA (Fig. 1). Positions of protection—no enzymatic cleavage—due to protein binding could then be detected by primer extension using a 5′-end-labeled oligo. Digestion of unbound RNA was compared to digestion of RNA pre-incubated with SRSF2 or hnRNP A1. Results showed that both SRSF2 and hnRNP A1 compete for binding at a similar region of the silencer site ESS2, located in the terminal loop of the B motif [35]. Binding of SRSF2 to the ESS2 was then confirmed by kethoxal chemical probing, which chemically modifies G residues that are unbound or unprotected. The binding of hnRNP A1 via its RNA binding domain (UP1) to the ESS2 was also confirmed by NMR that examined the proton exchange rate of G residues between the RNA and the ^{15}N-labeled UP1 [35].

2.4 RNA Affinity Chromatography

RNA affinity chromatography is an excellent way to determine in cellular extracts what RNA-binding proteins interact with a specific RNA sequence. Target RNAs are bound to agarose beads, which are then incubated in cellular extracts. Bound proteins are eluted,

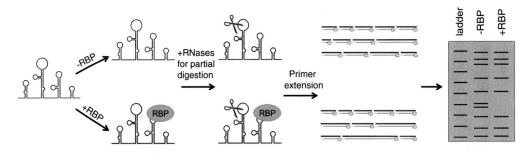

Fig. 1 Enzymatic footprinting assay. In vitro transcribed RNAs are incubated with or without purified RNA-binding protein (RBP). RNAs are then treated with RNases specific for single-stranded or double-stranded RNA. After a partial digestion, RNA fragments are converted to cDNA fragments using reverse transcriptase-mediate primer extension of ^{32}P-labeled primers. Resulting cDNA is ran on a gel for fragment size analysis

for example with SDS buffer, ran on a polyacrylamide gel, transferred to membrane, and probed with antibodies. This method was used to characterize the involvement of SR proteins in the aberrant splicing of HIV exon 6D that occurs with a naturally occurring U to C mutation within the exon. RNA affinity chromatography using both the wildtype and mutant HIV-1 exon 6D sequences showed that the C to U mutation increases binding of both SRSF1 and SRSF2 [36]. Splicing assays that used the wildtype and mutant exonic sequences confirmed that increased SRSF1 and SRSF2 binding results in increased inclusion of the exon [36]. RNA affinity chromatography has also been used in identifying the SRSF1 and SRSF2 binding to the chicken β-tropomyosin 6B exon [37] as well as the SRSF2 binding to HIV-1 tat exon 2 [38].

RNA affinity chromatography has also been used in molecular mechanistic studies of splicing-related diseases involved SR proteins. One example is a naturally occurring mutation downstream of exon 7 of pyruvate dehydrogenase gene that results in mitochondrial encephalopathy. RNA affinity chromatography showed that this mutation causes increased binding of SRSF2, but not SRSF1, despite binding of SRSF1 to exon 7 also occurring [39]. These results were also confirmed with UV crosslinking. Another example is the aberrant exclusion of exon 3 of human growth hormone, which results in a dominant-negative protein isoform that causes growth hormone deficiency. This autosomal dominance is often caused by a mutation in ESE2 of exon 3. RNA affinity chromatography showed that SRSF1 and SRSF2 bind to ESE2, and both prefer the mutant ESE2 [40]. However, only the overexpression of SRSF2, but not SRSF1, causes hGH exon 3 skipping in the cell. UV-crosslinking experiments showed that not only did the disease-related mutation increase SRSF2 binding of ESE2, SRSF2 also bound to another region downstream of ESE2. Binding of SRSF2 to both sites synergistically increases skipping of exon 3 [40].

Interaction Present   Interaction Not Present

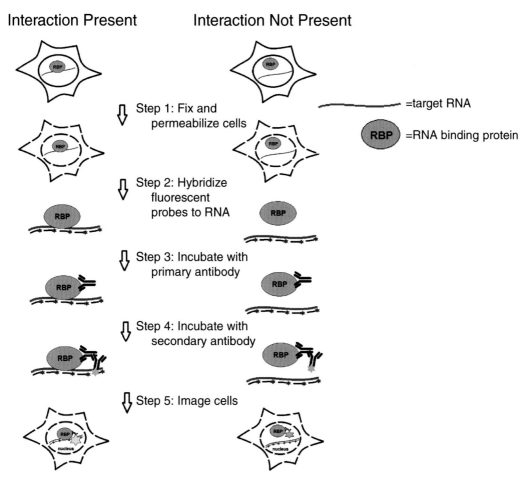

Step 1: Fix and permeabilize cells

Step 2: Hybridize fluorescent probes to RNA

Step 3: Incubate with primary antibody

Step 4: Incubate with secondary antibody

Step 5: Image cells

———— =target RNA

RBP =RNA binding protein

Fig. 2 Quantitative single cell recruitment assay. Cells are fixed and permeabilized. After washing, fluorescent-labeled anti-sense RNA probes (*red*) are hybridized to the target RNA within the fixed cells. The RNA-binding protein of interest is then labeled with a primary antibody, followed by a fluorescent-labeled secondary antibody (*green*). During imagining with either a fluorescence or confocal microscope, direct interaction of the protein and RNA will be observed by overlapping fluorescent signals (*yellow*)

2.5 Quantitative Single Cell Recruitment Assay

The quantitative single cell recruitment assay uses RNA fluorescence in situ hybridization (FISH) and indirect Immunofluorescence (IF) against splicing factors to image their co-localization in single cells (Fig. 2). A specific RNA inside a fixed and permeabilized cell is imaged by hybridization to fluorescent-labeled anti-sense RNA probe. Protein localization in the cell was detected using primary antibody targeting the RNA-binding protein of interest followed by a fluorescent-labeled secondary antibody that will detect the primary antibody. Co-localization of the FISH signal and antibody signal is indicative of interaction of the RNA-binding protein with the RNA of interest.

This method has been used to investigate the in vivo interaction of SRSF1/SRSF2 with the exon 10 of the tau mRNA that encodes a microtubule binding protein. The exon 10 of tau is alternatively spliced in normal individuals, while a parkinsonism-like disorder mutation causes exon 10 to be included nearly 100 %. To investigate which splicing factors are differentially recruited to this alternatively spliced exon, minigenes with the wild-type or the mutant sequence were subjected to the aforementioned FISH-IF analysis. Mutant tau exon 10, which has nearly 100 % exon 10 inclusion, exhibited higher recruitment of both SRSF1 and SRSF2 in the FISH-IF experiment than WT tau exon 10, which exhibits >50 % exon 10 exclusion [41]. The study provides the first in vivo evidence for physical recruitment of splicing regulators to the alternative spliced exon.

3 Biophysical and Deep Sequencing Methods of Studying SRSF1 and SRSF2 Protein–RNA Interactions

3.1 Solution NMR of RRMs with RNA Oligos

More recently, solution NMR has been used to study the structural basis of SR protein RRM–RNA interactions. Most previous methods are specific for defining the nucleotide sequences where SR proteins bind. In contrast, solution NMR is used to determine the structural motifs of the protein itself that are responsible for binding to RNA. Solution NMR studies have shown that the general SR RRM structure consists of 4 β sheets against 2 α helices [42–44]. In 2012, the structure of RNA-bound RRM from SRSF2 was solved [43]. When compared to the SRSF1 RRM1 (1X4A) in the Protein Data Bank, and other published SR protein RRMs, SRSF2 has a longer L3 loop, along with a larger region of exposed hydrophobic residues, between the β2 and β3 beta sheets [43]. SRSF2 also has less hydrophobicity in β1, and a missing tyrosine residue in β1 that is conserved in most other SR proteins [43]. NMR chemical shift mapping showed that SRSF2 binds to the 5-mer GAGUA previously found with SELEX with a K_d of 0.5 mM; however, lengthening the oligo to a 9-mer increases binding efficiency to a K_d of ~61 nM, as determined by isothermal titration calorimetry (ITC), as a result of increasing protein–RNA contact residues [43]. Mutation assays within SRSF2's unique L3 loop followed by UV-crosslinking and NMR showed that unique L3 loop is important for specific protein–RNA binding, but does not change the overall structure of SRSF2's RRM [43].

Although the structure of RRM1 of SRSF1 with RNA has not been published, the structure of the RRM2 of SRSF1 has been studied extensively. The structure of SRSF1 RRM2 was studied in conjunction with the 8-mer AGGAGAAC that had been determined using SELEX of the combined SRSF1 RRM1 and RRM2 domains rather than the canonical RRM1 alone [42]. ITC was used to show that RRM2 is the domain primarily responsible for binding the purine-rich oligo [42]. ITC of the 8-mer UGAAGGAC with SRSF1's RRM2 determined that the K_d is 0.8 µM, and NMR studies of this binding determined that the primary regions responsible for binding are α1 and β2 [42]. In contrast, usually only the β sheets of SR protein canonical RRMs are responsible for direct binding of RNA, not α helices [42]. Another NMR study of SRSF1 RRM2 with RNA determined that residues in α1, β2, loop 3, and loop5 are all important for RNA–protein binding, and mutation of any of these residues reduced protein–RNA binding as indicated by UV-crosslinking and NMR titration [44]. These NMR studies are consistent with the previously determined RNA sequence specificities for SRSF1 and SRSF2.

3.2 CLIP-seq

The second technique that has been more recently developed to study protein–RNA interactions of RNA-binding proteins is an adaptation of the CLIP assay. UV-crosslinking is used at the whole cell level to covalently link RNA binding proteins to the RNA sequences with which they are in contact in vivo. Antibodies specific to a protein of interest are then used to isolate its respective protein–RNA complexes. Enzymatic digestion is then used to shorten the bound RNA to small fragments, and the resulting RNA is identified by deep sequencing. In 2013, a paper was published that showed global CLIP-seq analysis for both SRSF1 and SRSF2. This study was able to identify some general binding characteristics of SRSF1 and SRSF2. Although ~10 % of sequences bound by each are found to be in unannotated regions, the majority of bound regions are within introns (~50 %) and exons (~40 %) [45], which mimics the previously known actions of SR proteins. There is an extensive overlap in SRSF1 and SRSF2 binding regions for both constitutive and alternative exons, though unique binding sites are also observed [45]. Both SRSF1 and SRSF2 exhibit higher levels of binding near weak splice sites or long introns [45]. CLIP-seq also showed binding motifs of SRSF1 and SRSF2 similar to those found with physical interaction SELEX and functional SELEX. The most common SRSF1 binding motifs are GAKGASG, CAGGRGG, and CGGAGG, while the most common SRSF2 binding motifs are CUUCWSS, CGMVG, and CHGG (Table 1) [45]. Furthermore, SRSF1 and SRSF2 were shown to compete for binding because strong SRSF2 binding sites showed an increase in SRSF1 binding upon SRSF2 knockdown [45].

4 Concluding Remarks

The studies of the two prototypical SR proteins, SRSF1 and SRSF2, illustrate how important it is to use multiple methods to determine how RNA-binding proteins interact with RNA. For some experiments, SRSF1 and SRSF2 appear to have redundant function by binding the same pre-mRNA region and promoting splicing, while

CLIP-seq, SELEX, and enzymatic footprinting elucidate that SRSF1 and SRSF2 bind unique motifs, though sometimes overlapping regions.

UV-crosslinking using ^{32}P-labeled RNA was the first to show differences in RNA-binding capabilities of SRSF1 and SRSF2, but UV-crosslinking can only show interaction with known RNAs, has no predictive power, and does not provide the exact RNA motif that the RNA-binding protein interacts with. This is exemplified in how UV-crosslinking shows that both SRSF1 and SRSF2 bind the Pu2 sequence of the adenovirus E1A pre-mRNA, but not how SRSF1 and SRSF2 are likely binding different motifs near, and possibly overlapping, each other. RNA affinity chromatography and single cell recruitment assays have similar strengths and drawbacks to UV crosslinking in that there are excellent for determining the binding of an RNA-binding protein to a known transcript in vitro (RNA affinity chromatography) or in vivo (single cell recruitment assay), but neither can be used to determine exact binding motifs or predict whether the same RNA-binding protein can bind a different transcript that was not assayed. On the other hand, enzymatic footprinting improves upon UV crosslinking by further honing in on the motifs and secondary structures that RNA-binding proteins bind.

SELEX was the first technique able to define conserved motifs that SRSF1 and SRSF2 bind, providing excellent predictive power for potential binding sites in transcripts. Solution NMR goes even once step further by determining what residues of the protein are involved in the protein–RNA interaction of these conserved motifs. However, as in vitro models, SELEX and solution NMR fail to provide insight on which motifs are actually used by SRSF1 and SRSF2 in vivo and do not provide functional analysis regarding whether binding is directly linked to functional splicing.

CLIP-seq has revolutionized the ability to study protein–RNA interactions in vivo. It has added in the determination of global SR protein binding to mRNA regions and provided a large amount of data that can be further mined for other experiments regarding RBP function in the context of specific alternative splicing events. Further modifications of the original CLIP-seq protocol, such as photoactivatable ribonucleoside-enhanced CLIP (PAR-CLIP) [46] and individual nucleotide resolution CLIP (iCLIP) [47], are even capable of determining exact binding sites of the RNA-binding protein being assessed by either a base transition during reverse transcription at the site of crosslinking or an alternative adaptor ligation step that identifies cDNAs that become truncated at the crosslinking site, respectively [48]. However, easy-to-use algorithms have not yet been developed to allow quick searches of individual genes for sites of RNA binding protein interaction based on previously published RNA-seq data. Furthermore, similar to the previous problem with traditional SELEX assays, knowledge of the regions where RBPs bind does not provide direct knowledge of the consequences of these protein–RNA interactions. Both assays for confirmation of RNA binding by RBPs, such as traditional UV-crosslinking and EMSA, and assays for functional consequences of this binding, including in vitro splicing assays, and ESE mutation assays followed by reanalysis of splicing and RNA binding, will all continue to be indispensable for studying RNA binding proteins.

The summary above outlines how important it is to use multiple methods to determine how RNA-binding proteins interact with RNA and differ from each other. For some experiments, SRSF1 and SRSF2 appear to have redundant function by binding the same pre-MRNA region and promoting splicing, while CLIP-seq, SELEX and enzymatic footprinting further elucidate that SRSF1 and SRSF2 bind unique motifs, though sometimes overlapping regions, compared to each other. Therefore, it is imperative that a variety of methods are used in conjunction with each other to study the precise biological functions of RNA binding proteins in the future.

Acknowledgment

L.S. was supported in part by H.N. & Frances Berger Foundation Fellowship and the Norman and Melinda Payson Fellowship. This work was supported by grants from the Beckman Research Institute to R.-J.L.

References

1. Bandziulis RJ, Swanson MS, Dreyfuss G (1989) RNA-binding proteins as developmental regulators. Genes Dev 3:431–437

2. Kenan DJ, Query CC, Keene JD (1991) RNA recognition: towards identifying determinants of specificity. Trends Biochem Sci 16:214–220

3. Fu XD, Maniatis T (1990) Factor required for mammalian spliceosome assembly is localized to discrete regions in the nucleus. Nature 343:437–441

4. Fu XD, Maniatis T (1992) The 35-kDa mammalian splicing factor SC35 mediates specific interactions between U1 and U2 small nuclear ribonucleoprotein particles at the 3′ splice site. Proc Natl Acad Sci U S A 89:1725–1729

5. Ge H, Manley JL (1990) A protein factor, ASF, controls cell-specific alternative splicing of SV40 early pre-mRNA in vitro. Cell 62:25–34

6. Krainer AR, Conway GC, Kozak D (1990) Purification and characterization of pre-mRNA splicing factor SF2 from HeLa cells. Genes Dev 4:1158–1171

7. Spector DL, Fu XD, Maniatis T (1991) Associations between distinct pre-mRNA splicing components and the cell nucleus. EMBO J 10:3467–3481

8. Fu XD (1993) Specific commitment of different pre-mRNAs to splicing by single SR proteins. Nature 365:82–85

9. Kohtz JD, Jamison SF, Will CL et al (1994) Protein-protein interactions and 5′-splice-site recognition in mammalian mRNA precursors. Nature 368:119–124

10. Wu JY, Maniatis T (1993) Specific interactions between proteins implicated in splice site selection and regulated alternative splicing. Cell 75:1061–1070

11. Fu XD, Maniatis T (1992) Isolation of a complementary DNA that encodes the mammalian splicing factor SC35. Science 256:535–538

12. Krainer AR, Maniatis T (1985) Multiple factors including the small nuclear ribonucleoproteins U1 and U2 are necessary for pre-mRNA splicing in vitro. Cell 42:725–736

13. Fu XD, Mayeda A, Maniatis T et al (1992) General splicing factors SF2 and SC35 have equivalent activities in vitro, and both affect alternative 5′ and 3′ splice site selection. Proc Natl Acad Sci U S A 89:11224–11228

14. Mayeda A, Screaton GR, Chandler SD et al (1999) Substrate specificities of SR proteins in constitutive splicing are determined by their RNA recognition motifs and composite pre-mRNA exonic elements. Mol Cell Biol 19:1853–1863

15. Zahler AM, Lane WS, Stolk JA et al (1992) SR proteins: a conserved family of pre-mRNA splicing factors. Genes Dev 6:837–847

16. Caceres JF, Krainer AR (1993) Functional analysis of pre-mRNA splicing factor SF2/ASF structural domains. EMBO J 12:4715–4726

17. Ge H, Zuo P, Manley JL (1991) Primary structure of the human splicing factor ASF reveals similarities with Drosophila regulators. Cell 66:373–382

18. Krainer AR, Mayeda A, Kozak D et al (1991) Functional expression of cloned human splicing factor SF2: homology to RNA-binding proteins, U1 70K, and Drosophila splicing regulators. Cell 66:383–394

19. Zuo P, Manley JL (1993) Functional domains of the human splicing factor ASF/SF2. EMBO J 12:4727–4737

20. Sun Q, Mayeda A, Hampson RK et al (1993) General splicing factor SF2/ASF promotes alternative splicing by binding to an exonic splicing enhancer. Genes Dev 7:2598–2608

21. Gallego ME, Gattoni R, Stevenin J et al (1997) The SR splicing factors ASF/SF2 and SC35 have antagonistic effects on intronic enhancer-dependent splicing of the beta-tropomyosin alternative exon 6A. EMBO J 16:1772–1784

22. Chandler SD, Mayeda A, Yeakley JM et al (1997) RNA splicing specificity determined by the coordinated action of RNA recognition motifs in SR proteins. Proc Natl Acad Sci U S A 94:3596–3601

23. Li X, Shambaugh ME, Rottman FM et al (2000) SR proteins Asf/SF2 and 9G8 interact to activate enhancer-dependent intron D splicing of bovine growth hormone pre-mRNA in vitro. RNA 6:1847–1858

24. Rooke N, Markovtsov V, Cagavi E et al (2003) Roles for SR proteins and hnRNP A1 in the regulation of c-src exon N1. Mol Cell Biol 23:1874–1884

25. Bourgeois CF, Popielarz M, Hildwein G et al (1999) Identification of a bidirectional splicing enhancer: differential involvement of SR proteins in 5′ or 3′ splice site activation. Mol Cell Biol 19:7347–7356

26. Arrisi-Mercado P, Romano M, Muro AF et al (2004) An exonic splicing enhancer offsets the atypical GU-rich 3′ splice site of human apolipoprotein A-II exon 3. J Biol Chem 279:39331–39339

27. Wang E, Huang Z, Hobson GM et al (2006) PLP1 alternative splicing in differentiating oligodendrocytes: characterization of an exonic splicing enhancer. J Cell Biochem 97:999–1016

28. Oliphant AR, Brandl CJ, Struhl K (1989) Defining the sequence specificity of DNA-binding proteins by selecting binding sites from random-sequence oligonucleotides: analysis of yeast GCN4 protein. Mol Cell Biol 9:2944–2949

29. Tuerk C, Gold L (1990) Systematic evolution of ligands by exponential enrichment: RNA ligands to bacteriophage T4 DNA polymerase. Science 249:505–510

30. Tacke R, Manley JL (1995) The human splicing factors ASF/SF2 and SC35 possess distinct, functionally significant RNA binding specificities. EMBO J 14:3540–3551

31. Liu HX, Zhang M, Krainer AR (1998) Identification of functional exonic splicing enhancer motifs recognized by individual SR proteins. Genes Dev 12:1998–2012

32. Liu HX, Chew SL, Cartegni L et al (2000) Exonic splicing enhancer motif recognized by human SC35 under splicing conditions. Mol Cell Biol 20:1063–1071

33. Schaal TD, Maniatis T (1999) Selection and characterization of pre-mRNA splicing enhancers: identification of novel SR protein-specific enhancer sequences. Mol Cell Biol 19:1705–1719

34. Schaal TD, Maniatis T (1999) Multiple distinct splicing enhancers in the protein-coding sequences of a constitutively spliced pre-mRNA. Mol Cell Biol 19:261–273

35. Hallay H, Locker N, Ayadi L et al (2006) Biochemical and NMR study on the competition between proteins SC35, SRp40, and heterogeneous nuclear ribonucleoprotein A1 at the HIV-1 Tat exon 2 splicing site. J Biol Chem 281:37159–37174

36. Caputi M, Zahler AM (2002) SR proteins and hnRNP H regulate the splicing of the HIV-1 tev-specific exon 6D. EMBO J 21:845–855

37. Expert-Bezancon A, Sureau A, Durosay P et al (2004) hnRNP A1 and the SR proteins ASF/SF2 and SC35 have antagonistic functions in splicing of beta-tropomyosin exon 6B. J Biol Chem 279:38249–38259

38. Zahler AM, Damgaard CK, Kjems J et al (2004) SC35 and heterogeneous nuclear ribonucleoprotein A/B proteins bind to a juxtaposed exonic splicing enhancer/exonic splicing silencer element to regulate HIV-1 tat exon 2 splicing. J Biol Chem 279:10077–10084

39. Gabut M, Mine M, Marsac C et al (2005) The SR protein SC35 is responsible for aberrant splicing of the E1alpha pyruvate dehydrogenase mRNA in a case of mental retardation with lactic acidosis. Mol Cell Biol 25:3286–3294

40. Solis AS, Peng R, Crawford JB et al (2008) Growth hormone deficiency and splicing fidelity: two serine/arginine-rich proteins, ASF/SF2 and SC35, act antagonistically. J Biol Chem 283:23619–23626

41. Mabon SA, Misteli T (2005) Differential recruitment of pre-mRNA splicing factors to alternatively spliced transcripts in vivo. PLoS Biol 3, e374

42. Clery A, Sinha R, Anczukow O et al (2013) Isolated pseudo-RNA-recognition motifs of SR proteins can regulate splicing using a non-canonical mode of RNA recognition. Proc Natl Acad Sci U S A 110:E2802–E2811

43. Phelan MM, Goult BT, Clayton JC et al (2012) The structure and selectivity of the SR protein SRSF2 RRM domain with RNA. Nucleic Acids Res 40:3232–3244

44. Tintaru AM, Hautbergue GM, Hounslow AM et al (2007) Structural and functional analysis of RNA and TAP binding to SF2/ASF. EMBO Rep 8:756–762

45. Pandit S, Zhou Y, Shiue L et al (2013) Genome-wide analysis reveals SR protein cooperation and competition in regulated splicing. Mol Cell 50:223–235

46. Spitzer J, Hafner M, Landthaler M et al (2014) PAR-CLIP (Photoactivatable Ribonucleoside-Enhanced Crosslinking and Immunoprecipitation): a step-by-step protocol to the transcriptome-wide identification of binding sites of RNA-binding proteins. Methods Enzymol 539:113–161

47. Huppertz I, Attig J, D'Ambrogio A et al (2014) iCLIP: protein-RNA interactions at nucleotide resolution. Methods 65:274–287

48. Konig J, Zarnack K, Luscombe NM et al (2011) Protein-RNA interactions: new genomic technologies and perspectives. Nat Rev Genet 13:77–83

Chapter 2

Identification of mRNA-Interacting Factors by MS2-TRAP (MS2-Tagged RNA Affinity Purification)

Je-Hyun Yoon and Myriam Gorospe

Abstract

Posttranscriptional gene expression is governed by the interaction of mRNAs with vast families of RNA-binding proteins (RBPs) and noncoding (nc)RNAs. RBPs and ncRNAs jointly influence all aspects of posttranscriptional metabolism, including pre-mRNA splicing and maturation, mRNA transport, editing, stability, and translation. Given the impact of mRNA-interacting molecules on gene expression, there is great interest in identifying mRNA-binding factors comprehensively. Here, we provide a detailed protocol to tag mRNAs with MS2 hairpins and then affinity-purify *trans*-binding factors (RBPs, ncRNAs) associated with the MS2-tagged mRNA. This method, termed MS2-TRAP, permits the systematic characterization of ribonucleoprotein (RNP) complexes formed on a given mRNA of interest. We describe how to prepare the mRNA-MS2 expression vector, purify the MS2-tagged RNP complexes, and detect bound RNAs and RBPs, as well as variations of this methodology to address related questions of RNP biology.

Key words RBP, Ribonucleoprotein complexes, lncRNA, RNA affinity pulldown

1 Introduction

Posttranscriptional gene regulation is mainly controlled by two classes of factors associating with mRNAs, RNA-binding proteins (RBPs) and noncoding RNAs (ncRNAs) [1, 2]. RBPs are involved in all steps of posttranscriptional gene regulation, including pre-mRNA splicing and maturation, mRNA export, degradation, editing, localization, storage, and translation [3]. Among the various small ncRNAs (miRNAs, piRNAs, siRNAs, tiRNAs, snoRNAs, etc.), microRNAs have been studied most extensively and function by promoting mRNA degradation and/or translation repression [4, 5]. Long (l)ncRNAs also control many posttranscriptional steps through their interaction with target mRNAs [6]. Together, these families of ncRNAs and RBPs regulate the posttranscriptional fate of the mRNA in a robust and dynamic manner, allowing gene expression patterns to respond adequately to developmental, metabolic, and environmental cues [2]. Accordingly, studying the

Ren-Jang Lin (ed.), *RNA-Protein Complexes and Interactions: Methods and Protocols*, Methods in Molecular Biology, vol. 1421, DOI 10.1007/978-1-4939-3591-8_2, © Springer Science+Business Media New York 2016

composition of mRNA RNPs is critical for understanding the ultimate outcome of mRNAs.

Interactions between RBPs and RNAs have been studied for decades (reviewed comprehensively in Ref. [7] and chapters therein). The binding of a specific RBP to a specific RNA has been studied using classic in vitro methods like RNA electrophoretic mobility shift, biotin pulldown, and agarose gel retardation assays [7–9]. Systematic methods to detect binding of one RBP to collections of target mRNAs are more recent and include binding assays to measure native complexes (ribonucleoprotein immunoprecipitation or RIP) and crosslinked complexes (crosslinking immunoprecipitation or CLIP) [10]. The physical and functional interactions of specific mRNAs and microRNAs have also been studied using a variety of analytic methods. However, few high-throughput approaches are available to identify the collection of factors that interact with a given mRNA in the cell. One such approach, affinity pulldown using antisense oligonucleotides complementary to the mRNA. is feasible in cell-free systems, but is not a dependable method for targeting intracellular RNP complexes in vivo.

MS2 hairpin sequences have long been utilized for studying mRNAs in eukaryotic cells [11]. The high-affinity interaction of the MS2 RNA hairpin and its binding protein, the viral coat protein MS2-BP, has long been exploited for visualizing and purifying RNP complexes without introducing purified proteins or RNAs into cells. The existence of many chimeric variants of MS2-BP, bearing different tags for imaging and for biochemical detection gives this approach additional advantages. Here, we describe a detailed protocol for a methodology we have named MS2-TRAP (MS2-tagged RNA affinity purification) to purify RNP complexes that include RBPs and ncRNAs bound to an mRNA of interest. Detailed directions to design target plasmids expressing MS2-tagged RNA, step-by-step procedures for complex purification, detection of co-purified proteins and RNAs, troubleshooting suggestions, and additional notes are provided.

2   Materials

2.1   Cell Culture and Transfection

1. DMEM.

2. 10 % FBS.

3. 2 mM l-Glutamine.

4. 100 U/ml penicillin/streptomycin.

5. Lipofectamine 2000 (Invitrogen).

6. Opti-MEM (Invitrogen).

7. Plasmids pMS2, pMS2-GST, pMS2-YFP (Addgene).

8. Plasmid pcDNA3-FLAG-MS2 [11–13].

2.2 Immunopre-cipitation

1. Glutathione agarose beads (GE Healthcare).
2. Protein-A or -G sepharose beads (GE Healthcare).
3. Anti-YFP antibody (Santa Cruz Biotechnology).
4. Anti-Flag antibody (M2, Sigma).
5. Normalized IgG (Santa Cruz Biotechnology).
6. Proteinase K (Roche).
7. DNase I (Ambion): 2 U/µl.
8. Acid-phenol–chloroform, pH 4.5 (with IAA, 25:24:1) (Ambion).
9. GlycoBlue™ coprecipitant (15 mg/ml) (Ambion) ice-cold phosphate-buffered saline (PBS): 137 mM NaCl, 2.7 mM KCl, 10 mM Na_2HPO_4, and 1.8 mM KH_2PO_4.
10. NP-40 lysis buffer: 20 mM Tris–HCl at pH 7.5, 100 mM KCl, 5 mM $MgCl_2$, 0.5 % NP-40, 1X protease inhibitors (Roche), RiboLock RNase inhibitor (Fermentas, 40 U/ml), and 1 mM DTT.
11. NT2 buffer: 50 mM Tris–HCl at pH 7.5, 150 mM NaCl, 1 mM $MgCl_2$, and 0.05 % NP-40.
12. RIPA buffer: 10 mM Tris–HCl, pH 7.4, 150 mM NaCl, 1 % (v/v) Nonidet P-40, 1 mM EDTA, 0.1 % (w/v) SDS, and 1 mM DTT.
13. 4X SDS sample buffer: 40 % glycerol, 240 mM Tris–HCl pH 6.8, 8 % SDS, 0.04 % bromophenol blue, and 5 % β-mercaptoethanol.

2.3 Reverse Transcription and Real-Time Quantitative (q) PCR

1. dNTP mix (10 mM).
2. Random hexamers (150 ng/µl).
3. Reverse transcriptase (Fermentas): Maxima Reverse Transcriptase (200 U/µl).
4. Gene-specific primer sets.
5. KAPA SYBR® FAST qPCR Kits QuantiMir™ RT kit (System Biosciences).

3 Methods

3.1 Plasmid Construction, Cell Culture, and Transfection

1. Plasmid pMS2 has a pcDNA3 backbone containing 24 repeats of MS2 sequences between the *Eco*RI and *Eco*RV sites. Cloning sites at *Hin*dIII, *Kpn*I, and *Eco*RI are available for attaching complementary DNA (cDNA, prepared from the mRNA of interest) upstream of the 24 MS2 repeats (Fig. 1). Alternatively, MS2 can be transferred to the parental plasmid expressing the mRNA of interest by standard enzymatic digestion or PCR amplification (*see* **Note 1**).

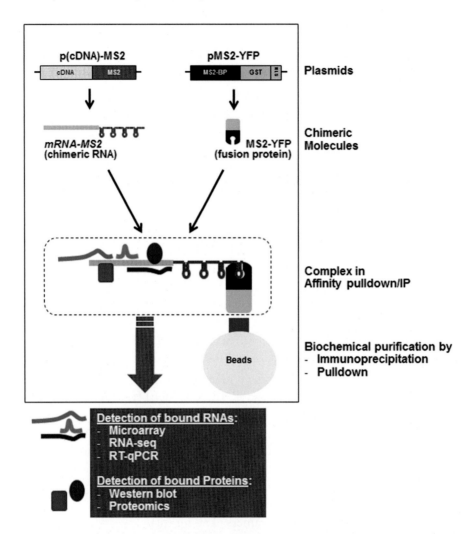

Fig. 1 Schematic representation of the MS2-TRAP methodology, including the plasmids required ("Plasmids"), the fusion RNA and protein expressed from the plasmids ("Chimeric Molecules") and the resulting RNP complex ("Complex in Affinity pulldown/IP"). After pulldown or immunoprecipitation using the appropriate conjugated beads, RNAs and proteins bound to the chimeric transcript can be identified using are methodologies listed (*gray box*)

2. Once the p(cDNA)-MS2 plasmid construct is ready, prepare a mammalian cell culture (e.g., human embryonic kidney 293 cells, HeLa cells, or mouse embryonic fibroblasts) at 80 % confluence in 100-mm culture plates in DMEM supplemented with 10 % fetal bovine serum and appropriate antibiotics.

3. Transfect 3 μg of p(cDNA)-MS2 or pMS2 empty vector, together with 1 μg of pMS2-GST (alternatively pMS2-YFP or pcDNA3-FLAG-MS2) diluted in 200 μl of Opti-MEM, mixed with 5 μl lipofectamine 2000 diluted with 200 μl of Opti-MEM (*see* **Note 2**).

4. After incubation for 20 min at room temperature, rinse cells with phosphate-buffered saline (PBS) once, replace with Opti-MEM, add the transfection mixture.

5. Incubate cells for an additional 5 h at 37 °C in 5 % CO_2, rinse cells once with PBS, and replace DMEM supplemented with 10 % FBS without antibiotics.

6. Forty-eight hours later, wash the cells with PBS once and lyse them in 500 μl NP-40 lysis buffer for 10 min on ice.

7. Scrape the cells, collect the lysate, and centrifuge them at 10,000 g for 30 min at 4 °C.

8. Transfer the supernatant to a new tube and use 2 mg of lysate in 1 ml for pulldown or immunoprecipitation.

3.2 Complex Purification and RNA Isolation

1. Prepare the beads (glutathione agarose, Protein-A or -G sepharose) by washing them with ice-cold PBS three times and resuspending them with an equal volume of PBS. The 40-μl slurry of glutathione beads can be used directly for pulldown analysis.

2. For coating anti-Flag or anti-YFP antibody, incubate 40 μl of the slurry with 3 μg of normalized IgG or anti-Flag (or anti-YFP) antibody for 3 h at 4 °C in NT2 buffer, centrifuge the beads at 2000 × g for 2 min at 4 °C, and wash five times using NT2 buffer.

3. Add 1 ml of cell lysate to the beads (GSH agarose or antibody-coated sepharose) and incubate for 2 h at 4 °C (*see* **Note 3**).

4. Centrifuge samples at 2000 × g for 2 min at 4 °C, and wash three times with NP-40 lysis buffer.

5. Add 20 units of RNase-free DNase I (Ambion) in 100 μl NP-40 lysis buffer and incubate them for 15 min at 37 °C.

6. Add 700 μl NT2 buffer and centrifuge at 2000 × g for 2 min at 4 °C.

7. For protein analysis, go to Subheading 3.4 (protein detection).

8. For RNA analysis, add 0.1 % SDS/0.5 mg/ml Proteinase K and incubate for 20 min at 55 °C.

9. Centrifuge at 10,000 × g for 10 min at 4 °C and collect the supernatant.

10. Add 500 μl RNase-free water/500 μl of acidic phenol and vortex them for 3 min.

11. Centrifuge at 10,000 × g for 30 min at 4 °C and transfer 400 μl of the supernatant.

12. Add 800 μl of 100 % ethanol/40 μl of 3 M sodium acetate/2 μl of glycoblue and incubate at −80 °C for 1 h or overnight.

13. Centrifuge at $10,000 \times g$ at 4 °C for 30 min, wash with 400 µl of 70 % ethanol, and centrifuge at $10,000 \times g$, for 15 min 4 °C.

14. Dry the pellets and resuspend in 12 µl of RNase-free water.

3.3   RNA Detection (mRNAs, lncRNAs, and miRNAs)

1. For detecting mRNA or lncRNA in the pulldown/IP sample, add 1 µl dNTP mix (10 mM) and 1 µl of random hexamers (150 ng/µl) with 12 µl RNAs.

2. Incubate at 65 °C for 5 min and 4 °C for 5 min.

3. Add 1 µl reverse transcriptase (200 U/µl), 1 µl RNase Inhibitor (40 U/µl), and 4 µl 5×reverse transcription reaction buffer.

4. Incubate reactions at 25 °C for 10 min, at 50 °C for 30 min, and at 85 °C for 5 min in thermo cycler.

5. Take 2.5 µl of cDNAs and add 2.5 µl forward and reverse gene specific primers (2.5–10 µM) and 5 µl of SYBR green master mix.

6. Once PCR is performed, calculate Ct values of each sample. For GSH pulldown, Ct values from pMS2 empty vector sample can be utilized for normalization. For immunoprecipitation, normalization can be performed with Ct values from pMS2 empty vector sample or from p(cDNA)-MS2 sample with IgG immunoprecipitation. Normalization of nonspecific binding can be performed using Ct values of mRNAs encoding housekeeping proteins such as GAPDH, ACTB, UBC, or SDHA.

7. For detecting miRNAs, take 5 µl of RNAs from precipitation and add 2 µl of 5×Poly(A) buffer, 1 µl of 25 mM $MnCl_2$, 1.5 µl of 5 mM ATP, and 0.5 µl Poly(A) polymerase included in the QuantiMir kit.

8. Incubate reaction for 30 min at 37 °C and add 0.5 µl of Oligo(dT) adaptor (QuantiMir kit).

9. Incubate reaction for 5 min at 60 °C and cool it down at room temperature for 2 min.

10. Add 4 µl of 5×RT buffer, 2 µl of dNTP mix, 1.5 µl of 0.1 M DTT, 1.5 µl of RNase-free water, and 1 µl of reverse transcriptase (QuantiMir kit).

11. Incubate reactions for 60 min at 42 °C and heat for 10 min at 95 °C.

12. Mix 2.5 µl of cDNAs with 2.5 µl of miRNA-specific primer (2.5 µM), universal primer (10 µM from QuantiMir Kit), and SYBR green master mix.

13. Perform PCR and calculate Ct values of each sample for normalization as described above for mRNA and lncRNA (*see* **Note 4**). Normalization of nonspecific binding can be performed using Ct values of housekeeping small RNAs such as *U6*.

3.4 Protein Detection

1. Forty-eight hours after transfection, whole-cell lysates can be prepared using RIPA buffer (500 μl) by incubating cells for 10 min on ice, scraping the cells, collecting the lysate, centrifuging it at $10,000 \times g$ for 30 min at 4 °C, and transferring the supernatant to a new tube.

2. After Subheading 3.2, **step 7**, the proteins present in the mRNA pulldown can be mixed with SDS sample buffer for further protein analysis.

3. After electrophoresis through SDS-containing polyacrylamide gels (SDS–PAGE), transfer samples onto nitrocellulose membranes (Invitrogen iBlot Stack) and probe membranes with primary antibodies of interest.

4. Use HRP-conjugated secondary antibodies to probe primary antibodies and detect the signals using chemiluminescence (Pierce) (*see* **Note 5**).

4 Notes

1. If the chimeric mRNA-MS2 transcript is not enriched relative to MS2 transcript alone, consider transfecting larger molar ratios of plasmid expressing the chimeric mRNA-MS2 than the detection plasmid (pMS2-GST, pMS2-YFP); for example 5:1 or 10:1. With this modification, nonspecific binding of abundant cellular RNAs can be minimized.

2. If target mRNAs, lncRNAs, or miRNAs are not enriched in the pulldown or immunoprecipitation materials, reduce the time allowed for coating with antibody or for binding with beads. The amount of beads, antibody, and lysates can be reduced too.

3. If the construction of pMS2-mRNA plasmid is difficult, MS2 can be inserted before the mRNA. Challenges may arise from limited availability of restriction sites in the pMS2 plasmid and/or parental plasmid. Despite successful construction of the plasmid to express MS2-tagged mRNA, the chimeric transcript may not behave like the endogenous mRNA because of the 24 MS2 repeats at the 5′end. In this case, additional methods to pull down the fusion RNA can be considered by using biotinylated antisense DNA or RNA oligomers.

4. If plasmid DNA contamination is suspected during RNA precipitation, RT-minus PCR amplification can be performed. In order to remove residual plasmid, it may be necessary to increase the amount of DNase and/or lengthen the incubation time.

5. If the goal is to identify *trans*-factors interacting with a lncRNA (which also interact extensively with RBPs and ncRNAs [14–16]) or with a partial RNA fragment, the same general

procedure can be adopted by tagging the lncRNA or RNA fragment with MS2. Additional RNA pulldown or IP can be performed as explained above.

Acknowledgments

J.H.Y. and M.G. were supported by the National Institute on Aging Intramural Research Program, National Institutes of Health and start-up fund from Medical University of South Carolina.

References

1. Morris AR, Mukherjee N, Keene JD (2010) Systematic analysis of posttranscriptional gene expression. Wiley Interdiscip Rev Syst Biol Med 2:162–180

2. Moore MJ (2005) From birth to death: the complex lives of eukaryotic mRNAs. Science 309:1514–1518

3. Glisovic T, Bachorik JL, Yong J, Dreyfuss G (2008) RNA-binding proteins and post-transcriptional gene regulation. FEBS Lett 582:1977–1986

4. Bartel DP (2009) MicroRNAs: target recognition and regulatory functions. Cell 136:215–233

5. Fabian MR, Sonenberg N, Filipowicz W (2010) Regulation of mRNA translation and stability by microRNAs. Annu Rev Biochem 79:351–379

6. Yoon JH, Abdelmohsen K, Gorospe M (2013) Posttranscriptional gene regulation by long noncoding RNA. J Mol Biol 425:3723–3730

7. Lin RJ (ed) (2008) RNA-protein interaction protocols. Methods Mol Biol 488:5–7. ISBN: 978-1-58829-419-7

8. Ryder SP, Recht MI, Williamson JR (2008) Quantitative analysis of protein-RNA interactions by gel mobility shift. Methods Mol Biol 488:99–115

9. Abdelmohsen K, Tominaga K, Lee EK, Srikantan S, Kang MJ, Kim MM, Selimyan R, Martindale JL, Yang X, Carrier F, Zhan M, Becker KG, Gorospe M (2011) Enhanced translation by nucleolin via G-rich elements in coding and non-coding regions of target mRNAs. Nucleic Acids Res 39:8513–8530

10. König J, Zarnack K, Luscombe NM, Ule J (2012) Protein-RNA interactions: new genomic technologies and perspectives. Nat Rev Genet 13:77–83

11. Bertrand E et al (1998) Localization of ASH1 mRNA particles in living yeast. Mol Cell 2:437–445

12. Dutko JA et al (2005) Inhibition of a yeast LTR retrotransposon by human APOBEC3 cytidine deaminases. Curr Biol 15:661–666

13. Gong C, Maquat LE (2011) lncRNAs transactivate STAU1-mediated mRNA decay by duplexing with 3′ UTRs via Alu elements. Nature 470:284–288

14. Chu C, Qu K, Zhong FL, Artandi SE, Chang HY (2011) Genomic maps of long noncoding RNA occupancy reveal principles of RNA-chromatin interactions. Mol Cell 44:667–678

15. Quinn JJ et al (2014) Revealing long noncoding RNA architecture and functions using domain-specific chromatin isolation by RNA purification. Nat Biotechnol 32:933–940

16. Engreitz JM et al (2014) RNA-RNA interactions enable specific targeting of noncoding RNAs to nascent pre-mRNAs and chromatin sites. Cell 159:188–199

Chapter 3

Biotin–Streptavidin Affinity Purification of RNA–Protein Complexes Assembled In Vitro

Shuai Hou, Lei Shi, and Haixin Lei

Abstract

RNA–protein complexes are essential for the function of different RNAs, yet purification of specific RNA–protein complexes can be complicated and is a major obstacle in understanding the mechanism of regulatory RNAs. Here we present a protocol to purify RNA–protein complexes assembled in vitro based on biotin–streptavidin affinity. In vitro transcribed RNA is labeled with ^{32}P and biotin, ribonucleoprotein particles or RNPs are assembled by incubation of RNA in nuclear extract and fractionated using gel filtration, and RNP fractions are pooled for biotin–streptavidin affinity purification. The amount of RNA–protein complexes purified following this protocol is sufficient for mass spectrometry.

Key words Biotin, Streptavidin, Affinity purification, RNA–protein complexes

1 Introduction

Recent years have witnessed rapid progresses in RNA research, and a major finding is the emerging of large quantity of long noncoding RNAs, but their functions are still largely unknown. Since RNA–protein interactions are the fundamental bases for RNA functions, purifying RNA–protein complexes and revealing the main components in the complexes would provide deep insights into the mechanisms on how these RNAs achieve their regulatory roles [1, 2].

However, purification of RNA–protein complexes formed in vivo or assembled in vitro remains to be a daunting technical challenge in many labs. Major obstacles encountered are degradation of RNAs, optimization of RNP assembly conditions, effective separation of the RNP from nonspecific binding, and the amount of RNPs available for downstream analysis. Current strategies in RNA–protein complex purification include biotin–streptavidin system [3], MS2 coat protein and modified MS2 hairpin [4], PP7 coat protein and PP7 hairpin, aptamer streptoTag and streptomycin [5], or S1m aptamer and streptavidin [6].

Ren-Jang Lin (ed.), *RNA-Protein Complexes and Interactions: Methods and Protocols*, Methods in Molecular Biology, vol. 1421, DOI 10.1007/978-1-4939-3591-8_3, © Springer Science+Business Media New York 2016

Here we describe the most traditional yet effective system using biotin and streptavidin. Biotin ($C_{10}H_{16}N_2O_3S$, M.W. 244.31) is a water soluble vitamin, also known as vitamin B_7, vitamin H, coenzyme R or biopeiderm, whereas streptavidin is a protein of 52.8 kD purified from the bacterium *Streptomyces avidinii*. Tetramers of streptavidin have extremely high affinity to biotin with the dissociation constant of 10^{-14} mol/L, which is one of the strongest non-covalent interactions found in nature so far. The high binding affinity between biotin and streptavidin is due to extensive network of hydrogen bonds, hydrophobic interactions, conformation complementarity and stabilization.

The biotin–streptavidin system is widely used in protein purification [7] and nano-biotechnology as it is resistant to many harsh conditions such as denaturants, detergents, organic solvents, proteolytic enzymes, extreme pH and temperature. In addition, uniform labeling is used instead of end labeling for biotin incorporation in this protocol, which would enhance the RNA recovery efficiency. Finally, the small size of biotin may minimize the interference of the biological activity of labeled substrate. For instance, pre-mRNA labeled with optimized amount of biotin could be efficiently spliced in vitro.

This protocol requires a minimum of 4 days, which is seemingly time and labor consuming compared to some fast protocols. However, the amount of RNP purified following the protocol is considerably higher and is quite sufficient for mass spectrometry and other downstream analysis.

2   Materials (*See* Note 1)

2.1   *In Vitro Transcription*

1. 10xNTPs: 4 mM ATP and CTP; 2 mM UTP and GTP.
2. Cap analog: 20 mM (New England Biolabs).
3. Biotin-16-UTP: dilute to 0.3 mM (Roche).
4. UTP-[α-^{32}P] (Perkin Elmer).
5. RNasin Inhibitor: 40 u/µl (Promega).
6. T7 RNA polymerase: 50 u/µl (New England Biolabs).
7. 5x Transcription buffer: Mix 2 ml of 1 M Tris-HCl 7.6, 300 µl of 1 M MgCl$_2$, 1 ml of 100 mM spermidine, and autoclaved milli-Q water for a total volume of 10 ml; aliquot after filter with a syringe filter (*see* **Note 2**).

2.2   *RNA Purification and Gel Separation*

1. RQ1 RNase-free DNase (Promega).
2. Glycogen (Roche).
3. Phenol–chloroform–IAA (25:24:1, pH 6.6)(Ambion).
4. 5 M ammonium acetate: dissolve 19.25 g in water and bring the volume to 50 ml, filter.

5. 100 % ethanol.

6. 70 % ethanol.

7. Formamide dye: mix 0.4 ml 0.5 M EDTA, 0.8 ml 2.5 % xylene cyanol, 0.8 ml 2.5 % bromophenol blue, 16 ml formamide, 2 ml sterile H_2O; aliquot and store at –20 °C.

8. 6.5 % denaturing PAGE gel solution: combine 81.25 ml 40 % acryl–bis (29:1), 215 g urea, 50 ml 10x TBE, and add H_2O to 500 ml. Filter and store at 4 °C.

9. RNA ladder: for example, φX174 DNA/HinfI dephosphory-lated markers (Promega).

10. ATP-[γ-^{32}P] (Perkin Elmer).

11. TEMED(Sigma).

12. 10 % ammonium persulfate (APS): dissolve 1 g in 10 ml H_2O, filter, aliquot and store at –20 °C.

13. 3 MM Whatman filter paper.

14. Plastic membrane (Saran Wraps).

2.3 In Vitro RNP Assembly

1. 12.5 mM ATP: add 250 μl 0.5 M ATP to 9.75 ml water.

2. 0.5 M Creatine Phosphate diTris salt (Sigma).

3. 80 mM $MgCl_2$: add 1.6 ml 1 M $MgCl_2$ to 18.4 ml water.

4. RNP assembly buffer: Mix 1 ml 1 M HEPES pH 7.6, 1.65 ml 3 M KCl, and add water to 50 ml, filter.

5. Nuclear extract (*see* **Note 3**).

2.4 Fractionation of the RNP by Gel Filtration

1. Sephacryl S-500 HR, store at 4 °C(GE Healthcare).

2. 10xColumn buffer: 200 mM HEPES pH 7.6, 600 mM KCl, 25 mM EDTA, 1 % Triton X-100, store at 4 °C.

3. 1xColumn buffer: dilute 100 ml 10x column buffer in 890 ml water, then add 10 ml freshly made 10 % sodium azide, store in the cold room.

4. 1.5 ml capless tube.

5. 3 MM filter paper.

6. Saran membrane.

2.5 Biotin–Streptavidin Affinity Purification

1. Low salt buffer: Mix 10 ml HEPES pH 7.6, 25 ml 3 M KCl, 2.5 ml 0.5 M EDTA, 500 μl Triton X-100, 1.25 ml 2 M DTT, and add water to 100 ml. Filter and store at 4 °C.

2. 3 M KCL: dissolve 112 g in H_2O and add H_2O to 500 ml, filter and store at 4 °C.

3. Streptavidin agarose resin (Thermal Scientific).

4. Protease inhibitor, EDTA free: dissolve 1 mini-tablet in 500 μl low salt buffer (Roche).

5. 4x stacking gel buffer: dissolve 30.3 g Tris-base in water, add 20 ml 10 % SDS, adjust pH to 6.8, bring up to 500 ml with water, filter and store at 4 °C.

6. PGB (Protein gel buffer) : Mix 12.5 ml 4x stacking gel buffer, 10 ml glycerol, 20 ml 10 % SDS, 0.5 ml 1 % bromophenol blue, and add water to 45 ml; aliquot 990 μl in each tube and store at –20 °C. Add 10 μl 2 M DTT before use.

7. Proteinase K: resuspend 100 mg (Roche) in 10 ml autoclaved milli-Q water, aliquot and store at –20 °C.

8. 2x PK buffer: Mix 10 ml 1 M Tris pH 8.0, 2.5 ml 0.5 M EDTA, 3 ml 5 M NaCl, 10 ml 10 % SDS, and add H_2O to 50 ml; filter and store at room temperature. Warm up in 37 °C water bath before use if precipitates are observed.

2.6 Sample Preparation for Mass Spectrometry

1. NuPAGE Novex 4–12 % Bis-Tris protein gels (Life Technologies).

2. The NuPAGE MOPS SDS buffer (Life Technologies).

3. NuPAGE antioxidant (Life Technologies).

4. SilverQuest staining kit (Life Technologies).

5. Trichloroacetic acid (TCA) (Sigma).

6. Acetone (HPLC grade).

3   Methods (*See* Note 4)

3.1 In Vitro Transcription

1. Assemble the following reaction at room temp:

DNA template (*see* **Note 5**)	1.0 ug
10x NTPs	2.5 μl
1 %Triton	2.5 μl
20 mM Cap analog	2.5 μl
200 mM DTT	1.5 μl
0.3 mM Biotin-16-UTP	0.8 μl
RNasin Inhibitor	0.5 μl
5xTranscription buffer	5.0 μl
UTP-[α-³²P]	1.0 μl
T7 RNA polymerase	1.5 μl

Add H_2O to a total volume of 25 μl.

2. Incubate at 37 °C in a water bath for 1 ~ 2 h.

3. Add 1 μl RQ1 RNase-free DNase, incubate for 10 min at 37 °C.

4. Add H_2O to a final of 200 μl, remove 2 μl for measuring radioactivity.

3.2 RNA Purification and Gel Separation

1. Add equal volume of phenol/chloroform, vortex and then spin at 13,500 g (12,000 rpm) for 5 min, transfer 160 μl supernatant to another Eppendorf tube.

2. Add 2 μl 20 mg/ml glycogen, 200 μl 5 M Ammonium Acetate and 950 μl 100 % ethanol, mix and centrifuge at 13,500 g (12,000 rpm) for 20 min.

3. Wash the pellet once with 70 % ethanol, resuspend in 20 μl H_2O, remove 1 μl for counting.

4. Count the samples aliquoted before and after purification using a liquid scintillation counter for 2 min, calculate the incorporation rate and the concentration of RNA (see **Note 6**).

5. Aliquot 1 μl of the RNA sample and dilute to 1 ng/μl, then mix 1–3 μl of the RNA dilution with formamide dye, incubate in 70–80 °C water for 10 min, vortex, spin down, and load the sample(s) on 6.5 % denaturing PAGE gel.

6. After gel separation, peel off the gel using filter paper and cover with saran membrane, dry the gel with a vacuum drier. Expose the gel to X-ray film or phosphoimager screen (Fig. 1) (see **Note 7**).

3.3 In Vitro RNP Assembly

1. In each of the four 2.0 ml Eppendorf tubes (see **Note 8**), add:

12.5 mM ATP	20 μl
0.5 M Creatine phosphate	20 μl
80 mM $MgCl_2$	20 μl
Dilution buffer	150 μl
Nuclear extract	150 μl
RNA	500 ng
H_2O	140 μl
Total	500 μl

2. Mix well and incubate at 30 °C for 30–60 min, put on ice and then load on the gel filtration column.

3.4 Fractionation of the RNP by Gel Filtration

1. Wash the gel filtration column thoroughly, soak Sephacryl S-500 HR resin in the column buffer (see **Note 9**).

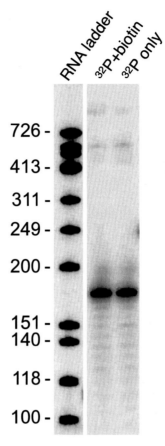

Fig. 1 In vitro T7 transcripts separated on 6.5 % denaturing PAGE. ^{32}P + biotin: transcripts were double labeled with ^{32}P and biotin; ^{32}P: transcripts were labeled with ^{32}P only

2. Attach stop-cock and tubing to the column outlet and put the column on a ring stand, inject milli-Q water into the column from the tubing using syringe to remove air bubbles.

3. In the cold room, attach the column to peristaltic pump, pour the resin into the column against a glass rod steadily.

4. Turn on the pump to start packing and keep adding the resin before it is fully packed (*see* **Note 10**).

5. When the column is packed to about 2.5 in. from the top, swirl the rod above the resin to create a completely flat surface for loading.

6. Mix 250 μl nuclear extract with 250 μl dilution buffer, block the column by running the mixture.

7. Load the assembled RNP complexes onto the column very carefully and collect fractions using capless tubes (*see* **Note 11**).

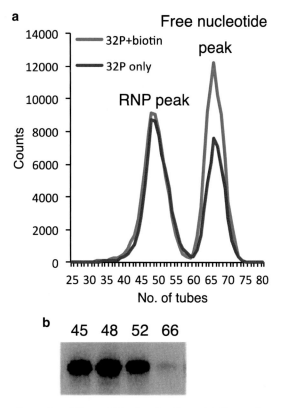

Fig. 2 Chromatography of the gel filtration fractions. (**a**) Chromatography showing the RNP peak and free nucleotide peak; (**b**) RNA purified from an aliquot of the designated fraction

8. Remove 50 µl from fractions 25 to 80 for counting and draw the chromatography (Fig. 2a) (*see* **Note 12**).

9. Remove an aliquot from randomly selected fractions of both peaks, add 5 µl proteinase K, 100 µl 2xPK buffer and H_2O to the final volume 200 µl, vortex and then incubate at 37 °C for 10 min.

10. Add 2 µl glycogen, 200 µl RNA phenol, vortex and then spin for 5 min at 13,500g (12,000 rpm) at room temperature.

11. Transfer 160 µl supernatant to another tube, add 600 µl 100 % ethanol, spin at room temperature for 20 min after vortex.

12. Wash once with 70 % ethanol, air dry briefly and add 10 µl of formamide dye, incubate in 70–80 °C water for 10 min, vortex and spin down, samples are ready for loading on RNA gel.

13. Run the gel and expose, RNA should be observed in samples taken from the RNP peak but not in samples from the free nucleotide peak (Fig. 2b).

3.5 Biotin–Streptavidin Affinity Purification

1. Pool the fractions from the RNP peak together and adjust the salt concentration to 150 mM with 3 M KCL, transfer to 15 ml tube (*see* **Note 13**).

2. Add 100–150 μl of streptavidin agarose resin and 1/100 V of protease inhibitor, Rotate at 4 °C overnight.

   The amount of streptavidin agarose resin added in the affinity purification should be optimized. To calculate the recovery efficiency, take equal aliquots from pooled fractions before streptavidin agarose resin is added and the supernatant after biotin–streptavidin binding, purify RNA and run RNA gel, quantify the bands.

   Recovery efficiency = 1-(Band intensity after/band intensity before)

   For example, the recovery efficiency before and after optimization is 43 % (1–57 % = 43 %) vs 84 % (1–16 % = 84 %), whereas there is almost no recovery without biotin (Fig. 3).

3. Wash with 10 ml 150 mM low salt buffer for four times, rotate at 4 °C for 10 min for each wash.

4. Transfer the agarose resin to 1.5 ml Eppendorf tube, Add 150 μl PGB and rotate at room temp for 10 min, incubate in hot water (70–80 °C) for 2 min.

5. Spin down the resin at 2000 *g* (4500 rpm), carefully transfer the supernatant to another tube and spin again, transfer the supernatant to the same tube. The supernatant is ready for gradient gel after boiling or mass spectrometry after precipitation (*see* **Note 14**).

3.6 Sample Preparation for Mass Spectrometry

1. Aliquot about 10 % elute, boiling and run gradient gel, stain the gel with silver staining kit (Fig. 4) (*see* **Note 15**).

2. For mass spectrometry, dilute the elute to 1 ml and add 100 μl 100 % TCA, vortex for 15 s, place on ice for 15 min (*see* **Note 16**).

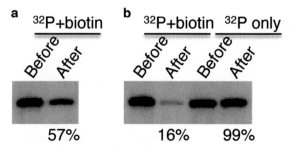

Fig. 3 Opitimization of the amount of streptavidin agarose resin enhances RNA recovery efficiency. (**a**) RNA recovery efficiency before optimization; (**b**) RNA recovery efficiency with optimized amount of agarose resin

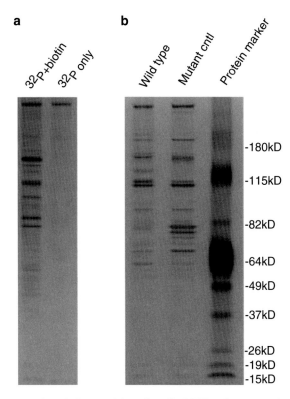

Fig. 4 Two examples of silver staining of purified RNPs after separation on gradient SDS-PAGE. (**a**) Silver staining showing RNP was purified only from biotin-labeled sample; (**b**) RNPs assembled on wild type transcripts and mutant control

3. Spin at full speed in a table-top microcentrifuge for 15 min, 4 °C, remove the supernatant.

4. Wash pellet with 100 μl ice-cold acetone twice, spin for 30 min each time.

5. Remove acetone and air dry the pellet, samples are ready for mass spectrometry (*see* **Note 17**).

3.7 Day-to-day Task   Day 1:

Pack the column

In vitro transcription

RNA purification

Run the RNA gel

Day 2:

In vitro RNP assembly

Gel filtration fractionation

Day 3–4:

Fraction counting/chromatography

RNA purification from selected fractions

Affinity purification

Day 4:

Run the gradient gel

Silver staining

TCA precipitation.

4 Notes

1. The supplier information is provided for convenience rather than a requirement.

2. All the reagents and chemicals should be of the highest grade. We routinely use autoclaved milli-Q water instead of DEPC-treated water. All the reagents should be stored at –20 °C.

3. Store nuclear extract at –80 °C and aliquot all the other reagents and store at –20 °C. Nuclear extract can be from bulk or small scale preparation. For small scale nuclear extract preparation, see the reference by Folco EG, et al. 2012 [8]. For RNPs present in the cytoplasm, replace nuclear extract with S100 [9].

4. Before the experiments, keep in mind that solution I from the miniprep kit is the major source for RNase contamination since large amount of RNase A is added to the solution. Other sources for RNase contamination include dust and hands. Therefore, DO NOT perform any RNA-related experiments on the same bench where miniprep is carried out and always keep the RNA bench CLEAN!

5. DNA template could be linearized plasmid or PCR products containing T7 promoter sequence. Transcription buffer should be added last to avoid precipitates. Follow instructions on working with isotope and have proper protections. For negative control, omit Biotin-16-UTP.

6. Calculation the incorporation rate and the concentration of RNA:
   Incorporation rate (%) = Counts after × 2 × 10/(counts before)
   For example, if the counts before purification is 12,500 and counts after purification is 60,000, then the incorporation rate is 96 %.
   Incorporation rate (%) = 60,000 × 2 × 10/12,500 = 96%
   Concentration (ng/µl) = Incorporation rate × 1000 × 6.25/20

(for two reactions, replace 6.25 by 12.5; for four reactions, replace 6.25 by 25)

For example, if the incorporation rate is 96 %, then the concentration of the RNA is 300 ng/μl.

Concentration $(ng/\mu l) = 96\% \times 1000 \times 6.25/20 = 300$ ng/μl

7. It is critical to run the T7 transcripts on RNA gel since this will provide the information about the length and the quality of the transcripts and whether there is any unspecific transcript. The RNA ladder should be end labeled using ATP-$[\gamma$-$^{32}P]$.

8. You may use a 50 ml Falcon tube if you prefer to do single tube incubation. The RNP assembly conditions may need to be optimized.

9. A 0.75 m × 1.5 cm column is recommended, the bed volume is ~90 ml. The column should be clean and RNase free; soak and wash with 0.5 % sodium azide if necessary. You may need to use other Sephacryl resin depending on the size of the RNP assembled for best resolution.

10. Do not let the packed resin dry out during packing! It will ruin the column.

11. Protect yourselves properly during loading and shield off the fractionation area since it is radioactive! The fractionation speed should be adjusted so that fractions of 1 ml are collected in each tube.

12. There is no need to count fractions 1 to 24 since they do not contain any RNA. The chromatography should be a smooth curve, if jigsaw peaks are observed, the column may be ruined or the order of the fractions for counting is mixed up.

13. The concentration of KCl in fractions is 60 mM, which should be taken into account when adjusting the salt concentration. The salt concentration can be adjusted up to 250 mM if more stringent binding and wash condition is preferred.

14. Keep the samples warm in winter to avoid precipitates. Be very careful not to transfer any agarose resin after the second spin. It may be helpful not to transfer all the supernatant at this point.

15. It is critical and essential to perform the control that only labeled with ^{32}P but without bibotin (^{32}P only) since it will help to determine whether the affinity purification is successful or not. As showed in Fig. 4a, this lane should be blank or only faint bands could be observed.

16. Add TCA directly without dilution of the elute may lead to significant amount of cloudy precipitates which will ruin the experiment.

17. You should NOT be able to see the pellet or only tiny pellet can be seen. Large pellet means bad sample preparation and you may not get results from mass spectrometry.

Acknowledgements

This work was supported by Innovative Research Team in University, Ministry of Education, China (No. IRT13049), NSFC Grant 81472491 and Liaoning Pandeng Scholar Program.

References

1. Zhou Z, Licklider LJ, Gygi SP, Reed R (2002) Comprehensive proteomic analysis of the human spliceosome. Nature 419(6903):182–185

2. Hogg JR, Collins K (2007) RNA-based affinity purification reveals 7SK RNPs with distinct composition and regulation. RNA 13:868–880

3. Lei H, Zhai B, Yin SY, Gygi S, Reed R (2013) Evidence that a consensus element found in naturally intronless mRNAs promotes mRNA export. Nucleic Acids Res 41(4):2517–2525

4. Zhou Z, Reed R (2003) Purification of functional RNA-protein complexes using MS2-MBP. Curr Protoc Mol Biol. Chapter 27: Unit 27.3

5. Bachler M, Schroeder R, von Ahsen U (1999) StreptoTag: a novel method for the isolation of RNA binding proteins. RNA 5(11):1509–1516

6. Leppek K, Stoecklin G (2014) An optimized streptavidin-binding RNA aptamer for purification of ribonucleoprotein complexes identifies novel ARE-binding proteins. Nucleic Acids Res 42(2), e13

7. Chodosh A (2001) Purifiation of DNA-binding proteins using biotin/streptavidin affinity systems. Curr Protoc Mol Biol. Chapter 12: Unit 12.6

8. Folco EG, Lei H, Hsu JL, Reed R (2012) Small-scale nuclear extracts for functional assays of gene-expression machineries. J Vis Exp (64). pii: 4140

9. Abmayr SM, Reed R, Maniatis T (1988) Identification of a functional mammalian spliceosome containing unspliced pre-mRNA. Proc Natl Acad Sci U S A 85(19):7216–7220

Chapter 4

Detecting RNA–Protein Interaction Using End-Labeled Biotinylated RNA Oligonucleotides and Immunoblotting

Xuexiu Zheng, Sunghee Cho, Heegyum Moon, Tiing Jen Loh, Ha Na Jang, and Haihong Shen

Abstract

RNA–protein interaction can be detected by RNA pull-down and immunoblotting methods. Here, we describe a method to detect RNA–protein interaction using RNA pull down and to identify the proteins that are pulled-down by the RNA using immunoblotting. In this protocol, RNAs with specific sequences are biotinylated and immobilized onto Streptavidin beads, which are then used to pull down interacting proteins from cellular extracts. The presence of a specific protein is subsequently verified by SDS- polyacrylamide gel electrophoresis and immunoblotting with antibodies. Interactions between the SMN RNA and the PSF protein and between the caspase-2 RNA and the SRSF3 protein (SRp20) in nuclear extract prepared from HeLa cells are illustrated as examples.

Key words Spliceosome, RNA binding protein, RNA–protein interaction, Biotinylation, Streptavidin, RNA pull-down, Immunoblotting

1 Introduction

The method for detecting RNA–protein interaction by RNA pull-down and immunoblotting is one of many methods in studying RNA–protein interaction [1–4]. The main principle of the method described here is to use streptavidin agarose beads to pull-down the biotin-labeled RNA and its associated proteins under the specified conditions [5, 6]. The streptavidin agarose can be used to effectively and specifically pull down biotinylated RNA–protein complexes because of the strong affinity between biotin and streptavidin [7–10]. Here we describe how to pull-down spliceosomal RNA-binding proteins using biotin labeled RNA and streptavidin agarose beads, and how to detect the RNA interacting proteins by immunoblotting.

Ren-Jang Lin (ed.), *RNA-Protein Complexes and Interactions: Methods and Protocols*, Methods in Molecular Biology, vol. 1421, DOI 10.1007/978-1-4939-3591-8_4, © Springer Science+Business Media New York 2016

2   Materials

All the reagents applied in this experimental protocol are molecular biology grade (unless indicated otherwise), and dissolved in ultrapure water (unless indicated) which is obtained through three steps of purification: firstly, impurity particle depletion procedure with 10 μm and 5 μm filters, secondly, impurity compounds depletion procedure through carbon filter, thirdly and finally, deionizing procedure through reverse osmosis system. The ultrapure water has an electrical resistivity of 18.2 million ohm-cm at 25 °C. All reagent solutions are prepared and stored in stock at room temperature (unless indicated otherwise). The diethylpyrocarbonate (DEPC) treated water is not applied in the experiments unless indicated otherwise.

All the procedures in RNA pull down protocol are carried out at 4 °C unless otherwise specified, and all the solutions applied to this experiment are kept 4 °C overnight before the experiment. All the procedures in protein immunoblotting are carried out at room temperature otherwise specified.

2.1   RNA Pull-Down

1. Biotin-labeled RNAs of 10 nucleotides in length, the biotin residue is at the 5′ end (obtained commercially; ours are from the Bioneer Corporation (Fig. 1).

2. Streptavidin agarose washing buffer: 20 mM Tris–HCl, pH 7.5, 100 mM KCl, 2 mM EDTA, 0.5 mM DTT, and 0.5 mM PMSF.

3. RNA–streptavidin interaction buffer: 20 mM Tris–HCl, pH 7.5, 300 mM KCl, 0.2 mM EDTA, 0.5 mM DTT, and 0.5 mM PMSF.

4. Strepavidin agarose beads.

5. Nuclear extracts prepared from HeLa cells as the source of the RNA-binding proteins.

SMN-WT:	biotin-AAGAAGGAAG
SMN mutant:	biotin-AAGAAUUAAG
Casp2 WT:	biotin-UUCUUCAUCC
Casp2 mutant:	biotin-UGAAGCGUCC

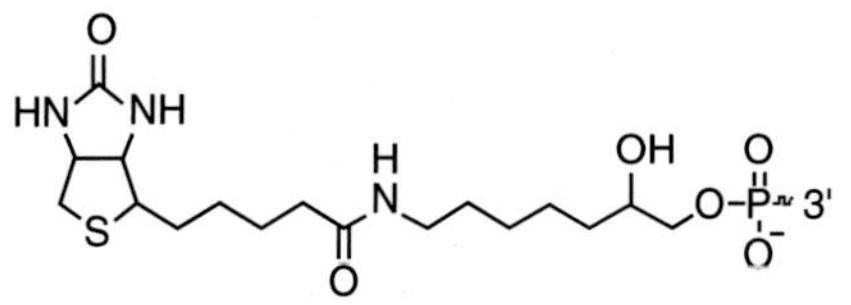

Fig. 1 Synthetic, biotin-labeled RNA. The sequences of the four 5′ end-biotin-labeled RNA decamers are shown. SMN is Survival of Motor Neuron, Casp2 is Caspase-2, and WT is wild-type. These sequences are from the messenger RNA of the respective gene. The structure of the biotin moiety is depicted

2.2 SDS-Polyacrylamide Gel Electrophoresis (SDS-PAGE) and Immunoblotting

1. 30 % acrylamide with 29:1 ratio of acrylamide and bis-acrylamide.

2. 1.5 M Tris–HCl, pH 8.8.

3. 10 % ammonium persulfate (APS).

4. *N,N,N′,N′*-tetramethylethane-1,2-diamine (TEMED).

5. 5× protein loading buffer: 60 mM Tris–HCl (pH 6.8), 25 % glycerol, 2 % SDS, 14.4 mM β-mercaptoethanol, 0.1 % bromophenol blue.

6. SDS PAGE gel running buffer: 25 mM tris base, 192 mM glycine, 0.1 % SDS (*see* **Note 1**).

7. SDS PAGE gel transfer buffer: 25 mM tris base, 192 mM glycine.

8. Nitrocellulose membrane blocking buffer: 5 % skim milk, 10 mM Tris–HCl, pH 7.5, 150 mM NaCl, 0.1 % Tween 20.

9. Immunoblotting buffer: 5 % skim milk, 10 mM Tris–HCl, pH 7.5, 150 mM NaCl, 0.1 % Tween 20.

10. Antibodies against the RNA-binding proteins (the antibody against PSF protein is from Santa Cruz and the antibody against SRSF3 protein is from Invitrogen).

11. ECL Western Blotting substrate (Thermo Scientific Pierce).

12. X-ray film (FUJI Incorporated).

3 Methods

3.1 Preparation of Biotin Labeled RNA

RNA can be obtained through two different ways, chemical synthesis or in vitro transcription. The shorter RNAs are practically feasible to be synthesized chemically while longer ones are suitable to be transcribed from a DNA template by in vitro transcription. There are also two different ways to label RNA with biotin, end labeling and body incorporation. In this protocol, end labeled biotin RNAs are used.

3.2 Preparation of Streptavidin Agarose Beads

1. Commercially obtained 50 % streptavidin agarose beads are mixed well by shaking in a cap-sealed tube.

2. 5 μl of streptavidin agarose beads is taken out into 1.5 ml Eppendorf (EP) tubes using end-cut yellow tip (*see* **Note 2**).

3. 500 μl ice-chilled streptavidin washing buffer is added into streptavidin agarose bead containing 1.5 ml EP tubes, and pipetting the slurry 4–5 times with end-cut blue tip.

4. The EP tubes are centrifuged at 5000 ($470 \times g$) revolutions per minute (RPM) for 1 min at 4 °C in a bench top centrifuge.

5. The supernatant is discarded using gentle pipetting with blue tip (*see* **Note 3**).

3.3 Binding Biotin Labeled RNA onto Streptavidin Agarose Beads

1. The EP tubes containing prepared streptavidin agarose are refilled with 500 μl of streptavidin washing buffer and gently pipetted 4–5 times with end cut blue tip.

2. 1 μl (100 pmol) of biotin labeled RNA is applied to above EP tubes except one of these EP tubes as negative control (*see* **Note 4**).

3. These EP tubes are incubated at 4 °C for 1–4 h under 100 RPM rotation in a rotator (*see* **Note 5**).

4. The EP tubes are then centrifuged at 5000 RPM for 1 min at 4 °C to precipitate the biotin labeled RNA bound streptavidin agarose bead.

5. The supernatant (streptavidin washing buffer) from EP tubes above are discarded with pipetting gently with blue tip (*see* **Note 3**).

6. 500 μl of the RNA–streptavidin interaction buffer is applied to the EP tubes with pellets of the biotin labeled RNA bound streptavidin agarose beads, and suspend the pellets with pipetting for 4–5 times with end cut blue tips on (*see* **Note 6**).

7. The EP tubes with biotin labeled RNA bound streptavidin agarose beads are centrifuged at 5000 RPM for 1 min at 4 °C in bench top centrifuge.

8. The supernatant of EP tubes above are discarded with gentle pipetting with blue tip.

9. The processes from **steps 6–8** listed above are repeated two more times.

3.4 Binding of Proteins to the Immobilized Biotin Labeled RNA

1. 500 μl of the RNA–streptavidin interaction buffer is applied to the EP tubes containing biotin labeled RNA bound streptavidin agarose bead, and suspend the agarose slurry with gentle pipetting with end cut blue tip for 4–5 times.

2. 100 μl of nuclear extract with a protein concentration of 3–5 μg/μl from HeLa cells are applied to each of above EP tubes.

3. The EP tubes with nuclear extract and biotin labeled RNA bound streptavidin slurry are kept at 4 °C overnight (about 12 h) under 100 RPM rotation in a rotator.

4. The EP tubes with nuclear extract and biotin labeled RNA bound streptavidin agarose bead are centrifuged at 5000 RPM for 1 min at 4 °C in bench top centrifuge.

5. The supernatant of EP tubes above are discarded with gentle pipetting with blue tip.

6. 500 μl of the RNA–streptavidin agarose interaction buffer is applied to the EP tubes with pellets of the nuclear extract and biotin labeled RNA bound streptavidin agarose bead, and suspend the pellets with pipetting for 4–5 times with end cut blue tips on (*see* **Note 6**).

7. The EP tubes with nuclear extract and biotin labeled RNA bound streptavidin agarose bead are centrifuged at 5000 RPM for 1 min at 4 °C in bench top centrifuge.

8. The supernatant of EP tubes above are discarded with gentle pipetting with blue tip.

9. The processes from **steps 6–8** listed above are repeated two more times.

3.5 Elution of the Bound Proteins from the Streptavidin Agarose Beads

1. 16 μl of 1× protein loading dye are applied into each EP tube obtained from above procedure, in which there is the complex of RNA binding proteins and biotin labeled RNA bound streptavidin slurry.

2. The pellet in each EP tube containing 1× protein loading dye, RNA bind proteins and biotin labeled RNA bound streptavidin slurry is suspended with gentle pipetting for 4–5 times with yellow tip (*see* **Note 7**).

3. The EP tubes are heated in heat block at 80 °C for 5 min (*see* **Note 8**).

4. The EP tubes are centrifuged at 11,000 RPM for 10 min at room temperature.

5. The supernatant (about 10 μl) of the EP tubes above is gently transfer to new fresh EP tubes with matched marker on the top of caps of EP tubes (*see* **Note 9**).

3.6 Analysis of the RNA Interacting Proteins by Immunoblotting

1. The mini gel plates are cleaned with 70 % ethanol and assembled with spacers according the manufacturer's instructions.

2. 12 ml of 10 % SDS-PAGE gel solution are prepared by adding 4 ml of 30 % acrylamide with 29:1 ratio of acrylamide and bis-acrylamide, 3 ml of 1.5 M Tris–HCl, pH 8.8, and 4.95 ml of double distilled H_2O. Just before pouring SDS-PAGE gel preparation solution into assembled mini gel plates, adding 50 μl of 10 % APS, 2.5 μl of TEMED, and mixing with a magnetic stirrer or shaking by hand (*see* **Notes 10** and **11**).

3. Immediately, the 10 % SDS PAGE gel preparation solution prepared above is poured into the space of pre-assembled mini gel plates followed by the comb is inserted, and keep at room temperature for about an hour to get gel formation completely. In our study, the 1.0 mm spacer and combs are preferred while this is completely case dependent (*see* **Note 12**).

4. The prepared mini gel plate is assembled to the gel running apparatus according the manufacturer's instructions followed by fulfilling running buffer to both anode side and cathode side space to have the electricity go through anode, running buffer, wells of gel, gel body, running buffer, and cathode.

5. Eluted 10 μl of the biotin labeled RNA bound proteins samples from Subheading 3.5 are applied to the SDS-PAGE gel and electrophoresed at 120 voltage (V) about 2 h until the bromophenol blue dye is at the bottom of the gel.

6. The mini gel is removed from mini gel plate and set into mini gel nitrocellulose transferring apparatus according the manufacturer's instruction.

7. The protein bands in mini gel are transferred into nitrocellulose membrane at constant electric current of 300 mA for about 1 h (*see* **Note 13**).

8. The nitrocellulose membrane is removed from transferring apparatus and appropriately trimmed before put into a container for blocking (*see* **Note 14**).

9. The nitrocellulose membrane is blocked in 10–20 ml blocking solution (*see* **Note 15**), for 1 h with gentle shaking at room temperature.

10. The blocking solution is exchanged with 10 ml fresh blocking plus 1 μl first antibody (*see* **Note 16**), and incubated at room temperature for another 1 h (*see* **Note 17**).

11. The first antibody incubating solution is discarded and the nitrocellulose membrane is washed with 20 ml TBS-T buffer for 5 min with gentle shaking at room temperature.

12. The nitrocellulose membrane is washed for 5 times by repeating above procedure.

13. The nitrocellulose membrane is then incubated with 10 ml of TBS-T plus 0.5 μl of HRP conjugated secondary antibody that recognize the C fragment of first antibody for 1 h at room temperature.

14. The secondary antibody incubating solution is discarded and the nitrocellulose membrane is washed with 20 ml TBS-T buffer for 5 min with gentle shaking at room temperature.

15. The nitrocellulose membrane is washed for 5 times by repeating above procedure.

16. The nitrocellulose membrane is then spread with 0.5 ml of mixture of ECL solution A and solution B according to the manufacturer's instruction for 1 min in the dark room.

17. The nitrocellulose membrane is taken out and put onto a transparent overhead projector sheet, and suck any leftover liquid from the edge of nitrocellulose membrane with a paper tower.

18. The nitrocellulose membrane is covered with one layer of wrap, and set into X-ray film cassette.

19. One sheet of X-ray film is exposed to the nitrocellulose membrane for 1–15 min until the optimized results are obtained. Figure 2 shows a couple of representative results from this protocol which have already been published in the journal of

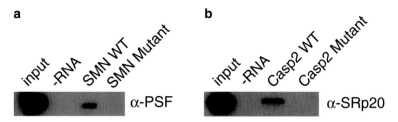

Fig. 2 Immunoblots of two RNA-binding proteins pulled down with biotinylated RNA. Nuclear extracts were incubated with biotinylated RNAs immobilized on Streptavidin agarose beads and the bound proteins were eluted and probed with specific antibodies: (**a**) an anti-PSF antibody used at a 1:5000 dilution [11] and (**b**) an anti-SRp20 antibody used at a 1:1000 dilution [12]. The PSF protein is only pulled down by the SMN wild-type RNA fragment, while the SRp20 protein is only pulled down by the Caspase-2 RNA fragment with the wild type sequence. The proteins were not pulled down with RNAs with a mutant sequence of without RNA on the beads. Both figures have already been published in the journal of Biochimica et Biophysica Acta [11, 12], and the permission documents for this reprint were obtained from the publishing company Elsevier

Biochimica et Biophysica Acta [11, 12]. The permission documents for the reprint here were obtained from the publishing company Elsevier ahead.

4   Notes

1. There are a couple of references that use different recipes, for example, some protocols suggest 250 mM glycine. We compared the results using different recipes and found no significant difference among different concentrations of glycine. We choose the 192 mM concentration of glycine because the time needed to dissolve is more reasonable and the higher concentration of glycine often needs heating during preparation [13, 14].

2. In order to collect the 50 % streptavidin agarose beads evenly, the end of yellow tip must be cut to enlarge the opening for passing streptavidin agarose beads, the number of EP tubes depends on the experimental design, in any case, one of the EP tubes must designated as negative control and no biotin labeled RNA will be added in this EP tube.

3. The supernatant should not be drained out fully, because the pellet is easy to suspend by any slight disturbs.

4. The molar ratio of biotin labeled RNA and streptavidin should be more than 2:1, more exactly, the streptavidin binding sites must be fully occupied by biotin labeled RNA, otherwise non-specific biotin containing proteins will contaminate the pull-down sample and irrelevant bands may show up in the final assay.

5. The duration of incubating time for making biotin labeled RNA bound streptavidin agarose beads depends on the length of biotin labeled RNA, if the biotin labeled RNA is less than 10 nucleotides, then the duration of incubating time is about 1 h; if the length of the RNA is more than 40 nucleotides, then the duration of incubating time is about 4 h.

6. The blue tips must be changed whenever they are used for EP tubes with different components, for example, the tubes with pellet of streptavidin, or the tubes with the pellet of biotin labeled RNA bound streptavidin.

7. The pellets shall not be suspended by vortexing, otherwise, there would be a lot of bubbles and the beads would stick to the side wall of EP tubes. Even with additional centrifugation, the bubbles would not disappear completely, which makes it difficult to separate the supernatant and the pellet cleanly.

8. The reason to set the heating temperature at 80 °C is that 80 °C is sufficient for denaturing most proteins in the presence of 1× protein loading buffer. Many references suggest boil protein samples at 100 °C in 1× protein loading dye before loading to SDS-polyacrylamide gel for electrophoresis. The protein denaturing efficacy may be improved at 100 °C, however, this will increase the risk of losing protein samples due to spilling or explosion [13–15].

9. The precipitates should never be disturbed when taking out the supernatant by pipetting. Otherwise the immunoblotted bands will be not sharp, especially when the quality of the antibody you work with is not high.

10. The percentage of SDS-PAGE gel depends on the protein bands you want to separate. The relationships between protein molecular weight and SDS-PAGE gel percentage have been described [13]. There are two different protocols to prepare SDS-PAGE gel, one of them has the separating gel at the bottom part and the stacking gel on the upper part, another one has no stacking gel at all, the whole gel are composed only of separating gel. We compared results from these two different protocols and found no significant differences. Therefore, it is not necessary to use SDS-PAGE gels with an upper stacking gel.

11. There is another difference in SDS-PAGE gel preparation recipes, one is with 0.1 % SDS in SDS-PAGE gel preparation solution and another one is without 0.1 % SDS. We have also compared the results from these two different recipes and found no significant differences. The reason is probably that SDS runs faster compared with proteins so that the PAGE gel becomes saturated with 0.1 % SDS before the proteins migrate into SDS-PAGE gel from the loading well. The handling of SDS-PAGE gel preparation without 0.1 % SDS is much sim-

pler because it is less likely to form bubbles during SDS-PAGE gel preparation.

12. The comb width and well numbers are decided according to the volume of loading sample. 10-well, 1 mm thickness comb can be used for loading about 25–30 μl of sample, while 15-well ones about 15–20 μl of sample. 10-well, 1.5 mm thickness comb can be applied to loading about 35–40 μl of sample, while 15-well ones about 25–30 μl of sample [13].

13. The efficacy of protein transfer in this step must be checked with the transfer efficacy of pre-stained protein size marker. Generally, the protein transfer efficacy depends on mini gel percentage, molecular weight of protein, and the electric current applied. After transfer protein bands from mini gel to nitrocellulose, the mini gel must be carefully checked to see if there are pre-stained protein size markers left in mini gel or not; the percentage of mini gel or the transferring electric current may need to be adjusted accordingly.

14. The membrane must be kept in the container in a position in which protein transferred side is always on the top side.

15. There are at least three different recipes for blocking solution, one 3–5 % skim milk in TBS-T buffer, another one is 1 % BSA in TBS-T, and the last one is the combination of the first two, 3–5 % skim milk plus 1 % BSA in TBS-T. The efficacy of these different blocking solutions for immunoblotting shall be checked for each application.

16. The fold of dilution to use for the first antibody can range from 100× to 20,000×; it must be tested as soon as the antibody is obtained commercially, even there is a recommended dilution.

17. The incubation condition of nitrocellulose membrane with first antibody is case dependent. Some antibodies bind specifically to the target protein (antigen) with very little non-specific binding, but some others could have a lot of non-specific bands. To decrease non-specific bands, one can try various blocking solutions—3 % skim milk, 1 % BSA, or the combination of the two. One can also apply the blocking solution in all first antibody incubating solution.

References

1. Ascano M, Hafner M, Cekan P, Gerstberger S, Tuschl T (2012) Identification of RNA-protein interaction networks using PAR-CLIP. Wiley Interdiscip Rev RNA 3(2):159–177. doi:10.1002/wrna.1103

2. Shen H, Zheng X, Shen J, Zhang L, Zhao R, Green MR (2008) Distinct activities of the DExD/H-box splicing factor hUAP56 facilitate stepwise assembly of the spliceosome. Genes Dev 22(13):1796–1803

3. Kyung Oh H, Lee E, Jang HN, Lee J, Moon H, Sheng Z, Jun Y, Loh TJ, Cho S, Zhou J (2013) hnRNP A1 contacts exon 5 to promote exon 6 inclusion of apoptotic Fas gene. Apoptosis 18(7):825–835

4. Moon H, Cho S, Loh TJ, Oh HK, Jang HN, Zhou J, Kwon YS, Liao DJ, Jun Y, Eom S, Ghigna C, Biamonti G, Green MR, Zheng X, Shen H (2014) SRSF2 promotes splicing and transcription of exon 11 included isoform in

Ron proto-oncogene. Biochim Biophys Acta 1839(11):1132–1140. doi:10.1016/j.bbagrm.2014.09.003

5. Moritz B, Wahle E (2014) Simple methods for the 3′ biotinylation of RNA. RNA 20(3):421–427. doi:10.1261/rna.042986.113

6. Shi M, Han W, Spivack SD (2013) A quantitative method to identify microRNAs targeting a messenger RNA using a 3′UTR RNA affinity technique. Anal Biochem 443(1):1–12. doi:10.1016/j.ab.2013.08.002

7. Li H, Chen W, Zhou Y, Abidi P, Sharpe O, Robinson WH, Kraemer FB, Liu J (2009) Identification of mRNA binding proteins that regulate the stability of LDL receptor mRNA through AU-rich elements. J Lipid Res 50(5):820–831. doi:10.1194/jlr.M800375-JLR200

8. Lopez de Silanes I, Galban S, Martindale JL, Yang X, Mazan-Mamczarz K, Indig FE, Falco G, Zhan M, Gorospe M (2005) Identification and functional outcome of mRNAs associated with RNA-binding protein TIA-1. Mol Cell Biol 25(21):9520–9531. doi:10.1128/MCB.25.21.9520-9531.2005

9. Hall-Pogar T, Liang S, Hague LK, Lutz CS (2007) Specific trans-acting proteins interact with auxiliary RNA polyadenylation elements in the COX-2 3′-UTR. RNA 13(7):1103–1115. doi:10.1261/rna.577707

10. Butter F, Scheibe M, Morl M, Mann M (2009) Unbiased RNA-protein interaction screen by quantitative proteomics. Proc Natl Acad Sci U S A 106(26):10626–10631. doi:10.1073/pnas.0812099106

11. Cho S, Moon H, Loh TJ, Oh HK, Williams DR, Liao DJ, Zhou J, Green MR, Zheng X, Shen H (2014) PSF contacts exon 7 of SMN2 pre-mRNA to promote exon 7 inclusion. Biochim Biophys Acta 1839(6):517–525. doi:10.1016/j.bbagrm.2014.03.003

12. Jang HN, Lee M, Loh TJ, Choi SW, Oh HK, Moon H, Cho S, Hong SE, Kim do H, Sheng Z, Green MR, Park D, Zheng X, Shen H (2014) Exon 9 skipping of apoptotic caspase-2 pre-mRNA is promoted by SRSF3 through interaction with exon 8. Biochim Biophys Acta 1839(1):25–32. doi:10.1016/j.bbagrm.2013.11.006

13. Bollag DM, Edelstein SJ, Rozycki MD (1996) Protein methods, 2nd edn. Wiley-Liss, New York

14. Sambrook J, Russell DW (2001) Molecular cloning: a laboratory manual, 3rd edn. Cold Spring Harbor Laboratory Press, Cold Spring Harbor, NY

15. Penna A, Cahalan M (2007) Western blotting using the invitrogen NuPage Novex Bis Tris minigels. J Vis Exp 7:264. doi:10.3791/264

Chapter 5

Purification of RNA–Protein Splicing Complexes Using a Tagged Protein from In Vitro Splicing Reaction Mixture

Naoyuki Kataoka

Abstract

In eukaryotes, pre-mRNA splicing is an essential step for gene expression. Splicing reactions have been well investigated by using in vitro splicing reactions with extracts prepared from cultured cells. Here, we describe protocols for the preparation of splicing-competent extracts from cells expressing a tagged spliceosomal protein. The whole-cell extracts are able to splice exogenously added pre-mRNA and the RNA–protein complex formed in the in vitro splicing reaction can be purified by immunoprecipitation using antibodies against the peptide tag on the splicing protein. The method described here to prepare splicing-active extracts from whole cells is particularly useful when studying pre-mRNA splicing in various cell types, and the expression of a tagged spliceosomal protein allows one to purify and analyze the RNA–protein complexes by simple immunoprecipitation.

Key words Splicing, Whole-cell extracts, Sonication, Immunoprecipitation, Anti-Flag antibody

1 Introduction

In eukaryotic cells, most of the nuclear-encoded genes contain introns. Thus, the excision of introns from pre-mRNAs by RNA splicing is an essential step for gene expression [1, 2]. The pre-mRNA splicing reaction has been extensively studied by in vitro splicing reaction using cell extracts. In vitro analyses demonstrated that splicing occurs via a two-step reaction. As the first-step reaction, cleavage at the 5′ splice site and the formation of the phosphodiester bond between the branch point and the 5′ end of the intron take place. This results in the formation of two splicing intermediates: the 5′ exon and the intron lariat with downstream exon. In the second step, cleavage at the 3′ splice site occurs, resulting in ligation of both the 5′ and 3′ exons to produce the mRNA and release of the intron lariat.

Analyses of the splicing reaction also revealed that splicing occurs in a large RNA–protein complex, termed spliceosome, which consists of five Uridine-rich small nuclear ribonucleoproteins

Ren-Jang Lin (ed.), *RNA-Protein Complexes and Interactions: Methods and Protocols*, Methods in Molecular Biology, vol. 1421, DOI 10.1007/978-1-4939-3591-8_5, © Springer Science+Business Media New York 2016

(U snRNPs) and a number of non-snRNP splicing factors [1, 2]. Each U snRNP contains a small nuclear RNA (snRNA) and several specific proteins, including the Sm/Lsm proteins. It is estimated that at least 100 proteins are required for splicing reaction [3, 4].

In order to study the splicing reaction in mammals, cultured HeLa cells and HEK293T cells have been useddas the most common source of extracts, since those cells are easy to grow, contain relatively low amounts of ribonucleases, and procedures had been previously developed for preparation of extracts from them for transcription studies [5–7]. Immunoprecipitation with specific antibodies from in vitro splicing reaction is a powerful tool in order to test whether the target proteins are associated with spliceosomes, which contain pre-mRNA, spliced RNA, and splicing intermediates [6, 8, 9].

Here, I describe a detailed protocol for preparation of efficient splicing extracts from HEK293T cells expressing transfected tagged proteins [6, 7]. This whole-cell extract system can be used for in vitro splicing followed by immunoprecipitation [6, 10–15]. This simple procedure for a small-scale whole-cell extract preparation is also useful for rapidly checking and studying pre-mRNA splicing in many different cell types.

2 Materials

All reagents should be prepared with high-quality water, e.g., Milli Q (Millipore) or equivalent. Sterilization of all the reagents used for extract preparation is carried out by autoclaving wherever possible. Vendor information is provided simply for convenience and many common reagents can be obtained from various reliable sources.

2.1 Cell Culture

1. HEK293T cells.
2. Dulbecco's modified Eagle's medium (DMEM) with low glucose (Life Technologies). Add sterile fetal bovine serum (Life Technologies) and 100X Antibiotic Antimycotic solution (Life Technologies) to 100 mL/L and 10 mL/L, respectively.
3. 10-cm cell culture plates (Falcon).

2.2 Transfection

1. Lipofectamine 2000 Transfection Reagent (Life Technologies).
2. Opti-MEM I Reduced Serum Medium (Life Technologies).

2.3 Preparation of Extracts

Reagents 1 and 2 are stock solutions and they are added to the specified solution immediately before use. These stock solutions should be stored at –20°C.

1. 1 M Dithiothreitol (DTT).
2. 20 mg/mL Phenylmethanesufonyl fluoride (PMSF) dissolved in ethanol.

3. Phosphate-buffered saline (PBS): 137 mM NaCl, 2.7 mM KCl, 8 mM Na_2HPO_4, 1.5 mM KH_2PO_4, 0.5 mM $MgCl_2$.

4. Buffer E: 20 mM HEPES-KOH, pH 7.9, 100 mM KCl, 0.2 mM EDTA, 10 % (v/v) glycerol, 1 mM DTT.

5. Ultrasonic processor Model VC130PB (Sonics & Materials Inc., Newtown, PA, USA).

2.4 In Vitro Splicing Reaction

All reagents should be prepared with high-quality water, e.g., Milli Q (Millipore) or equivalent.

1. m^7GpppG (New England Biolabs).

2. 10X SP mixture: 5 mM ATP, 200 mM creatine phosphate, 16 mM $MgCl_2$. Make small aliquots (0.2 mL) in microcentrifuge tubes and store at –20 °C from the following stocks:

   (a) 100 mM ATP, pH 7.5 (GE Healthcare).

   (b) 1 M Creatine phosphate, prepared from reagent with disodium salt (Calbiochem).

   (c) 1 M $MgCl_2$.

3. 2X PK buffer: 0.2 mM Tris–HCl, pH 7.5, 25 mM EDTA, 0.3 M NaCl, 2 % SDS. Do not autoclave. Sterilize by filtration through a 0.2 μm filter and store at room temperature.

4. 20 mg/mL proteinase K solution: dissolve proteinase K in water. Sterilization is not required if high-quality water is used. Aliquot 200 μL in microcentrifuge tubes and store at –20 °C.

5. Tris-saturated phenol (e.g., UltraPure Buffer-Saturated Phenol, Life Technologies).

2.5 Immuno-precipitation

1. IP buffer: 20 mM HEPES/KOH, pH 7.9, 150 mM NaCl, 0.05 % (v/v) Triton X100.

2. Anti-Flag M2 Affinity Gel (Sigma-Aldrich).

3. Elution Buffer: Buffer E plus 40 ng/μL 3X Flag peptide (Sigma-Aldrich).

3 Methods

3.1 HEK293T Cell Culture

HEK293T cells are grown on 10-cm plates at 37 °C up to 50–80 % confluency for transfection (*see* **Note 1**). The cells should not be overgrown.

3.2 Transfection of Plasmids into HEK293T Cells

1. Change cell culture medium to DMEM without antibiotics at least 30 min prior to transfection.

2. Dilute 10 μg plasmid DNA in 500 μL of Opti-MEM I Reduced Serum Medium. Mix gently by pipetting.

3. Dilute 25 µL of Lipofectamine 2000 reagent with 500 µL of Opti-MEM I Medium. Incubate it for 5 min at room temperature.

4. After the 5 min incubation, combine the diluted DNA solution with diluted Lipofectamine 2000 solution. The total volume will be 1000 µL. Mix gently by pipetting and incubate for 20 min at room temperature (*see* **Note 2**).

5. Add the 1000 µL of mixture to each dish containing cells and medium. Mix gently by rocking the plate back and forth.

6. Incubate cells at 37 °C in a CO_2 incubator for 18–48 h for transgene expression. Medium may be changed 4–6 h after transfection.

3.3 Preparation of HEK293T Whole-Cell Extract

1. The cells are washed three times with 10 mL of ice-cold PBS each time on the plates. Since HEK293T cells can be easily detached from the plates, add PBS to cells and wash them gently.

2. The cells are detached from the plates by pipetting, and resuspended in 10 mL PBS (*see* **Note 3**). Harvest cells by low-speed centrifugation at ~$500 \times g$ (e.g., Beckman JA25.50 rotor, 1000 rpm) for 5 min at 4 °C, and remove PBS completely.

3. The resulting pellet is resuspended in ice-cold buffer E. Typically, 200 µL of buffer E is used for cells isolated from one 10-cm plate.

4. Disrupt cells by sonication with an Ultrasonic processor Model VC130PB or equivalent. Sonication is carried out at 30 % continuous for 5 s; the 5-s sonication is done three times with a 30-s incubation on ice between bursts (*see* **Note 4**).

5. Transfer the sonicated lysate into a microcentrifuge tube and centrifuge at ~$15,000 \times g$ (e.g., TOMY microcentrifuge MX-300, 13,000 rpm) for 20 min at 4 °C. Carefully save the supernatant, which is the whole-cell extract, into a new microfuge tube placed on ice.

6. Aliquot the extract in 20–50 µL portions into microcentrifuge tubes as desired. Quickly freeze the tubes by dropping them into liquid nitrogen and store at –80 °C or lower. The extract remains active for at least 1 year if stored at –80 °C and never thawed.

3.4 In Vitro Splicing Assays Using Whole-Cell Extracts

The pre-mRNA substrate for in vitro splicing is usually prepared by in vitro run-off transcription with a bacteriophage RNA polymerase, such as T7 or SP6, by using a linearized plasmid. The pre-mRNAs are capped by priming the transcription with a dinucleotide primer, m^7GpppG and labeled internally with $\alpha^{32}P$-UTP or $\alpha^{32}P$-GTP [9, 16, 17]. After in vitro splicing reaction, RNAs are purified and analyzed by electrophoresis on a urea-denaturing polyacrylamide gel (PAGE) and autoradiography.

1. Prepare in vitro splicing mixture by mixing pre-mRNA and reagents. A recipe for a typical splicing reaction of 40 µL is as follows (*see* **Note 5**).

   (a) Pre-mRNA 2 µL (40 fmol).

   (b) 10X SP mixture 4 µL.

   (c) Cell extracts 24 µL.

   (d) Milli-Q Water 10 µL.

2. Incubate at 30 °C for 30 min to 4 h. The kinetics and efficiency of splicing depends on which kind of pre-mRNA substrate is used for reaction.

3. Stop splicing reaction by adding 100 µL of 2X PK buffer and 60 µL of Milli-Q water.

4. Add 4 µL of 20 mg/mL proteinase K solution and incubate at 37 °C for 30 min.

5. Extract with 200 µL of Tris-saturated phenol, the supernatant (~180 µL) is transferred to a new tube, and the RNA is precipitated with ethanol.

6. Analyze recovered RNAs on denaturing PAGE.

RNA recovered from in vitro splicing reactions using whole-cell extracts from HEK293T is shown in Fig. 1a (lanes marked as input).

3.5 Immuno-precipitation of RNA–Protein Complexes from In Vitro Splicing Reactions

1. Take 30 µL of Flag-M2 resin (bed volume is ~25 µL) and transfer to a microcentrifuge tube.

2. Add 500 µL of ice-cold PBS and mix by inverting the tube. Centrifuge at ~1000×*g* (e.g., TOMY MX-300 microcentrifuge, 3300 rpm) briefly at 4 °C. Carefully remove the supernatant.

3. Add 500 µL of ice-cold IP buffer and mix by inverting the tube. Centrifuge at ~1000×*g* briefly at 4 °C. Carefully and completely remove the supernatant. Repeat this step once more.

4. Dilute 40 µL of in vitro splicing reaction mixture (Subheading 3.4, **step 2**) with 160 µL of ice-cold IP buffer and mix gently with pipetting.

5. Add 200 µL of in vitro diluted splicing reaction mixture with the Flag M2 resin prepared by the procedure described above. Incubate them at 4 °C, 30 min to 1 h by rotating the tube.

6. Centrifuge at ~1000×*g* briefly at 4 °C. Completely remove the supernatant.

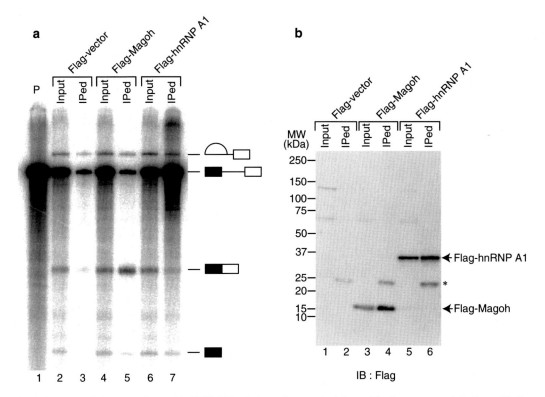

Fig. 1 In vitro splicing reactions with HEK293T whole-cell extracts followed by immunoprecipitation with the anti-Flag M2 antibody. (**a**) Immunoprecipitation of RNAs from splicing reaction mixtures. ^{32}P-labeled chicken δ-crystallin pre-messenger RNA (pre-mRNA, lane 1, marked as P) [17] was incubated with HEK293T whole-cell extracts prepared from cells transfected with Flag-vector, Flag-Magoh, or Flag-hnRNP A1 plasmid. The input lanes contain 10 % of total. Flag-Magoh preferentially precipitates spliced mRNA (lanes 4 and 5), while Flag-hnRNP A1 precipitates pre-mRNA efficiently (lanes 6 and 7). Immunoprecipitation from Flag-vector-transfected HEK293T whole-cell extracts show some nonspecific precipitation with anti-Flag M2 antibody (lanes 2 and 3). (**b**) Western blotting of the immunoprecipitated proteins from the splicing reactions with the anti-Flag M2 antibody. The immunoprecipitated Flag-tagged proteins (Flag-Magoh and Flag-hnRNP A1) were detected as the expected size (lanes 3, 4 and 5, 6). The *lanes marked input* contain 10 % of total. The *asterisk* shows the band corresponding to the immunoglobulin light chain from the Flag-M2 resin (lanes 2, 4, and 6)

7. Add 500 μL of ice-cold IP buffer and mix by inverting the tube. Centrifuge at ~1000×*g* briefly at 4 °C. Completely remove the supernatant. Repeat this step three more times.

8. Add 500 μL of ice-cold IP buffer and transfer the suspension into a new microcentrifuge tube (*see* **Note 6**). Centrifuge at ~1000×*g* briefly at 4 °C. Completely remove the supernatant.

9. In order to recover RNA–protein complexes, add 50 μL of Elution buffer to the resin and incubate at 4 °C for 30 min to 1 h by rotating the tube. Briefly centrifuge at ~1000×*g* at 4 °C. Carefully save the supernatant.

10. Repeat elution step once again and combine the supernatant (~100 μL in total).

11. (a) To recover RNAs, add 100 µL of 2X PK buffer to the eluate and perform proteinase K treatment, phenol extraction, and ethanol precipitation as described in the previous section (Subheading 3.4. **steps 4–6**). Recovered RNAs can be analyzed by urea-denaturing polyacrylamide gels. An example is shown in Fig. 1a.

(b) Immunoprecipitated proteins recovered by mixing and boiling with SDS-loading buffer can be analyzed by SDS-PAGE followed by either western blotting with anti-Flag antibody (Fig. 1b) or silver staining of the gel (*see* **Note 7**).

4   Notes

1. Using healthy, growing cells under optimal growth condition is required. Slowly growing or overgrown cells should be avoided as they produce extracts with much less splicing activity.

2. The transfection efficiency is affected by the amount of plasmid DNA and the ratio between the DNA and the transfection reagent. The optimal condition for transfection has to be determined for each cell line before using it for the preparation of whole-cell extracts.

3. For other cells that attach well to tissue culture plates, like HeLa cells, use Trypsin-EDTA (Life Technologies) to detach cells from the plates. Briefly, add 0.5 mL of Trypsin-EDTA with 2 mL PBS to the cells on a 10-cm plate after a 10 mL PBS wash. Incubate cells at 37 °C for 3 min and add 7.5 mL of culture medium to stop trypsinization. Resuspend the cells and transfer to a clean disposable tube to harvest cells.

4. The setting for sonication may differ between different sonicators. It is necessary to determine the optimal settings for cell disruption for each type of sonicator. The extent of cell lysis can be monitored by checking the cell lysate under microscope. Foaming the cell suspension during sonication should be avoided, since it causes inactivation of the extract.

5. The in vitro splicing condition described here is applicable for many kinds of pre-mRNAs [9, 16, 17]. However, optimal concentrations of magnesium and potassium vary with different pre-mRNAs, which must be determined empirically.

6. Transferring the resuspended beads with the IP buffer to a new tube greatly reduces the background for immunoprecipitation of both RNA and protein, since a microcentrifuge tube nonspecifically absorb RNA and protein, which can be recovered during elution step.

7. For western blotting by anti-Flag antibody, using anti-Flag rabbit polyclonal antibody (Sigma-Aldrich) is strongly recommended, since the eluate may contain some amount of mouse anti-Flag antibody from the beads that cross-reacts with anti-mouse secondary antibody.

References

1. International Human Genome Sequencing, C (2004) Finishing the euchromatic sequence of the human genome. Nature 431:931–945

2. Lander ES, Linton LM, Birren B et al (2001) Initial sequencing and analysis of the human genome. Nature 409:860–921

3. Hastings ML, Krainer AR (2001) Pre-mRNA splicing in the new millennium. Curr Opin Cell Biol 13:302–309

4. Jurica MS, Moore MJ (2003) Pre-mRNA splicing: awash in a sea of proteins. Mol Cell 12:5–14

5. Dignam JD, Lebovitz RM, Roeder RG (1983) Accurate transcription initiation by RNA polymerase II in a soluble extract from isolated mammalian nuclei. Nucleic Acids Res 11:1475–1489

6. Kataoka N, Dreyfuss G (2004) A simple whole cell lysate system for in vitro splicing reveals a stepwise assembly of the exon-exon junction complex. J Biol Chem 279:7009–7013

7. Kataoka N, Dreyfuss G (2008) Preparation of efficient splicing extracts from whole cells, nuclei, and cytoplasmic fractions. Methods Mol Biol 488:357–365

8. Hanamura A, Caceres JF, Mayeda A, Franza BR Jr, Krainer AR (1998) Regulated tissue-specific expression of antagonistic pre-mRNA splicing factors. RNA 4:430–444

9. Kataoka N, Yong J, Kim VN et al (2000) Pre-mRNA splicing imprints mRNA in the nucleus with a novel RNA-binding protein that persists in the cytoplasm. Mol Cell 6:673–682

10. Kashima I, Yamashita A, Izumi N et al (2006) Binding of a novel SMG-1-Upf1-eRF1-eRF3 complex (SURF) to the exon junction complex triggers Upf1 phosphorylation and nonsense-mediated mRNA decay. Genes Dev 20:355–367

11. Kataoka N, Diem MD, Kim VN, Yong J, Dreyfuss G (2001) Magoh, a human homolog of Drosophila mago nashi protein, is a component of the splicing-dependent exon-exon junction complex. EMBO J 20:6424–6433

12. Kataoka N, Diem MD, Yoshida M et al (2011) Specific Y14 domains mediate its nucleo-cytoplasmic shuttling and association with spliced mRNA. Sci Rep 1:92. doi:10.1038/srep00092

13. Kawano T, Kataoka N, Dreyfuss G, Sakamoto H (2004) Ce-Y14 and MAG-1, components of the exon-exon junction complex, are required for embryogenesis and germline sexual switching in *Caenorhabditis elegans*. Mech Dev 121:27–35

14. Kim VN, Kataoka N, Dreyfuss G (2001) Role of the nonsense-mediated decay factor hUpf3 in the splicing-dependent exon-exon junction complex. Science 293:1832–1836

15. Yoshimoto R, Okawa K, Yoshida M, Ohno M, Kataoka N (2014) Identification of a novel component C2ORF3 in the lariat-intron complex: lack of C2ORF3 interferes with pre-mRNA splicing via intron turnover pathway. Genes Cells 19:78–87

16. Pellizzoni L, Kataoka N, Charroux B, Dreyfuss G (1998) A novel function for SMN, the spinal muscular atrophy disease gene product, in pre-mRNA splicing. Cell 95:615–624

17. Sakamoto H, Ohno M, Yasuda K, Mizumoto K, Shimura Y (1987) In vitro splicing of a chicken delta-crystallin pre-mRNA in a mammalian nuclear extract. J Biochem 102:1289–1301

Chapter 6

Loading of Argonaute Protein with Small Duplex RNA in Cellular Extracts

Keith T. Gagnon

Abstract

Argonaute (Ago) proteins are the minimum core proteins required for executing RNA interference (RNAi) mechanisms of gene regulation. For Ago proteins to regulate gene expression through RNAi they must be loaded, or "programmed," with a single strand of small RNA. Natural small RNAs are typically double-stranded duplexes that require additional factors for efficient and specific loading into Ago proteins. Here, a protocol is described for investigating RNAi programming through loading of human Ago2 using radio-labeled small interfering RNA (siRNA) and HeLa cell extracts. This protocol provides an Ago loading assay to study RNAi programming when starting with crude or partially purified cell extracts. The Ago loading assay should prove useful for studying other Ago proteins using a variety of mammalian cell extracts.

Key words RNA interference, RNAi programming, Argonaute loading, Small duplex RNA, Cell extract, In vitro, Ago2, Human, Mammalian

---

1 Introduction

RNA interference (RNAi) can be divided into at least two distinct phases, generically called "programming" and "execution" [1, 2]. First, an Argonaute (Ago) protein is loaded, or programmed, with small RNAs [2, 3]. Once loading has occurred, the ribonucleoprotein complex and associated proteins can then be guided to complementary target RNA through base-pairing with the small RNA that is bound to the Ago protein [4–6]. Interaction with the targeted RNA results in execution of RNAi by either target RNA cleavage and degradation or modulation of target RNA function, such as reduced translation of an mRNA [1, 2, 7–9]. Many Ago proteins have "slicer" activity, an RNase activity that breaks a specific phosphodiester bond of the targeted RNA [10–13]. In humans there are four Ago proteins, Ago1–4, but only the Ago2 protein is catalytic [14, 15].

The most common small RNAs in humans are microRNAs (miRNAs), which are imperfect duplexes that can vary in length from approximately 20–30 bases (Fig. 1a) [1, 16, 17]. RNAi not

Ren-Jang Lin (ed.), *RNA-Protein Complexes and Interactions: Methods and Protocols*, Methods in Molecular Biology, vol. 1421, DOI 10.1007/978-1-4939-3591-8_6, © Springer Science+Business Media New York 2016

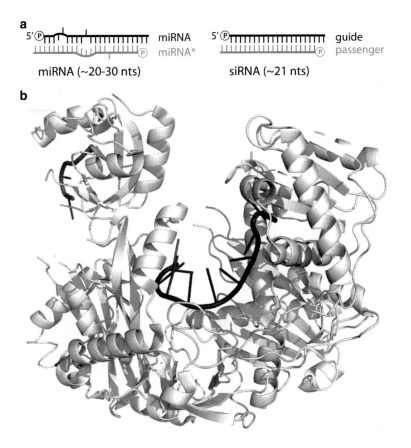

Fig. 1 Small RNAs and their loading into Argonaute protein. (a) Secondary structure of miRNAs and siRNAs. miRNAs are imperfect duplexes composed of a miRNA strand and a miRNA* strand, both of which can be loaded into Ago proteins and target different RNAs, although the miRNA* strand is of lower abundance [4]. siRNAs are perfect duplexes composed of guide and passenger strands, the latter of which is not usually loaded efficiently into the Ago protein. (b) Crystal structure of human Ago2 bound to miRNA miR20a (PDB ID 4F3T) [27]. Ago protein is shown in *light gray*. MiR20a was only partially resolved in this structure and is shown centered in *black*

only functions in natural regulation, but it can be redirected to regulate almost any cellular RNA by introducing small RNAs into cells [9, 18]. For biotechnology applications, small interfering RNAs (siRNAs) are commonly used, consisting of 19 nucleotide perfect duplexes with two nucleotide 3′ overhangs (Fig. 1a) [18–20]. RNAi has become a premier tool for genetic research and a promising future approach for therapeutic treatment of human disease [19, 21, 22]. RNAi continues to be intensely studied with the aim of improving biotechnology applications and understanding the cellular function and evolution of RNAi.

To initiate RNAi, Ago proteins are loaded with a single strand of short RNA [3, 12, 23]. Under in vitro conditions, purified Ago protein alone can usually bind and load small single-stranded nucleic acids of the appropriate size range [24, 25]. Once bound to Ago, the small RNA is displayed in a manner which presents its bases for specific recognition of complementary RNAs through Watson–Crick base-pairing (Fig. 1b) [26–28]. Naturally occurring small RNAs are typically double-stranded duplexes. Thus, loading is a regulated and multistep process whereby small RNA duplexes are unwound and one strand is selected for loading into the Ago protein [11, 29, 30]. The loaded strand is often referred to as the "guide" strand whereas the discarded strand is called the "passenger" strand [11].

Ago proteins alone do not appear to possess strong helicase activity for removing the passenger strand nor a clear mechanism for choosing the appropriate strand as the guide [11, 29, 31]. Therefore, these activities are primarily catalyzed and directed by other factors [30, 32]. One protein implicated in this process is heat shock protein 90 (Hsp90). Previous studies have shown that the ATP-dependent chaperoning activity of Hsp90 is important for efficient RNAi [33]. In addition to Hsp90, other factors are also implicated in Ago loading [30, 32, 34]. Currently, it is unclear whether all of the factors involved in Ago loading activity are known [6, 32, 34–36]. In addition, production of active, recombinant proteins is resource-intensive. Thus, reconstitution of efficient loading activity of Ago proteins in vitro using purified components is challenging [30, 33, 34].

Experimental approaches which assay the binding of targeted RNA by Ago complexes [6], the cleavage of the targeted RNA [12], or the reduced translation of targeted mRNAs [8], are not direct assays for Ago loading. While connected, the efficiency of the execution phase of RNAi does not necessarily correlate with efficiency of the programming phase of RNAi [33]. These distinct phases must be separated and studied individually to understand the function and regulation of RNAi and to further improve its biotechnology applications.

To facilitate investigation of Ago loading activity and discovery of Ago loading factors in the absence of a defined biochemical system with purified components, a protocol using cellular extracts and synthetic small RNAs was developed (Fig. 2a) [37, 38]. This protocol provides an activity assay for small duplex RNA loading of Ago proteins. The Ago2 loading assay presented here was originally developed to compare loading activities in nuclear and cytoplasmic extracts (Fig. 2b) [37]. No loading activity was observed in nuclear extracts, which do not contain Hsp90 protein [37]. Thus, the Ago2 loading assay enabled the discovery of an unexpected layer of RNAi regulation, which appears to restrict Ago2 loading to the cytoplasmic compartment.

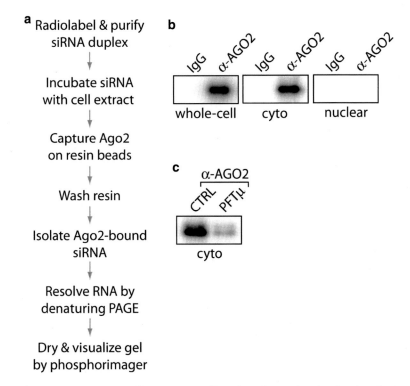

Fig. 2 In vitro Ago2 loading assay. (**a**) Flowchart for performing in vitro Ago2 siRNA loading assay with human cell extracts. (**b**) Typical results from Ago2 loading assays using whole-cell, cytoplasmic (cyto) or nuclear extracts prepared from HeLa cells. The antibody used for immunoprecipitation is indicated above the gel. (**c**) Treatment of HeLa cells with pifithrin-μ (PFTμ), a known Hsp70/90 inhibitor, prior to extract preparation results in reduced Ago2 loading activity in vitro

2   Materials

2.1   Chemicals and Reagents

2.1.1   Preparation of Radiolabeled RNA

1. RNase-free distilled and deionized water (ddH$_2$O).

2. siRNA strands dissolved in ddH$_2$O at 100 μM:

    (a) siLuc guide (siLuc_as): rUrGrUrUrCrArCrCrUrCrGrArUA rUrGrUrGrCTT.

    (b) siLuc passenger (siLuc_ss): rGrCrArCrArUrArUrCrGrArGr GrUrGrArArCrATT.

3. [γ]-^{32}P-ATP (7000 Ci/mmol) (MP Biomedicals) (*see* **Note 1**).

4. SUPERase-In RNase inhibitor (Ambion) (*see* **Note 1**).

5. T4 polynucleotide kinase (PNK) and 10× PNK buffer: 0.7 M Tris–HCl, pH 7.6, 100 mM MgCl$_2$, 50 mM dithiothreitol.

6. Redistilled phenol, water-saturated.

7. Chloroform:isoamyl alcohol (24:1).

8. Acetone.

9. Black India ink (local art supply store).

10. Yeast transfer RNA (tRNA) at 10 mg/mL in ddH$_2$O.

11. Acetone.

12. Lithium perchlorate (LiClO$_4$).

13. Urea, molecular biology grade.

14. 10 % ammonium persulfate (APS), prepared fresh.

15. *N,N,N,N'*-Tetramethyl-ethylenediamine (TEMED).

16. 10× TBE: 0.89 M Tris base, 0.89 M boric acid, 20 mM EDTA.

17. 4× Native loading buffer: 4× TBE, 40 % glycerol (v/v), 0.2 mg/mL xylene cyanol, 0.2 mg/mL bromophenol blue.

2.1.2 Ago Loading Assay   HeLa cell extracts (*see* **Note 2**).

1. 40 % acrylamide solution, 19:1 (acrylamide:bisacrylamide) crosslinking.

2. Protein G Plus/Protein A agarose (EMD Millipore, IP05-1.5ML) (*see* **Note 1**).

3. Anti-Ago2 antibody (Abcam, ab57113) (*see* **Note 1**).

4. Mouse IgG antibody (Millipore, 12-371) (*see* **Note 1**).

5. 100 mM adenosine triphosphate (ATP).

6. 1 M phosphocreatine.

7. Creatine kinase at 4U/μL.

8. 0.5 M ethylendiaminetetraacetic acid (EDTA).

9. Denaturing loading buffer: 90 % formamide, 1× TBE, 5 mM EDTA, 0.2 mg/mL bromophenol blue, 0.2 mg/mL xylene cyanol.

10. IP Equilibration Buffer (IPEB): 20 mM Tris, pH 7.4, 0.15 M NaCl, 2 mM MgCl$_2$, 0.05 % NP-40 (Igepal CA-630) (v/v), 0.1 % Polyvinylpyrrolidone (PVP) (w/v).

11. IP Binding Buffer (IPBB): 20 mM Tris, pH 7.4, 0.15 M NaCl, 3 mM MgCl$_2$.

12. IP Wash Buffer (IPWB): 20 mM Tris, pH 7.4, 0.5 M NaCl, 4 mM MgCl$_2$, 0.05 % NP-40 (Igepal CA-630) (v/v).

2.2 Equipment and Supplies

1. Temperature-controlled heating block.

2. Automatic pipettor and pipettes.

3. Clear plastic wrap (Saran™ wrap or Glad™ wrap).

4. 3MM Whatman filter paper.

5. Temperature-controlled vacuum gel drier.

6. Phosphorimager and phosphorimager screen.

7. 1.5 mL microfuge tubes.

8. Room temperature bench-top centrifuge (1.5 mL tube rotor).

9. Liquid and solid ^{32}P radioactive waste containers.

10. Geiger counter.

11. Scintillation counter.

12. Electrophoresis power supply (with constant current setting).

13. Denaturing polyacrylamide gel electrophoresis apparatus.

14. Water-cooled native polyacrylamide gel electrophoresis apparatus.

15. Glass plates ($19 \times 20 \times 0.3$ and $16 \times 20 \times 0.3$ cm) and gel spacers and combs (0.75 cm).

16. Large binder clips (2 in wide) (local business supply store).

17. Clear scotch tape.

18. Razor blades.

19. Bunsen burner.

20. Microcentrifuge tube rotator.

21. RNase-free mini-spin filtration columns (<10 μm) (Pierce® Spin Cups-Paper Filter) (*see* **Note 1**).

22. Vortex mixer.

3   Methods

3.1   General Methods

3.1.1   Tissue Culture and Cell Extracts

It is expected that the user will have some experience with tissue culture and existing protocols for how to grow and maintain healthy cell cultures. For the protocol described here, we recommend HeLa cells as they are straightforward to culture. HeLa cells were grown in Dulbecco's Modified Eagle's Medium (DMEM) supplemented with 5 % fetal bovine serum (FBS) and 0.5 % non-essential amino acids (NEAA). Cells were grown in sterile incubators at 37 °C in 5 % CO_2. Extracts were prepared according to methods described previously (*see* **Note 2**) [37, 38].

3.1.2   Handling RNA

General guidelines for working with RNA solutions include the use of baked glassware or DEPC-treated plastic-ware, wearing of gloves, practicing single-use of pipette tips, keeping work surfaces clean and free of dust, and using reagents that are certified RNase-free by the manufacturer. RNA should generally be kept cold in neutral buffers that lack divalent metal ions and stored long-term at −80 °C, preferably in a dry form when not being actively used.

3.2   Preparation of Radiolabeled siRNA for the Ago Loading Assay

RNA is commonly "end-labeled," where a radioactive nucleotide or phosphate is placed at the terminal end of the RNA sequence. Here we use 5′-end labeling, which requires that the RNA not have a 5′-phosphate (*see* **Note 3**). Although 5′-end radiolabeling of RNA is a standard procedure, it is described here because of special considerations, including differential labeling of the siRNA guide strand, purification on a native gel, high specific activity labeling, and precipitation with an uncommon reagent. The success of the subsequent Ago2 loading assay in cell extracts requires purified siRNA of very high specific radioactivity. The siRNA used in this protocol targets a sequence of the *Luciferase* mRNA [18]. There are no special considerations when choosing an siRNA or duplex RNA for this method. However, standard siRNA designs should be used initially or as controls, such as a duplex RNA of 19 nucleotides that contains 3′ dTdT overhangs [18, 20].

3.3 5′-End Labeling With T4 Polynucleotide Kinase (PNK)

1. Mix the reaction components below in a 1.5 mL microcentrifuge tube:

   (a) 1 µL of siLuc_as (guide strand) (100 µM stock).

   (b) 2 µL 10× PNK buffer.

   (c) 2.5 µL [γ]-^{32}P-ATP (~0.3 mCi).

   (d) 2 µL PNK (10 U/µL).

   (e) 1 µL SUPERase-In (40 U/µL).

   (f) ddH$_2$O to 20 µL.

2. Incubate at 37 °C for 2.5 h. Phenol/chloroform extract (*see* **Note 4**). Keep on ice or store at

   (a) −20 °C until ready to proceed.

CAUTION: Work behind a shield and follow safety regulations when handling radioactivity.

3.4 Preparing and Purifying Radiolabeled Duplex siRNA

Prior to use in the Ago loading assay, the radiolabeled guide strand must be annealed to a complementary passenger strand and gel-purified to remove unincorporated [γ]-^{32}P-ATP, single-stranded RNAs that did not anneal, and RNA that may have partially degraded during labeling.

3.4.1 Annealing Radiolabeled siRNA Guide and Passenger Strands

1. Add 1.2 µL (120 pmol) of siLuc_ss (passenger strand) RNA to the radiolabeled siLuc_as guide strand phenol extracted from above.

2. Incubate at 90 °C for 3 min. Remove from heating block and let cool at room temperature (RT) for 15 min.

3. Mix 8 µL of Native loading buffer with siRNA sample. Place siRNA sample on ice or freeze at −20 °C until ready for gel-purification.

3.4.2 Casting a Native Polyacrylamide Gel for Purification of Radiolabeled siRNA

1. Prepare a 40 mL solution containing 15 % acrylamide (19:1 acrylamide:bisacrylamide), 1× TBE buffer.

2. Add 10 % APS (6 µL/mL) and TEMED (1 µL/mL) and mix by inverting or stirring.

3. Pour into an assembled glass sandwich ($19 \times 20 \times 0.75$ cm) and position a 15-well comb in the top of the gel (*see* **Note 5**).

4. Allow the gel to polymerize for 30 min. Remove the comb and place the gel in a water-cooled gel electrophoresis apparatus. Add 1× TBE buffer to anode and cathode tanks. Be sure to cover the electrodes and gel wells with sufficient buffer. Rinse the wells with 1× TBE buffer.

5. Pre-run the gel at 30–35 mA for 15 min. Make sure water is turned on and circulating to keep the gel cool.

3.4.3 Resolving
the Radiolabeled siRNA
by Native Gel
Electrophoresis

1. Turn off the current to the gel. Load siRNA sample into the wells. One sample should fit into a single well if possible. Otherwise split the sample equally into multiple wells.

2. Turn current to the gel back on and run at 30–35 mA until the bromophenol blue band has migrated about 2/3 of the way through the gel (1.5–2 h). Make sure water circulation is turned on (*see* **Note 6**). The glass plates should remain cool to the touch and not exceed 30–35 °C.

3. Turn off the current to the gel and carefully drain the anode tank (typically the lower tank) buffer into a radioactive liquid waste container (*see* **Note 7**). Rinse the gel apparatus to remove trace radioactivity from the surface of the lower buffer chamber.

CAUTION: Work behind a shield and follow safety regulations when handling radioactivity.

3.4.4 Purifying
Radiolabeled siRNA
by Native Polyacrylamide
Gel Electrophoresis

1. Remove the gel cassette from the apparatus, rinse the glass plates to remove trace radioactivity from the external surface of the gel cassette (*see* **Note 8**), then place the gel cassette behind a shield.

2. Separate the glass plates with a plastic wedge tool so that the gel sticks to one plate. Cover the exposed gel with clear plastic wrap (Saran™ wrap).

3. Place three small drops of radioactive dye (1 μL [γ]-^{32}P-ATP in 30 μL black India ink) on the Saran™ wrap at three corners of the gel and allow them to air dry. Cover them with clear tape.

4. Expose in darkroom to autoradiography film placed on top of the gel. Expose for 15–60 s and develop.

5. Slide the film underneath the glass that the gel is on. Orient and align the spots at the three corners of the gel to locate the RNA bands. Carefully peel back the Saran wrap and cut out the band with a flamed (RNase-free) razor.

6. Crush the gel slice into a paste with a 1 mL plastic pipette tip (RNase-free and flame-sealed at the small end) in a 1.5 mL microcentrifuge tube.

7. Pipette 300 μL of nuclease-free water down the side of the tip and into the microcentrifuge tube so as to collect all the gel bits. Rotate at 4 °C for 4–16 h to elute RNA.

8. Spin down sample to pellet elution and gel bits. Cut ~2 mm off the end of a 1 mL pipette tip with flamed scissors or a flamed razor and use the tip to move the entire sample, gel bits and elution, to an RNase-free filter spin column (~10 μm cutoff) (*see* **Note 9**). Spin at 2000×*g* for 2 min to collect eluted RNA. Discard gel bits.

9. Add 10 μg of tRNA to the elution and split the sample evenly into two microcentrifuge tubes to accommodate the total volume needed for precipitation. Precipitate RNA from solution by adding 9 vol of 2 % lithium perchlorate ($LiClO_4$) in acetone and incubating at –20 °C for >15 min (*see* **Note 10**).

10. Spin at 12,000×*g* for 10 min, wash pellet with ice-cold acetone, and spin again at 12,000×*g* for 2 min. Let the pellet air dry at room temperature.

11. Resuspend RNA pellet in 30 μL of RNase-free water. Measure the radioactivity by scintillation counting of 1 μL (do not use scintillation fluid). Label the date and CPM/μL on the side of the tube and store RNA frozen at –20 °C or –80 °C (*see* **Note 11**).

3.5 In Vitro Ago Loading Assay

3.5.1 Loading Ago2 and Capturing on Resin

1. Mix the reaction components in the order shown below in a 1.5 mL microcentrifuge tube and rotate at RT for 1 h (*see* **Note 12**).
   - 250 μL extract.
   - 1 μL siRNA ($1–2 \times 10^6$ CPMs/uL).
   - 2.5 μL ATP (100 mM).
   - 2.5 μL phosphocreatine (1 M).
   - 2.5 μL creatine kinase (4 U/μL).

2. While loading reactions rotate, equilibrate resin for immunoprecipitation. Resuspend Protein G Plus/Protein A resin, aliquot 40 μL into two 1.5 mL microcentrifuge tubes, and add 1 mL of IPEQ buffer. Rotate for 5 min at RT.

3. Pellet resin by centrifugation at 2000×*g* for 1 min at RT. Remove buffer from resin with a pipet. Repeat wash step with IPBB buffer.

4. Spin down loading reaction at 12,000×*g* for 5 min at RT to remove precipitation (*see* **Note 13**). Move supernatant to equilibrated resin. Add 2 μg of IgG to one tube and 2 μg of anti-Ago2 to the other tube. Rotate at RT for 1 h.

3.5.2 Isolating Ago2-Bound siRNA

1. Spin down loading reaction and resin mixtures at 2000×*g* for 1 min. Discard supernatant.

2. Wash the resin 3 times by adding 0.5 mL of IPWB, rotating at RT for 5 min, then spinning at $2000 \times g$ for 1 min. Perform a final wash with 0.5 mL of IPBB. On the final wash, move resin and buffer to a new tube before centrifugation (*see* **Note 14**).

3. Add 40 µL of ddH$_2$O, 0.5 µL of EDTA (0.5 M), 0.5 µL of tRNA (10 mg/mL), and 100 µL of phenol to the resin in each tube (*see* **Note 15**). Vortex for 2 min.

4. Spin down sample at $12,000 \times g$ for 5 min. Carefully collect the top aqueous layer by pipetting and move to a new tube. Add 700 µL of 2 % LiClO$_4$ in acetone, vortex briefly, and precipitate at 4 °C for >15 min.

5. Centrifuge samples at $12,000 \times g$ for 10 min, remove supernatant and wash pellet by adding 0.5 mL of acetone, inverting to mix, and spinning at $12,000 \times g$ for 2 min. Remove acetone and allow pellet to air dry on bench top.

3.5.3 Visualizing Ago2-Bound siRNA by Resolving Complexes on a Denaturing Gel

1. Prepare a 40 mL solution containing 15 % acrylamide (19:1 acrylamide:bisacrylamide), 7 M urea, 1× TBE buffer, and 2 % glycerol (*see* **Note 16**).

2. Add 10 % APS (6 µL/mL) and TEMED (1 µL/mL) and mix by inverting.

3. Pour into an assembled glass sandwich ($19 \times 20 \times 0.75$ cm) and position a 15-well comb in the top of the gel.

4. Allow the gel to polymerize for 30 min. Remove the comb and place in a denaturing gel electrophoresis apparatus. Add 1× TBE buffer to the anode and cathode tanks. Be sure to cover the electrodes and gel wells with sufficient buffer. Rinse the wells with 1× TBE buffer.

5. Pre-run the gel at 40 mA for 30 min. The outer glass plate of the gel should become very warm to the touch (*see* **Note 17**).

6. Add 6–10 µL of formamide loading buffer to the dried RNA samples. Heat to 90 °C for 3 min. Let cool to RT on bench top. Vortex briefly and centrifuge at $2000 \times g$ for 30 s.

7. Turn off the current to the gel. Load RNA sample into individual wells.

8. Turn current to the gel back on and run at 40 mA until the bromophenol blue band has migrated about 1/2 of the way through the gel (1 h). Monitor the gel so that it does not get too hot (*see* **Note 18**). The RNA will run between the two blue dye bands on the gel.

9. Stop the gel, separate the glass plates, and lay Saran™ wrap over the top of the gel. Flip the gel over and peel the glass plate off so that the gel is now adhered to the Saran™ wrap. Press a sheet of Whatman 3 M paper to the gel and dry in a vacuum gel dryer at 80 °C for 2 h.

10. Expose the dried gel to a phosphorimager screen overnight. Develop the screen to visualize radioactive RNA that co-eluted with Ago2 in the in vitro loading assay. Typical results for different HeLa cell extracts are shown in Fig. 2b.

3.6 Analysis and Application of the Ago2 Loading Assay

Ago 2 loading assay results should show a single band on a denaturing gel after immunoprecipitation of Ago2 from HeLa whole-cell or cytoplasmic extracts incubated with radiolabeled duplex (Fig. 2b). No band or a very faint band should appear in the samples immunoprecipitated with non-specific IgG antibody when used. For HeLa nuclear extracts, either no band or very faint bands are detected for Ago2 loading (Fig. 2b). This observation reflects the fact that known loading factors have been reported as absent from HeLa nuclear extracts, as well as nuclear extracts from other tested cell lines [37]. This assay can be used to test different loading conditions and the effect of compounds on loading activity. For example, the Hsp90 inhibitor pifithrin-μ has been reported to inhibit Ago2 loading and RNAi activity [33]. Here, we demonstrate that pifithrin-μ does inhibit Ago2 loading of duplex siRNA from a HeLa cytoplasmic extract in vitro (Fig. 2c). These results suggest that the Ago2 loading assay presented here depends on Hsp90 activity and recapitulates the loading mechanism of Ago2 inside of cells.

The Ago2 loading assay can be modified to investigate other aspects of loading. Whether or not an Ago protein is accessible for loading can be tested by first immunoprecipitating Ago protein, then incubating with a radiolabeled single-strand guide RNA [37, 38]. Another modification is to run the eluted RNA on a native, non-denaturing gel. Native gel electrophoresis might be helpful in discriminating between duplex loading and passenger strand removal. If the passenger strand has not been completely removed two distinct bands will be observed. The faster migrating band represents RNA that was bound to Ago2 whereas the slower migrating band represents duplex RNA containing both guide and passenger strands [37, 38].

When considering the use of other cell types or extracts for Ago2 loading, or investigating the loading of other Ago proteins, several factors should be considered. HeLa cells appear to have a robust RNAi system and other cell lines, such as primary fibroblasts, may not produce comparable Ago loading activities. While many cell types and extract will need to be individually optimized, general rules would include increasing the amount of extract used, increasing the concentration of the extract, and increasing the amount of radioactive siRNA used [38]. For various Ago proteins, the specificity and performance of the antibody in immunoprecipitation applications can affect the results. Antibodies would therefore need to be tested. A FLAG-HA Ago2 fusion has been successfully used in this assay before [37]. Therefore, affinity tagged versions of Ago proteins may provide a more universal and predictable immunoprecipitation step in this assay.

4   Notes

1. Vendor names are provided for convenience. The chemicals from the specific commercial suppliers named have been tested to work in this assay, the same reagents from other vendors may also be suitable.

2. HeLa cell extracts were prepared as in Gagnon et al. [37, 38]. For whole-cell extracts, cells were sonicated (same sonication conditions used for nuclear extract preparations) in ice-cold nuclear lysis buffer [20 mM Tris, pH 7.4, 150 mM NaCl, 2 mM $MgCl_2$, 0.05 % NP-40 (Igepal CA-630) (v/v), 10 % glycerol (v/v)] containing phosphatase and protease inhibitors. Insoluble cell debris was removed by centrifugation at $10,000 \times g$ at 4 °C and extracts were flash frozen in liquid nitrogen and stored at −80 °C for later use.

3. The siRNA strands used in this protocol are chemically synthesized and by default do not have 5′ phosphate groups. Thus, they are immediately ready for 5′ radiolabeling.

4. Phenol extraction is a standard method. Briefly, add 1 volume of redistilled water-saturated phenol to the sample. Vortex for 30 s to mix, then spin at top speed to separate phenol and aqueous phases. Collect the top aqueous phase by pipetting. Add 1 volume of chloroform:isoamyl alcohol (24:1), vortex for 30 s to mix, then spin at top speed to separate chloroform and aqueous phases. Collect top aqueous phase by pipetting.

5. Prior to assembling glass plates with spacers for pouring the gel, the plates should be clean and free of detergent. Glass plates can be wiped or rinsed with ethanol before assembly to remove dust. Glass plates can also be baked, although a thorough washing is often sufficient to remove most contaminating RNases.

6. The glass plates should remain cool to the touch and not get above 30–35 °C. If gel running temperatures exceed 40 °C, the integrity of the siRNA duplex may become compromised, depending on its GC content and thermal stability.

7. The anode tank buffer where the positive electrode is located is contaminated with high levels of radioactive ^{32}P. Transfer from the tank to an approved liquid waste container should be performed in a sink or over a large bucket to prevent lab contamination on bench surfaces or floors. Monitor this process carefully with a Geiger counter. Afterward, decontaminate the area and electrophoresis apparatus by washing with soapy water.

8. Radioactivity can adhere to the outside of the plates that contain the gel. Since the gel between the glass plates is radioactive, the surface of the gel plates cannot be checked for contamination directly. After rinsing, wipe the outside of the glass plates with a paper towel and use a Geiger counter to check the paper towel.

9. Spin column filters are useful for removing gel bits and allowing the eluted RNA in solution to flow through. Avoid filter units with low molecular weight cutoffs as these tend to get clogged by the gel bits. Prepare spin columns for RNA work by filtering RNase-free water or buffer over the filter prior to adding sample.

10. The radiolabeled RNA is extremely dilute and will not precipitate well by standard ethanol precipitation. Addition of tRNA acts as a co-precipitant and helps keep the precipitated RNA insoluble during the acetone wash step. $LiClO_4$ in acetone efficiently precipitates very dilute and small RNAs. $LiClO_4$ is a dangerous oxidizer—handle with caution.

11. Radiolabeled RNA-specific activity should range from several hundred thousand to a few million CPMs/μL. The half-life of ^{32}P is 14.2 d.

12. For in vitro Ago2 loading assays, loading needs to be performed in the extract before Ago2 is subsequently captured with immuno-affinity resin. If Ago2 is bound to the resin first, duplex siRNA loading is inefficient. Thus, resin and antibody should not be mixed with the siRNA and extract together, but only added after an initial time of Ago2 loading in solution. No buffer is added to the extract since the extract preparation already contains the necessary buffer and salt components.

13. It is important to centrifuge samples prior to addition of resin to remove any precipitation that has occurred during incubation. Otherwise, the precipitated material will co-pellet with the resin in later steps and increase contaminating background levels during analysis.

14. Moving resin to a new tube on the final wash can help reduce background levels, which are important for obtaining clean results for IgG controls.

15. Binding of Ago2 to small RNA is Mg^{2+}-dependent. Addition of excess EDTA helps disrupt this interaction. In addition, EDTA is not soluble in acetone, so it also acts as a co-precipitant to help visualize the precipitated RNA. Addition of tRNA also acts as a co-precipitant and helps keep the precipitated RNA insoluble during the acetone wash step.

16. Glycerol is an important component of the gel. Do not omit. During drying of the gel prior to visualization, glycerol prevents the gel from shrinking and cracking, which can make a gel unusable for publication.

17. Pre-running the gel is necessary to heat it up. It should be very warm or hot to the touch. The heat and the urea in the gel help to keep the RNA denatured.

18. The glass plates should be warm or hot to the touch during the run, but not unbearable. If the plates get too hot they will

crack. If samples "smile," where the middle samples run faster than the outer samples, this indicates uneven heating of the gel. Too much smiling should be avoided by reducing the current passing through the gel.

References

1. Gurtan AM, Sharp PA (2013) The role of miR-NAs in regulating gene expression networks. J Mol Biol 425:3582–3600

2. Wilson RC, Doudna JA (2013) Molecular mechanisms of RNA interference. Annu Rev Biophys 42:217–239

3. Lingel A, Simon B, Izaurralde E, Sattler M (2004) Nucleic acid 3′-end recognition by the Argonaute2 PAZ domain. Nat Struct Mol Biol 11:576–577

4. Bartel DP (2009) MicroRNAs: target recognition and regulatory functions. Cell 136:215–233

5. Berezhna SY, Supekova L, Supek F, Schultz PG, Deniz AA (2006) siRNA in human cells selectively localizes to target RNA sites. Proc Nat Acad Sci USA 103:7682–7687

6. Landthaler M, Gaidatzis D, Rothballer A, Chen PY, Soll SJ, Dinic L, Ojo T, Hafner M, Zavolan M, Tuschl T (2008) Molecular characterization of human Argonaute-containing ribonucleoprotein complexes and their bound target mRNAs. RNA 14:2580–2596

7. Chen PY, Meister G (2005) microRNA-guided posttranscriptional gene regulation. Biol Chem 386:1205–1218

8. Guo H, Ingolia NT, Weissman JS, Bartel DP (2010) Mammalian microRNAs predominantly act to decrease target mRNA levels. Nature 466:835–840

9. Valencia-Sanchez MA, Liu J, Hannon GJ, Parker R (2006) Control of translation and mRNA degradation by miRNAs and siRNAs. Genes Dev 20:515–524

10. Hammond SM, Bernstein E, Beach D, Hannon GJ (2000) An RNA-directed nuclease mediates post-transcriptional gene silencing in Drosophila cells. Nature 404:293–296

11. Matranga C, Tomari Y, Shin C, Bartel DP, Zamore PD (2005) Passenger-strand cleavage facilitates assembly of siRNA into Ago2-containing RNAi enzyme complexes. Cell 123:607–620

12. Meister G, Landthaler M, Patkaniowska A, Dorsett Y, Teng G, Tuschl T (2004) Human Argonaute2 mediates RNA cleavage targeted by miRNAs and siRNAs. Mol Cell 15:185–197

13. Zamore PD, Tuschl T, Sharp PA, Bartel DP (2000) RNAi: double-stranded RNA directs the ATP-dependent cleavage of mRNA at 21 to 23 nucleotide intervals. Cell 101:25–33

14. Ender C, Meister G (2010) Argonaute proteins at a glance. J Cell Sci 123:1819–1823

15. Liu J, Carmell MA, Rivas FV, Marsden CG, Thomson JM, Song JJ, Hammond SM, Joshua-Tor L, Hannon GJ (2004) Argonaute2 is the catalytic engine of mammalian RNAi. Science 305:1437–1441

16. Burroughs AM, Ando Y, de Hoon MJ, Tomaru Y, Suzuki H, Hayashizaki Y, Daub CO (2011) Deep-sequencing of human Argonaute-associated small RNAs provides insight into miRNA sorting and reveals Argonaute association with RNA fragments of diverse origin. RNA Biol 8:158–177

17. Leung AK, Young AG, Bhutkar A, Zheng GX, Bosson AD, Nielsen CB, Sharp PA (2011) Genome-wide identification of Ago2 binding sites from mouse embryonic stem cells with and without mature microRNAs. Nat Struct Mol Biol 18:237–244

18. Elbashir SM, Harborth J, Lendeckel W, Yalcin A, Weber K, Tuschl T (2001) Duplexes of 21-nucleotide RNAs mediate RNA interference in cultured mammalian cells. Nature 411:494–498

19. Watts JK, Corey DR (2010) Clinical status of duplex RNA. Bioorg Med Chem Lett 20:3203–3207

20. Yuan B, Latek R, Hossbach M, Tuschl T, Lewitter F (2004) siRNA selection server: an automated siRNA oligonucleotide prediction server. Nucleic Acids Res 32:W130–W134

21. Braasch DA, Jensen S, Liu Y, Kuar K, Arar K, White MA, Corey DR (2003) RNA interference in mammalian cells by chemically-modified RNA. Biochemistry 42:7967–7975

22. Burnett JC, Rossi JJ (2012) RNA-based therapeutics: current progress and future prospects. Chem Biol 19:60–71

23. Ma JB, Ye K, Patel DJ (2004) Structural basis for overhang-specific small interfering RNA recognition by the PAZ domain. Nature 429:318–322

24. Rand TA, Ginalski K, Grishin NV, Wang X (2004) Biochemical identification of Argonaute 2 as the sole protein required for RNA-induced silencing complex activity. Proc Nat Acad Sci USA 101:14385–14389

25. Rivas FV, Tolia NH, Song JJ, Aragon JP, Liu J, Hannon GJ, Joshua-Tor L (2005) Purified Argonaute2 and an siRNA form recombinant human RISC. Nat Struct Mol Biol 12:340–349

26. Song JJ, Smith SK, Hannon GJ, Joshua-Tor L (2004) Crystal structure of Argonaute and its implications for RISC slicer activity. Science 305:1434–1437

27. Wang Y, Juranek S, Li H, Sheng G, Tuschl T, Patel DJ (2008) Structure of an Argonaute silencing complex with a seed-containing guide DNA and target RNA duplex. Nature 456:921–926

28. Elkayam E, Kuhn CD, Tocilj A, Haase AD, Greene EM, Hannon GJ, Joshua-Tor L (2012) The structure of human Argonaute-2 in complex with miR-20a. Cell 150:100–110

29. Khvorova A, Reynolds A, Jayasena SD (2003) Functional siRNAs and miRNAs exhibit strand bias. Cell 115:209–216

30. Liu Y, Ye X, Jiang F, Liang C, Chen D, Peng J, Kinch LN, Grishin NV, Liu Q (2009) C3PO, an endoribonuclease that promotes RNAi by facilitating RISC activation. Science 325:750–753

31. Meister G (2013) Argonaute proteins: functional insights and emerging roles. Nat Rev Genet 14:447–459

32. Pare JM, LaPointe P, Hobman TC (2013) Hsp90 co-chaperones p23 and FKBP4 physically interact with hAgo2 and activate RNAi-mediated silencing in mammalian cells. Mol Biol Cell 24:2303–2310

33. Iwasaki S, Kobayashi M, Yoda M, Sakaguchi Y, Katsuma S, Suzuki T, Tomari Y (2010) Hsc70/Hsp90 chaperone machinery mediates ATP-dependent RISC loading of small RNA duplexes. Mol Cell 39:292–299

34. MacRae IJ, Ma E, Zhou M, Robinson CV, Doudna JA (2008) In vitro reconstitution of the human RISC-loading complex. Proc Nat Acad Sci USA 105:512–517

35. Lee HY, Zhou K, Smith AM, Noland CL, Doudna JA (2013) Differential roles of human Dicer-binding proteins TRBP and PACT in small RNA processing. Nucleic Acids Res 41:6568–6576

36. Martinez NJ, Chang HM, Borrajo Jde R, Gregory RI (2013) The co-chaperones Fkbp4/5 control Argonaute2 expression and facilitate RISC assembly. RNA 19:1583–1593

37. Gagnon KT, Li L, Chu Y, Janowski BA, Corey DR (2014) RNAi factors are present and active in human cell nuclei. Cell Rep 6:211–221

38. Gagnon KT, Li L, Janowski BA, Corey DR (2014) Analysis of nuclear RNA interference in human cells by subcellular fractionation and Argonaute loading. Nat Protoc 9:2045–2060

Chapter 7

In Vitro Analysis of Ribonucleoprotein Complex Remodeling and Disassembly

Hua-Lin Zhou and Hua Lou

Abstract

Ribonucleoprotein (RNP) complexes play essential roles in gene expression. Their assembly and disassembly control the fate of mRNA molecules. Here, we describe a method that examines the remodeling and disassembly of RNPs. One unique aspect of this method is that the RNA-binding proteins (RBPs) of interest are produced in HeLa cells with or without the desired modification and the RNP is assembled in cellular extracts with synthetic RNA oligonucleotides. We use this method to investigate how ubiquitination of an RBP affects its ability to bind its RNA target.

Key words RBP, RNP assembly/disassembly, Ubiquitination

1 Introduction

Ribonucleoprotein (RNP) complexes form when RNA-binding proteins (RBPs) interact with their cognate target sequences on RNA. RNPs play essential roles in gene expression, as they control many aspects of the life of an mRNA from its birth to maturation, localization, translation, and turnover [1, 2]. Remarkably, RNPs are assembled and disassembled in a highly dynamic and orderly manner, because a different array of RBPs is required to be loaded onto pre-mRNA/mRNA during every step of post-transcriptional mRNA metabolism. Thus, the precise regulation of RNP remodeling is essential in ensuring appropriate gene expression both spatially and temporally, and failure to remodel RNP complexes leads to disruption of downstream events such as mRNA export, translation, and decay.

In order to understand the mechanisms by which RNP remodeling is regulated, we developed an in vitro assay of RNP remodeling [3]. Using this assay, we investigated how a specific form of post-translational modification, ubiquitination, affects the ability of the RBP HuR to interact with its RNA target. To ensure that

Ren-Jang Lin (ed.), *RNA-Protein Complexes and Interactions: Methods and Protocols*, Methods in Molecular Biology, vol. 1421, DOI 10.1007/978-1-4939-3591-8_7, © Springer Science+Business Media New York 2016

the HuR protein is ubiquitinated in its native form, we used proteins prepared from transfected HeLa cells.

Figure 1 illustrates the general procedure of this assay. We first transfected HeLa cells with Myc-tagged HuR and HA-tagged ubiquitin. Forty-eight hours post-transfection, we lysed the cells and prepared total protein lysate. We then assembled RNP complexes on RNA oligonucleotides, either with or without biotin, that contain the HuR target sequence on p21 3′-UTR by incubating the protein lysate with the RNA. The assembled RNP complexes contain both ubiquitinated and non-ubiquitinated Myc-HuR. The central question of this particular study is how ubiquitination of HuR affects its ability to interact with its target RNA. We answered this question by incubating the RNP complexes with recombinant UBXD8 and p97, the complex-remodeling machine known to recognize ubiquitinated protein substrate and extract it from a macromolecular complex [4].

Following the incubation, we tested if the HuR-RNA-containing RNP is remodeled to release RNA from ubiquitinated HuR by conducting the following two assays. The first assay is the RBP-releasing assay. As shown in Fig. 1a, in this assay, biotinylated RNA oligonucleotides were used and the RNP complexes were collected using the streptavidin beads. We examined if ubiquitinated HuR is released from the RNP after incubation with UBXD8-p97 by separating the supernatant from the beads that contains biotinylated RNA and proteins assembled on the RNA, and conducting western blot analysis using the anti-Myc antibody. An increase in the level of ubiquitinated HuR in the supernatant after the incubation with UBXD8-p97 is an indication that ubiquitination of HuR reduces its ability to remain interacting with its target RNA. The second assay is the RNA-releasing assay. As shown in Fig. 1b, in this assay, RNA oligonucleotides without biotin were used and the RNP complexes were collected using anti-Myc agarose beads. We examined the level of RNA oligonucleotides that is complexed with HuR when ubiquitination of HuR is reduced. In this assay, following the incubation of p97 and UBXD8, we isolated RNA from the supernatant, labeled RNA with ^{32}P and detected it by electrophoresis followed by phosphoimaging.

The methods described here can be adapted to examine how RNP remodeling is affected by changes of RNA sequence or structure, as well as changes of RBP sequence, structure or post-translational modifications.

2 Materials

1. *For cell transfection:* HeLa cells, DMEM medium, fetal bovine serum (Life Technologies), plasmid DNA Myc-HuR and HA-Ub, PolyJet (SignaGen Laboratories); for protein

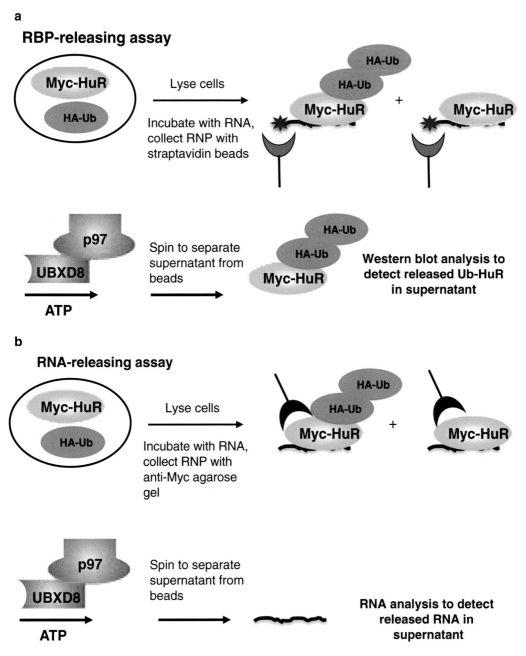

Fig. 1 Diagrams depicting the RNP remodeling assays. (**a**) RBP-releasing assay. HeLa cells are transfected with Myc-HuR and HA-Ub. The protein lysate is incubated with biotinylated p21 RNA. The RNP complexes are collected using streptavidin beads. The RNP complexes are then incubated with recombinant UBXD8 and p97. The bound and released Myc-HuR is detected by western blot analysis using anti-Myc antibodies; (**b**) RNA-releasing assay. HeLa cells are transfected with Myc-HuR and HA-Ub. The protein lysate is incubated with p21 RNA that is not biotinylated. The RNP complexes are collected with anti-c-Myc agarose affinity gel. The RNP complex was then incubated with recombinant UBXD8 and p97. The bound and released p21 RNA is ^{32}P-labeled, purified, run on a denaturing acrylamide gel and detected by phosphoimaging

extraction: proteinase inhibitor cocktail, dialysis cassette, protein concentration assay reagent; for RNP assay: RNA oligos (Dharmacon), TE, glycogen, glycerol, RNaseOUT, Streptavidin beads, anti-c-myc agarose, UBXD8 protein, p97 protein, ADP, AMP-PNP, gradient gel, SDS loading dye, TRIzol, ^{32}P-ATP, polynucleotide kinase.

2. *Cold lysis buffer:* 1 mM PMSF, 20 units/ml RNaseOut, 10 mM Tris–HCl, pH 7.4, 10 mM NaCl, 3 mM MgCl$_2$, 0.1 mM DTT, 0.5 % NP40.

3. *Dialysis buffer:* 20 % glycerol, 20 mM HEPES, pH 7.6, 1 mM EDTA, 100 mM KCl, 2 mM DTT, 0.1 mM PMSF.

4. *10X Binding buffer:* 200 mM HEPES, pH 7.5, 1 M NaCl, 30 mM MgCl$_2$.

5. *1X Wash buffer:* 10 mM Tris–HCl, pH 8.0, 1 mM EDTA, 0.5 M NaCl.

6. *10X Reaction buffer:* 200 mM HEPES, pH 7.5, 1 M NaCl, 30 mM MgCl$_2$.

7. *Elution buffer:* 10 mM Tris–HCl, pH 6.0, 2 M NaCl, 1 mM EDTA, 0.5 M MgCl$_2$.

8. Nutator-mixer.

3   Methods

3.1   Cell Culture and Transfection

1. Split 1 plate (10 cm) of HeLa cells that is 90 % confluent into 20 plates (10 cm) in DMEM complete medium containing 10 % Fetal Bovine Serum (FBS) and 1 % Penicillin-Streptomycin. After 2 days, the plates should be approximately 70 % confluent. At this point, transfection HeLa cells using PolyJet by following manufacturer's instructions (*see* **Note 1**). The procedure is briefly described below.

2. Feed cells with 10 ml of fresh DMEM 30–60 min before transfection.

3. For each plate, dilute 8 μg of plasmid DNA into 200 μl of serum-free DMEM. Gently pipette up and down to mix.

4. For each plate, dilute 16 μl of PolyJet™ reagent into 200 μl of serum-free DMEM. Gently pipette up and down 3–4 times to mix.

5. Add the diluted PolyJet™ reagent to the diluted DNA solution. Immediately pipette up and down 3–4 times to mix.

6. Incubate for 10–15 min at room temperature to allow PolyJet™/DNA complexes to form.

7. Add the 400 μl PolyJet™/DNA mixture drop-wise onto the medium in each plate and gently swirl the plate.

8. Remove PolyJet™/DNA complex-containing medium and replace with fresh DMEM-containing 10 % FBS 17–18 h post transfection.

3.2 Cell Lysis

1. Two days post transfection, wash cells on the plate with cold 1XPBS twice. Scrape cells in 2 ml cold 1XPBS. Move all of cells to a 50 ml conical tube. Collect cells by centrifugation for 1 min at $2000 \times g$ at 4 °C. Remove as much supernatant as possible.

2. To the cell pellet, add 1 ml lysis buffer, which is supplemented with proteinase inhibitor cocktail immediately before use (*see* **Note 2**). Rock the tube on a Nutator mixer for 10 min at 4 °C. Centrifuge for 5 min at $10,000 \times g$ at 4 °C. Transfer the supernatant to a 1.5 ml tube.

3. Add 500 µl lysis buffer with proteinase inhibitor cocktail to the pellet, resuspend and rock for 5 min at 4 °C. Centrifuge for 5 min at $10,000 \times g$ at 4 °C. Remove supernatant and combine with that from **step 2**. There should be 1.5 ml of total supernatant.

4. Dialyze the supernatant in the dialysis buffer for 4 h at 4 °C with Slide-A-Lyzer Dialysis Cassette. Change to fresh dialysis buffer after 2 h.

5. Measure protein concentration with the Bio-Rad Protein Assay Dye Reagent Concentrate. Dilute the protein lysate to 18 µg/µl and aliquot the lysate into 0.5 ml tube, 50 µl/tube. Flash freeze the aliquots in liquid nitrogen, and store them at –80 °C for future use (*see* **Note 3**).

3.3 RNA Oligonucleotides

1. Order wild type and mutant RNA oligonucleotides. In our case, the RNAs are 42 nucleotides in length and contain sequences of the p21 3′-untranslated region that have either wild type or mutated HuR binding sites (wild type: UCUUAAUUAUUAUUUGUGUUUUAAUUUAAACAC CUCCUCAUG; mutant: UCGGACGUGGGGAGUCGUG CGGUAGGGUGCACACCUCCUCACG). The RNAs used for RBP-releasing assay are biotinylated at their 5′ end. The RNAs used for RNA-releasing assay have a 2′-bis(2-acetoxyethoxy)methyl (ACE) protection instead of biotin at their 5′ ends. Order the oligonucleotides at the 0.01 µmol scale. Dissolve the oligonucleotides in 2 ml TE buffer to make 5 µM stock, aliquot RNA in 50 µl/tube and store at –80 °C. These RNA oligonucleotides are named as WT-BIO and Mut-BIO (for RBP-releasing assay) or WT-w/o-BIO and Mut-w/o-BIO (for RNA-releasing assay).

2. Deprotection of RNA oligonucleotides. The RNA oligonucleotides that come with the proprietary 2′-ACE protection are

deprotected before use. We use a modified version of the deprotection protocol from Dharmacon, briefly described as the following:

Precipitate RNA from one 50 μl aliquot. To the aliquot, add 2 μl glycogen (30 mg/ml), and 100 μl 100 % ethanol. Incubate at −80 °C for at least 30 min. Spin for 15 min at 14,000×g in cold room. Remove supernatant and wash pellet with 1 ml 100 % ethanol, spin for 1 min at 14,000×g, remove ethanol and wash with ethanol once. Air-dry pellet.

Resuspend RNA in 100 μl 2′-deprotection buffer (supplied with oligonucleotide). Vortex for 10 s and centrifuge for 10 s. Incubate at 60 °C for 30 min. Add 350 μl TE, 2.5 μl glycogen (30 mg/ml) and 1 ml ethanol. Incubate at −80 °C for 30 min. Spin down for 15 min at 14,000×g in cold room, wash 2X with 1 ml ethanol. Air-dry pellet. Resuspend RNA in 40 μl TE and store at −80 °C.

3.4 RNP Formation

3.4.1 Formation of RNP Complex for Protein-Releasing Assay

1. Dilute 5 μM WT-BIO or Mut-BIO to 500 nM by mixing 2 μl RNA stock with 18 μl TE. Heat the 20 μl WT-BIO (500 nM) or Mut-BIO (500 nM) to 70 °C for 5 min (*see* **Note 4**). Cool down to room temperature. Set up the following reaction:

    - 50 μl cell extract.
    - 10 μl biotinylated RNA oligonucleotide (500 nM).
    - 50 μl 10X binding buffer.
    - 100 μl 50 % w/v glycerol.
    - 25 μl RNaseOUT.
    - 50 μl 10X proteinase cocktail inhibitor.
    - Up to 500 μl ddH$_2$O.

2. Incubate for 3 h at 4 °C with gentle rocking on a Nutator.

3. Wash streptavidin agarose beads (~100 μl/sample 1:1 slurry). Add 1 ml 1X wash buffer to 100 μl beads (1:1 slurry) in a 1.5 ml microfuge tube (*see* **Note 5**). Spin down for 1 min at 1000×g. Remove supernatant. Repeat washing for five times.

4. Add streptavidin agarose beads into reaction sample. Incubate beads with RNP for 2 h at 4 °C with gentle rocking on a Nutator to form the beads-RNP.

5. Purification of RNP. Add 1 ml wash buffer to beads and rock for 5 min in cold room. Spin down for 1 min at 1000×g at 4 °C. Remove supernatant. Repeat washing five times. For the last wash, use 1X reaction buffer instead of washing buffer. After the last wash, remove supernatant as much as possible with a 30G needle. Save a small fraction of the supernatant after the last wash as the unbound fraction.

6. Resuspend beads with 100 μl 1X reaction buffer and store at 4 °C for future use. The resulting RNA/RBP complexes can be

used for the RNP remodeling assay. To detect if RNP complex is formed on the streptavidin agarose beads, 5 μl beads are removed and incubated with 100 μl of elution buffer at 65 °C for 5 min. Spin down for 1 min at $1000 \times g$ and remove supernatant to new 1.5 ml tube. Western blot analysis is carried out to detect the presence of the RBP of interest in the RNA/RBP complex.

3.4.2 Formation of RNP Complex for RNA-Releasing Assay

1. Dilute WT-w/o-BIO or Mut-w/o-BIO RNA oligonucleotide to 50 μM with TE (*see* **Note 4**). Incubate 20 μl diluted RNA at 70 °C for 5 min. Cool down at room temperature. Set up the following reaction:

   - 50 μl cell extract.
   - 10 μl RNA (50 μM).
   - 50 μl 10X binding buffer.
   - 100 μl 50 % w/v glycerol.
   - 25 μl RNaseOUT.
   - 50 μl 10X proteinase cocktail inhibitor.
   - Up to 500 μl ddH$_2$O.

2. Incubate for 3 h at 4 °C with gentle rocking on a Nutator.

3. Wash Anti-c-Myc Agarose Affinity Gel antibody (~100 μl/sample 1:1 slurry) (*see* **Note 5**). Add 1 ml 1X wash buffer to 100 μl beads in a 1.5 ml microfuge tube. Spin down for 1 min at $1000 \times g$. Remove supernatant. Repeat washing five times. In our experiment, as over-expressed RBP is Myc-tagged, we use the commercial Anti-c-Myc Agarose Affinity Gel antibody.

4. Same as the **steps 4** and **5** in preparation of RNP complex for the protein-releasing assay.

3.5 Releasing Assay

3.5.1 RBP-Releasing Assay

1. Resuspend the RNP complex prepared for the protein-releasing assay with a 200 μl cut-tip (tip cut to create bigger opening). Set up the following reaction:

   - 25 μl beads-RNP.
   - 25 μl 10X binding buffer.
   - 12.5 μl ATP (5 mM).
   - 5 μl recombinant p97 (ATPase) protein (12.5 μM) (*see* **Note 6**).
   - 5 μl UBXD8 (co-factor) protein (12.5 μM) (*see* **Note 6**).
   - 50 μl 50 % w/v glycerol.
   - 12.5 μl RNaseOUT.
   - 25 μl 10X proteinase cocktail inhibitor.
   - ddH$_2$O up to 250 μl.

2. Incubate the reaction mixture at 30 °C. At 0, 15, 30, and 60 min time points, resuspend the reaction solution and remove 50 μl of the mixture including beads to a new 1.5 ml tube.

3. Spin for 1 min at 1000×*g* and remove 40 μl supernatant containing released RBP into a new 1.5 ml tube (*see* **Note 7**). To examine the amount of RBP still bound to RNA after the releasing assay, spin for 1 min at 1000×*g* again and remove supernatant as much as possible with a 30G needle (*see* **Note 8**). Elude the bound RBP with 200 μl of elution buffer at 65 °C for 5 min.

4. To examine if the RNP remodeling is dependent on ATP, ADP. or AMP-PNP, an ATP analog can be used to replace ATP in the RBP-releasing reaction.

5. Run 20 μl released RBP and 20 μl bound RBP on a 4–15 % gradient Criterion Tris–HCl Gel. Add 6.6 μl 4x SDS loading dye, heat at 95 °C for 5 min and load the samples on the gel.

6. Detect the released RBP and bound RBP by western blot analysis using an appropriate antibody (*see* **Note 9**). Anti-Myc antibody was used in our assay. Quantification of the releasing assay was performed using the ImageJ software on the GE Healthcare Typhoon Trio Variable Mode Imager. The released RBP was normalized to the bound RBP.

3.5.2 RNA-Releasing Assay

1. Resuspend RNP complex prepared in RNP complex for RNA releasing with a 200 μl cut tip. Set up the following reaction:

   - 25 μl Beads-RNP.
   - 25 μl 10X binding buffer.
   - 12.5 μl ATP (5 mM).
   - 5 μl recombinant p97 (ATPase) protein (12.5 μM).
   - 5 μl UBXD8 (co-factor) protein (12.5 μM).
   - 50 μl 50 % w/v glycerol.
   - 12.5 μl RNaseOUT (Life Technologies, cat# 10777-019).
   - 25 μl 10X proteinase cocktail inhibitor.
   - ddH$_2$O up to 250 μl.

2. Incubate the reaction mixture at 30 °C. At 0, 15, 30. and 60 min time points, resuspend the reaction solution and remove 50 μl to a new 1.5 ml tube.

3. Spin for 1 min at 1000×*g* and remove 40 μl supernatant containing released RNA into a new 1.5 ml tube. To examine the amount of RNA still bound to the RBP after the releasing assay, spin for 1 min at 1000×*g* again and remove supernatant as much as possible with a 30G needle (*see* **Note 8**). Elute the bound RNA with 1 ml TRIzol.

4. Isolate the released and bound RNA following the TRIzol protocol. At the last step, dissolve the released RNA pellet in 10 μl TE and the bound RNA pellet in 50 μl TE.

5. Label the 5′-end of the RNA oligonucleotides with γ-^{32}P-ATP using T4 polynucleotide kinase. Set up the following reaction:

   - 5 μl RNA.
   - 3 μl 10X PNK buffer.
   - 1 μl T4 PNK.
   - 1 μl γ^{32}P-ATP (ICN Biomedical, cat# 38101X, 10 μCi/μl).
   - Up to 30 μl with ddH$_2$O.
   - Incubate for 30 min at 37 °C.

6. Add 370 μl TE and 400 μl phenol-chloroform (1:1) into the reaction tube, vortex, and spin for 5 min at 14,000×g. Remove supernatant to a new tube. Add 2.5 μl glycogen and 1 ml ethanol. Incubate at–80 °C for 1 h. Spin for 30 min at 14,000×g at 4 °C. Wash 2X with 1 ml ethanol. Air-dry.

7. Dissolve RNA pellet in 10 μl TE. Separate RNA on a denaturing 10 % polyacrylamide gel. The GE Healthcare Typhoon Trio Variable Mode Imager is used to calculate the released RNA amount.

4 Notes

1. High transfection efficiency is important for this assay. Use DMEM but not Opti-MEM to dilute PolyJet™ reagent and DNA. Do not vortex the mixture. Examine the RBP expression in the transfected cell by western blot analysis.

2. Lyse cells without SDS. In order to obtain high concentrations of cell extract, adjust the amounts of lysis buffer.

3. If post-translational modification of RBP is important for releasing, examine the post-translational modification of the RBP in the cell extract by western blot analysis as a control.

4. To reduce the potential RNP reassembly after the RBP-releasing assay, try not to use more RNA than RBP in this assay. Likewise, to reduce the potential RNP reassembly after the RNA-releasing assay, try not to use more RBP than RNA in this assay.

5. To reduce nonspecific binding, do not use too much beads and wash beads-RNP more than six times to remove nonspecific binding.

6. Carefully purify the recombinant ATPase and its co-factor protein to ensure that these recombinant proteins have activity.

7. After the beads-RNP disassociation reaction, remove up to only 2/3 supernatant to new tube to prevent from taking any beads.

8. 30G needles are used to remove as much as supernatant without accidentally taking any beads.

9. Gradient gels are used to better separate ubiquitinated proteins.

Acknowledgements

This work was supported by an NIH grant (NS-049103) and a DOD grant (NF060083) to Hua Lou. Hua-Lin Zhou was supported by post-doctoral fellowships from the American Heart Association (0725346B and 09POST2250749).

References

1. Moore MJ (2005) From birth to death: the complex lives of eukaryotic mRNAs. Science 309:1514–1518

2. Keene JD (2007) RNA regulons: coordination of post-transcriptional events. Nat Rev Genet 8:533–543

3. Zhou H-L, Geng C, Luo G, Lou H (2013) The p97-UBXD8 complex destabilizes mRNA by promoting release of ubiquitinated HuR from mRNP. Genes Dev 27:1046

4. Meyer H, Bug M, Bremer S (2012) Emerging functions of the VCP/p97 AAA-ATPase in the ubiquitin system. Nat Cell Biol 14:117–123

Chapter 8

Single-Turnover Kinetics of Methyl Transfer to tRNA by Methyltransferases

Ya-Ming Hou

Abstract

Methyl transfer from *S*-adenosyl methionine (abbreviated as AdoMet) to biologically active molecules such as mRNAs and tRNAs is one of the most fundamental and widespread reactions in nature, occurring in all three domains of life. The measurement of kinetic constants of AdoMet-dependent methyl transfer is therefore important for understanding the reaction mechanism in the context of biology. When kinetic constants of methyl transfer are measured in steady state over multiple rounds of turnover, the meaning of these constants is difficult to define and is often limited by non-chemical steps of the reaction, such as product release after each turnover. Here, the measurement of kinetic constants of methyl transfer by tRNA methyltransferases in rapid equilibrium binding condition for one methyl transfer is described. The advantage of such a measurement is that the meaning of kinetic constants can be directly assigned to the steps associated with the chemistry of methyl transfer, including the substrate binding affinity to the methyltransferase, the pre-chemistry re-arrangement of the active site, and the chemical step of methyl transfer. An additional advantage is that kinetic constants measured for one methyl transfer can be correlated with structural information of the methyltransferase to gain direct insight into its reaction mechanism.

Key words AdoMet-dependent methyl transfer, Rapid equilibrium binding

1 Introduction

Methyl transfer contributes to a broad spectrum of biological reactions in all living organisms, being involved in biosynthesis, metabolism, detoxification, signal transduction, protein sorting and repair, and nucleic acid processing and expression. The products of methyl transfer therefore play major roles from fetal development to brain function. For example, methylation of DNA is required to regulate gene expression, while methylation of mRNA is implicated to control the stability of message [1]. Methylation of hormones, neurotransmitters, and signal transduction pathways helps to regulate the action of each, while methylation of phospholipids keeps membranes fluid and receptors mobile. Methylation of the cobalamin cofactor of methionine synthase provides a repair

Ren-Jang Lin (ed.), *RNA-Protein Complexes and Interactions: Methods and Protocols*, Methods in Molecular Biology, vol. 1421, DOI 10.1007/978-1-4939-3591-8_8, © Springer Science+Business Media New York 2016

mechanism for this enzyme [2]. The most common form of methyl transfer uses AdoMet (Fig. 1a, also known as SAM) as the methyl donor, which is synthesized by condensation of methionine with ATP by methionine adenosyl transferase (or SAM synthetase) [3],

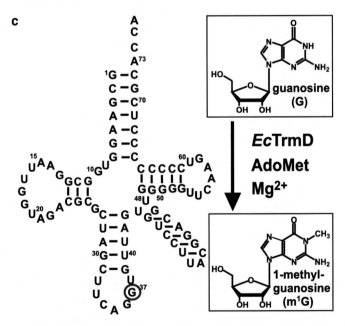

Forward primer 5′-GCT <u>TAA TAC GAC TCA CTA TAG</u> CGA AGG TGG CGG AAT TGG TAG ACG CGC TAG CTT CAG GTG T- 3′

Reverse primer 5′ -<u>mUmGG</u> TGC GAG GGG GGG GAC TTG AAC CCC CAC GTC CGT AAG GAC ACT AAC ACC TGA AGC- 3′

Fig. 1 AdoMet-dependent methyl transfer to G37-tRNA to synthesize m¹G37-tRNA. (**a**) The chemical structure of AdoMet. (**b**) Design of the former and reverse primers for synthesis of the template for transcription of *Ec*tRNA^Leu. The T7 promoter sequence is *underlined in red*, and the two 5′-terminal nucleotides in the reverse primer, which are 2′-*O*-methylated, are *underlined in red* as well. (**c**) The sequence and cloverleaf structure of *Ec*tRNA^Leu, showing the target base G37 (*circled in red*) and its chemical structure and the product m¹G37. The methyl transfer reaction is catalyzed by *Ec*TrmD, using AdoMet as the methyl donor and an Mg²⁺ ion as the catalyst [33]. The folding of the cloverleaf structure is driven by sequence complementarity among regions of the tRNA. The numbering of nucleotide sequence of *Ec*tRNA^Leu is based on the standard 76-nucleotide framework of tRNA [34]

and the product of methyl transfer is *S*-adenosyl homocysteine (AdoHcy). The ratio of AdoMet versus AdoHcy is known as the methylation potential and the level of this methylation potential can regulate the circadian length in mammals [4], indicating an even broader scope of regulation by AdoMet-dependent methyl transfer reactions. AdoMet is the second most widely used enzyme substrate [5], following ATP. The selective preference for AdoMet over other methyl donors, such as folate, reflects the favorable energetics resulting from nucleophilic attack on the positively charged methyl group of the sulfonium center (Fig. 1a). The energy release upon methyl transfer from AdoMet is ~17 kcal/mol [5], more than twice of the energy release upon hydrolysis of ATP to ADP and Pi. The target atoms of methyl transfer from AdoMet are also broadly ranged, including nitrogen, oxygen, carbon, and sulfur.

The widespread importance of AdoMet in biology has led to the discovery of at least 5 classes (class I–V) of structurally distinct methyltransferases [6]. The distinction among the 5 classes is in the topological fold of each class that binds AdoMet. This diversity is paralleled only by the diversification of multiple classes of ATP-dependent protein kinases and phosphoryl transferases. Why nature evolved 5 classes of AdoMet-fold is difficult to address, but the identification of different AdoMet conformations among these folds suggests mechanistic distinctions. For example, the class I fold is the most common of the 5 classes and it includes the greatest majority of AdoMet-dependent methyltransferases. The class I fold binds AdoMet in the open space of a dinucleotide-fold that is similar in structure to the NAD(P)-binding fold of the Rossmann-fold [6]. A major feature of the class I fold is that AdoMet adopts a straight conformation, where the adenosine and methionine moieties are extended in opposite direction from each other [6]. In contrast, other classes are much less common and they are characterized by positioning AdoMet in a bent conformation, where the two component moieties can face each other [6]. Of these, class IV stands out, because it binds AdoMet to the bottom of a topological knotted protein fold in a deep cleft. Proteins with a knotted fold are rare in the protein data bank and the class IV fold is distinguished because it is made up of three passages of the protein backbone in and out of a loop in a structure known as the trefoil-knot fold [7, 8]. Notably, the bent conformation of AdoMet in the trefoil-knot fold is the most pronounced, where the two component moieties are spatially facing each other almost at a right angle [6]. This is in a striking contrast from the straight conformation of AdoMet in class I methyltransferases.

An example of two different conformations of AdoMet leading to two distinct kinetic mechanisms of methyl transfer is found in the pair of the TrmD and Trm5 methyltransferases. These two enzymes are analogous to each other, because while they catalyze the same methyl transfer reaction, they do not use homologous motifs but instead unrelated motifs [9–13]. Specifically, TrmD is broadly conserved in the bacterial domain [14–16], and Trm5 is conserved in the eukaryotic and archaeal domain [9, 17], while both catalyzing methyl transfer from AdoMet to the N^1 atom of G37 on the 3′ side of the tRNA anticodon [9] (Fig. 1b). Because the synthesized m^1G37-tRNA is a critical determinant to prevent ribosomes from shifting into the +1-frame [16, 18, 19], both TrmD and Trm5 are essential for cell growth [14, 18]. However, despite their functional similarity, TrmD is a member of class IV and uses the trefoil-knot fold to bind AdoMet to the interface of an obligated dimer structure [10, 11, 20], whereas Trm5 is a member of class I and uses the Rossmann dinucleotide-fold to bind AdoMet to an active monomeric structure [12, 13]. Their structural distinction confers mechanistic distinction in virtually all aspects of the methyl transfer reaction. For recognition of AdoMet, TrmD binds it in the signature bent conformation using a lock-and-key mode, whereas Trm5 binds it in the characteristic straight conformation using an induced-fit mode [21]. For methyl transfer to N^1 of G37-tRNA, TrmD stabilizes only the anticodon stem-loop of the tRNA [22] but it carefully discriminates among all three major functional groups of the base with high stringency [23]. In contrast, Trm5 requires the entire tRNA L-shaped structure as the substrate [22] but it examines without high stringency only two of the three functional groups of the base [23]. Most importantly, TrmD catalyzes methyl transfer in a kinetic mechanism limited by the chemistry-associated steps, whereas Trm5 catalyzes methyl transfer in a mechanism limited by the non-chemical step involving release of the m^1G37-tRNA product [24]. Together, these observations show that the structural distinction between different conformations of AdoMet in TrmD and Trm5 has resulted in the kinetic distinction of methyl transfer of these two enzymes. This emphasizes the importance to perform kinetic analysis to correlate with structure as the basis to understand the mechanism of each methyl transfer.

For decades, kinetic analysis of methyl transfer, particularly to RNA substrates, has been dominated by steady-state assays. In these assays, the enzyme is in catalytic amounts, the RNA substrate is in excess, and the reaction proceeds over multiple rounds of turnover. Fitting the initial rate of methyl transfer (V_0) as a function of the RNA substrate to the Michaelis–Menten equation yields the catalytic turnover k_{cat} and the Michaelis constant K_m. However, because each parameter is a composite term of multiple turnovers, k_{cat} does not mean the rate constant of methyl transfer and K_m does not mean the binding affinity of the substrate to the enzyme, thus limiting insight into the reaction mechanism. The recent development of pre-steady-state assays has overcome this barrier [25]. In pre-steady-state assays, the enzyme is in excess, the RNA substrate is limiting, and the reaction proceeds only once. This affords two advantages. First, because the rate constant (k_{obs}) directly reports the kinetics of one methyl transfer, the extrapolation from the plot of k_{obs} vs. concentration leads to determination of kinetic parameters intrinsically associated with the methyl transfer. Second, because the enzyme is saturating relative to the RNA substrate, the enzyme–substrate affinity is measured under the condition of rapid binding equilibrium, leading to the determination of a true equilibrium binding affinity. It is with such pre-steady-state assays that TrmD and Trm5 are distinguished in their kinetic mechanism of methyl transfer [24].

This chapter describes the procedures to perform pre-steady-state assays to measure AdoMet-dependent methyl transfer to synthesize m^1G37-tRNA in one turnover by TrmD or Trm5. The representative enzymes are the bacterial *E. coli* TrmD (*Ec*TrmD) and the archaeal *Methanococcus jannaschii* Trm5 (*Mj*Trm5), each of which has been well characterized in our lab and has a high-resolution crystal structure in complex with AdoMet or the product AdoHcy [11–13, 20]. These assays have also been applied to study human Trm5 (*Homo sapiens* Trm5, *Hs*Trm5) to reveal similar kinetic parameters [17] to those of *Mj*Trm5, demonstrating the reliability of these assays for different members of the same methyltransferase family. The chosen assay temperature for each enzyme is 37 °C for *Ec*TrmD and *Hs*Trm5 and 55 °C for *Mj*Trm5. The G37-tRNA substrate for each enzyme is *Ec*tRNA^{Leu}, *Hs*tRNA^{Cys}, and *Mj* tRNA^{Cys}, respectively.

Two key features of these assays are worthy of mentioning. First, these assays use radioactive [^{3}H-methyl]-AdoMet as the methyl donor (where the methyl group is ^{3}H-labeled). Upon methyl transfer, the ^{3}H-methyl group is transferred to the tRNA substrate and becomes acid precipitable as an integral part of the tRNA, whereas the unincorporated ^{3}H-methyl group is not acid precipitated and is washed away. The product m^1G37-tRNA carrying the ^{3}H-methyl in m^1G37 is then quantified using scintillation counting and the fractional conversion from the substrate G37-tRNA is calculated. An alternative and also quantitative approach [26] is to prepare the G37-tRNA substrate with a site-specifically placed ^{32}P at the 5′ end of G37. After methyl transfer, both the G37 and m^1G37-tRNA are digested to single 5′-monophosphate nucleotides, which are resolved by one-dimensional TLC (thin layer chromatography). Due to their different chromatography mobility, m^1G37 is separated from G37 and their amounts can be quantified by analysis on a phosphorimager screen. Compared to the alternative method, the method using ^{3}H-AdoMet is simpler, at least with respect to the substrate G37-tRNA preparation.

Second, for methyl transfer that proceeds on the time scale around 0.1 s^{-1} or faster, such as the EcTrmD ($k_{obs} = 0.09 \pm 0.01$ s^{-1}) or MjTrm5 ($k_{obs} = 0.12 \pm 0.03$ s^{-1}) reaction [24], meaning that one turnover occurs in less than 10–11 s, the measurement is carried out on a rapid mixing and quench instrument. Our lab uses the KinTek RQF-3 model (KinTek Corp, Texas), which operates with a computer-controlled panel for rapid mixing the contents of two syringes and rapid quenching of the reaction over the time course of methyl transfer. For methyl transfer that proceeds on a slower time scale, the mixing and sampling can be performed without the instrument. We have found that data obtained from the same protocol performed with or without the instrument are reproducible, with the standard deviation being less than 20 %.

2   Materials

2.1 Preparation of the DNA Template for Synthesis of the G37-tRNA Substrate

All three enzymes, EcTrmD, MjTrm5, and HsTrm5, can use the transcript of G37-tRNA, made by in vitro transcription and devoid of any post-transcriptional modification, as the substrate for AdoMet-dependent methyl transfer. In the ^{3}H-based assay, the transcript is made according to a DNA template, using analytical grade reagents and autoclaved double deionized water (ddH$_2$O).

1. Sequenase, a T7 DNA polymerase lacking the editing domain. This enzyme can be purchased or purified from an overproducer strain [27].

2. DNA oligonucleotides for construction of the template sequence are synthesized by a commercial supplier and used directly without further purification.

3. Sequenase buffer (5×): 12.5 mM DTT, 250 mM NaCl, 100 mM MgCl$_2$, 200 mM Tris–HCl, pH 7.5.

4. dNTPs: 25 mM solution of each dNTP.

5. TE buffer: 10 mM Tris–HCl, pH 8.0, 1 mM EDTA.

6. 3 M ammonium acetate.

7. Ethanol, absolute and 70 %.

8. Speedvac system.

2.2 Synthesis of G37-tRNA by In Vitro Transcription and Purification

1. DNA template from Subheading 2.1 or prepared as a linearized plasmid.

2. T7 RNA polymerase, commercially available or purified from an over-expression clone such as pAT1219 in BL21(DE3) [28].

3. 1 M Tris–HCl, pH 8.0.

4. 1 M MgCl$_2$.

5. 0.1 M spermidine.

6. 0.5 M DTT.

7. 0.5 % Triton-X100.

8. 50 mM each of ATP, CTP, GTP, and UTP, pH 8.0.

9. 200 mM GMP, pH 8.0.

10. 0.5 M EDTA, pH 8.0.

11. **Items 5–7** from Subheading 2.1.

12. RNA loading dye solution: 7 M urea, 0.05 % xylene cyanol and 0.05 % bromophenol blue in 1x TBE buffer.

13. A hand-held UV lamp.

2.3 Denaturing Polyacrylamide Gel Analysis

1. 20× TBE buffer: dissolve 486.6 g of Tris-base, 37.2 g of EDTA (disodium salt) and 220 g of boric acid in water to a final volume of 2 L.

2. Stock solution of 12 % acrylamide gel solution with 7 M urea (12 % PAGE/7 M urea): mix 200 g of acrylamide:bis-acrylamide (29:1), 700 g of urea, 83.25 mL 20x TBE, and 1655 mL of water. Stir to dissolve all components at room temperature overnight. Adjust the final volume to 2 L with additional water, filter the solution through a 0.22 μm filtering unit and store the solution in a dark brown bottle at room temperature.

3. 1× TBE buffer: 89 mM Tris–HCl, pH 8.0, 89 mM boric acid, and 2 mM EDTA. Dilute 20× TBE 19:1 with water.

4. Mini gel electrophoresis system (e.g., from BioRad).

5. Preparative gel electrophoresis system with plates of dimensions ~400×200 mm and 2.0 mm spacers and comb. Depending on your system, heavy packaging tape and metal binder clips may be required to seal the plates when pouring the gel.

6. 10 % ammonium persulfate (APS).

7. N,N,N',N'-tetramethyl-ethane-1,2-diamine (TEMED).

2.4 Methyl Transfer to G37-tRNA to Synthesize m¹G37-tRNA

1. The substrate G37-tRNA synthesized by in vitro transcription and purified by denaturing 12 % PAGE/7 M urea.

2. Recombinant methyltransferases *Ec*TrmD (purified from an over-producer *E. coli* clone) [29], *Mj*Trm5 (purified from an over-producer *E. coli* clone) [9], and *Hs*Trm5 (purified from an over-producer *E. coli* clone) [17]. Each enzyme is expressed as a fusion to a His-tag and is purified from *E. coli* cells using a metal affinity resin (the cobalt resin Talon of CloneTech or the nickel resin of Promega). Preferably, each enzyme should be

stored at ~400 μM in 1x methyl transfer buffer and 40 % glycerol.

3. ^{3}H-AdoMet commercial solution (purchased from Perkin Elmer, NET155H, 60 Ci/mmol, 6.6 μM, 0.55 μCi/μL).

4. Unlabeled AdoMet (1 mM): dissolve 0.57 mg of AdoMet in 1.0 mL water with 1 μL concentrated H_2SO_4 (final concentration of the acid = 12 mM). Store the solution at –20 °C.

5. A working stock of AdoMet: mix 200 μL ^{3}H-AdoMet commercial solution (**step 3**) with 90 μL unlabeled AdoMet (1 mM, **step 4**) to give a final concentration of ~300 μM AdoMet with a specific activity of 2650 dpm/pmole (*see* **Note 1**).

6. TE buffer.

7. 1 M Tris–HCl, pH 8.0.

8. 3 M NH$_4$Cl.

9. 2 M KCl.

10. 1 M MgCl$_2$.

11. 1.0 M DTT.

12. 50 mM EDTA.

13. 10 mg/mL bovine serum albumin (BSA).

14. Heat-cool buffer: 10 mM Tris–HCl, pH 8.0, and 20 mM MgCl$_2$.

15. Whatman 3MM filter pads.

16. 5 % (w/v) trichloroacetic acid (TCA).

17. 95 % ethanol.

18. Ether.

19. A liquid scintillation counter.

20. Centri-Spin-20 spin columns (Princeton separations).

3   Methods

3.1 Preparation of DNA Template for Transcription of G37-tRNA

1. Design two DNA oligonucleotide primers with an overlapping sequence, such that upon hybridization the hybrid encodes the promoter sequence of T7 RNA polymerase and the coding sequence of a G37-tRNA substrate. For example, based on the sequence of *Ec*tRNA^{Leu}, the forward primer encodes a 5′ anchor of 3 nucleotides (5′-GCT), followed by the T7 RNA polymerase promoter sequence (underlined by red), and by the sequence for nucleotides G1 to U46, while the reverse primer encodes the sequence complements for nucleotides A76 to U33, followed by a 3′ anchor of 3 nucleotides (AGC-3′). In this design, the 5′-terminal two nucleotides of the

reverse primer contain a 2′-O-methyl ribose, which help to terminate transcription by T7 RNA polymerase without non-template nucleotide addition [30] (Fig. 1b). The hybrid of the two primers provides the template sequence for synthesis of the transcript of EctRNA^Leu (Fig. 1c).

2. Mix the two primers at a final concentration of 4 μM each in a final volume of 1 mL. Add 200 μL of a 5x Sequenase buffer and water to bring the volume to 970 μL.

3. Divide the solution containing the two primers into two equal volumes. Incubate for 2.5 min at 80 °C to denature any secondary structures of the primers. Remember to secure the top of each tube with a plastic clamp.

4. Spin the tubes briefly in a microfuge and place them in an ice bath to allow the two primers to anneal. Reserve one aliquot of 3.0 μL as a control for subsequent electrophoretic analysis of the Sequenase reaction.

5. Add 10 μL of 25 mM dNTPs and 2.5–10 μL of Sequenase to each solution at 37 °C. The amount of Sequenase to use is determined empirically by adding gradual amounts of the enzyme to each 100 μL reaction, followed by analysis for product formation on a mini-size 12 % PAGE/7 M urea gel for 30–40 min at 200 V in 1x TBE. The extension of primers by Sequenase will synthesize a double-stranded DNA that migrates slower than either of the two starting primers.

6. Incubate the rest of the Sequenase reaction overnight at 37 °C. Check for completion of the reaction by analysis of a 3 μL aliquot of one of the reactions on a mini-size 12 % PAGE/7 M urea gel (see **Note 2**).

7. Divide the total 1 mL reaction into 3 × 333 μL aliquots in separate tubes.

8. Add 33 μL of 3 M ammonium acetate (1/10 volume) and cold 1.0 mL ethanol (3 volumes) to each tube. Keep the solution at –20 °C for 30 min and then spin for 15 min at maximum speed in a refrigerated microfuge. Pour off the ethanol supernatant and wash the pellets with 1.0 mL cold 70 % ethanol. Dry pellets for 5 min in a Speedvac.

9. Dissolve each set of pellets in 100 μL of TE buffer.

10. Determine the concentration of the DNA by measuring the absorbance at 260 nm (1 OD unit = 50 μg/mL), which is usually in the range of 25 μM.

11. Store the DNA template at –20 °C, which should be stable for several months. For each transcription reaction, a 40 μL aliquot of the synthesized template is used for 1 mL of T7 transcription reaction.

3.2 Synthesis of G37-tRNA by In Vitro Transcription

1. In a 1 mL T7 RNA polymerase transcription reaction, add the following reagents in order:

	Stock concentration	Volume	Final concentration
DNA template	25 μM	40 μL	1.0 μM
Tris–HCl, pH 8.0	1 M	40 μL	40.0 mM
MgCl₂	1 M	24 μL	24.0 mM
Spermidine	0.1 M	10 μL	1.0 mM
DTT	0.5 M	10 μL	5.0 mM
Triton X-100	0.5 %	10 μL	0.005 %
ATP	50 mM	150 μL	7.5 mM
CTP	50 mM	150 μL	7.5 mM
GTP	50 mM	150 μL	7.5 mM
UTP	50 mM	150 μL	7.5 mM
GMP (see **Note 3**)	200 mM	150 μL	30.0 mM
T7 RNA polymerase (see **Note 3**)		10 μL	
ddH₂O		to 1000 μL	

2. Allow the transcription to continue for 3–5 h at 37 °C.

3. Spin down pyrophosphates that have formed for 10 min in a microfuge at the maximum speed. Transfer the supernatant containing the tRNA transcript to a new tube.

4. Adjust the solution by adding EDTA to a final concentration of 0.1 M and ammonium acetate to a final concentration of 0.3 M. Add two volumes of cold ethanol and precipitate the tRNA transcript at –20 °C for 15 min.

5. Spin the tRNA in a microfuge at the maximum speed for 15 min, 4 °C. Resuspend the pellet in 100 μL TE at 37 °C for 10 min. Not all pyrophosphate precipitation goes into solution.

6. Add 100 μL RNA loading dye to the resuspended pellet. The sample is ready for gel isolation of the tRNA transcript.

3.3 Isolation of tRNA Transcript from the Transcription Reaction

1. Make a large preparative 12 % PAGE/7 M urea gel by assembling two glass plates and spacers.

2. Seal the short-edge side of the glass plates with heavy packaging tape, and clamp along the two long-edge sides with metal binder clips.

3. Mix a 10 mL solution of 12 % PAGE/7 M urea with 120 μL of 10 % APS and 6 μL of TEMED and pour the solution to the bottom of the assembled glass plates.

4. Upon fast polymerization, the 10 mL gel serves as a seal at the bottom.

5. Mix 160 mL of 12 % PAGE/7 M urea with 960 μL of 10 % APS and 48 μL of TEMED and pour the gel solution to fill the rest of the assembled glass plates. Insert combs or spacers to form the sample-loading wells.

6. After 10–20 min, the gel should be polymerized and ready to use. Remove the combs, clamps, and tape from the glass plates, and place the gel in an appropriate electrophoresis apparatus. Fill both the top and bottom chambers with 1× TBE buffer.

7. Wash each well on the 12 % PAGE/7 M urea gel with 1× TBE buffer several times. When all wells are cleared of the gel material, immediately load the 200 μL solution of tRNA in RNA loading dye solution (from Subheading 3.2, **step 6**). Run the gel at 700 V for 15 h until xylene cyanol is about 8 cm above the bottom of the gel.

8. Remove the glass plates from the apparatus and transfer the gel to a saran wrap and place the gel and saran wrap on top of a silica gel fluorescent TLC plate.

9. Use a hand-held UV lamp to localize a UV shadow, indicating the site of migration of the tRNA transcript. Cut off the portion of the gel that exhibits UV shadow and crush the gel into pieces using a sterile glass rod.

10. Add 5 mL TE to the crushed gel pieces and shake the suspension on a rotator shaker at room temperature for 4–6 h to elute the tRNA.

11. Spin down gel pieces using a tabletop centrifuge. Remove the supernatant to a clean tube.

12. Add more TE back to the gel solution and continue elution for another 4–6 h. Again, spin and keep the supernatant.

13. Combine the two supernatants and precipitate the tRNA transcript by adding 1/10 volume of 3 M ammonium acetate and 3 volumes of ethanol.

14. Spin down the tRNA transcript from ethanol precipitation and wash the pellet two times with 70 % ethanol.

15. Dry the pellet and resuspend the tRNA in 100 μL of TE buffer. Determine the tRNA concentration by measuring the absorbance at 260 nm (1 OD unit = 40 μg/mL).

3.4 Assay for Methyl Transfer in One Turnover

1. Take an aliquot of the unmodified transcript of G37-tRNA (200 pmole) and adjust the volume with TE to 15 μL.

2. Heat the tRNA solution for 3 min at 80 °C, a temperature that is above the estimated melting temperature of the tRNA transcript.

3. Quickly spin the solution and add 5 μL of the heat–cool buffer. Anneal the tRNA at 37 °C for 15 min. The resulting stock concentration of tRNA is 100 μM.

4. Prepare a 5× EcTrmD buffer as an example (see Note 4):

EcTrmD component	Volume	5× concentration	1× concentration
1 M Tris–HCl, pH 8.0	500.0 μL	0.5 M	0.1 M
3 M NH$_4$Cl$_2$	40.0 μL	0.12 M	0.024 M
1 M MgCl$_2$	30.0 μL	30.0 mM	6.0 mM
1 M DTT	20.0 μL	20.0 mM	4.0 mM
50 mM EDTA	10.0 μL	0.5 mM	0.1 mM
10 mg/mL BSA	12.0 μL	0.12 mg/mL	0.024 mg/mL
ddH$_2$O		388.0 μL	
Total		1000.0 μL	

5. Prepare the tRNA solution for syringe #1 of the RQF-3 instrument (see Note 5):

    Mix G37-tRNA and ddH$_2$O first and denature the tRNA by heating at 85 °C for 3 min. Briefly spin down the heated solution and then add the 5x buffer to allow annealing of the tRNA at 37 °C for 15 min. Add the working stock of AdoMet and place the solution on ice until loading onto syringe #1.

Syringe 1	Volume	2× concentration	1× concentration
G37-tRNA transcript (100 μM)	1.5 μL	0.5 μM	0.25 μM
ddH$_2$O	209.5 μL		
5× buffer	60.0 μL		1x
Working stock of AdoMet (315 μM)	29.0 μL	30.0 μM	15.0 μM
Total	300.0 μL		

6. Prepare a series of EcTrmD solutions (ranging from 2 to 32 μM as the 2×) for syringe #2 of the RQF-3 instrument (see Note 6). For each concentration, prepare a 300 μL solution. A total of 6 concentrations are prepared.

    Mix ddH$_2$O and the 5× buffer, add an appropriate amount of the enzyme. Place on ice until loading onto syringe #2.

	1	2	3	4	5	6
2x EcTrmD	2 μM	4 μM	8 μM	16 μM	24 μM	32 μM
1x EcTrmD	1 μM	2 μM	4 μM	8 μM	12 μM	16 μM
EcTrmD stock (400 μM)	1.5 μL	3.0 μL	6 μL	12 μL	18 μL	24 μL
5x buffer	60 μL	60 μL	60 μL	60 μL	60 μL	60 μL
ddH₂O	238.5 μL	237 μL	234 μL	228 μL	222 μL	216 μL
Total	300 μL	300 μL	300 μL	300 μL	300 μL	300 μL

7. On the RQF-3 instrument, fill the large syringe on each side with 1x buffer and the middle syringe with 5 % TCA. Fill syringe #1 with 300 μL of the tRNA solution to one sample loop and syringe #2 with 300 μL of an enzyme solution to the second sample loop. Enter time points on the control panel. Upon hitting the start button, the 1x buffer pushes 15 μL from syringe #1 and 15 μL from syringe #2 into the reaction loop. After the specified time lapse, the reaction is quenched with 54 μL of 5 % TCA. Repeat the reaction with a different time point up to 17 time points (e.g., 0, 1, 3, 5, 7, 9, 12, 15, 20, 25, 30, 40, 50, 60, 80, 100, and 120 s).

8. Collect the quenched solution of each time point in an Eppendorf tube. Spot 20 μL of the quenched solution onto a 1 cm² Whatman 3 MM paper pad, which should be labeled for the reaction by a #2 HB pencil.

9. Drop all filter pads into a beaker containing 100–200 mL cold 5 % TCA. Estimate the volume of TCA as 5 mL per filter pad.

10. After all pads are in TCA, shake the solution for 10 min at 4 °C to wash off un-incorporated AdoMet, while allowing the synthesized m¹G37-tRNA to precipitate in TCA. Decant and repeat the 5 % TCA wash.

11. Wash all filter pads with 95 % ethanol by shaking for 10 min at 4 °C in the beaker. Repeat the ethanol wash one more time.

12. Wash all filter pads with ether. Agitate gently by hand and let the ether solution sit at room temperature for 5 min under a fume hood. Decant ether and dry the filter pads under the fume hood for 15 min.

13. Transfer each filter pad to a scintillation solution in a vial and measure the amount of radioactivity using a liquid scintillation counter.

14. Calculate m¹G37-tRNA synthesis based on the specific activity of AdoMet (see **Note 1**).

15. Correct the ³H counting by measuring the quenching factor using the following procedure: Take a 5 μL aliquot at the final

time point of a reaction and pass it through a quick spin column (*see* Subheading 2.4, **item 20**) to remove unincorporated ^{3}H-AdoMet, which stays with the column. Directly transfer the eluate 5 μL (which contains counts only associated with the methylated tRNA) into the liquid scintillation fluid and measure the counts. The ratio of the direct measurement of this count over the count on the TCA precipitated filter pad at the same time point reveals the quenching factor, which should be used to correct the fraction of methylation. For the protocol described here, the quenching factor is usually 4.

3.5 Data Analysis

1. Data points for each time course are fit to the single exponential equation:

$$y = y_o + A \times \left(1 - e^{-k_{app} \times t}\right) \qquad (1)$$

where y_o is the y intercept, A is the scaling constant, k_{app} is the apparent rate constant, and t is the time in seconds to determine k_{obs} [31, 32] (*see* **Note 7**). The data of k_{obs} vs. enzyme concentration for single-turnover analysis of m^{1}G37-tRNA synthesis are fit to the hyperbolic equation:

$$y = k_{chem} \times \frac{E_o}{E_o + K_d} \qquad (2)$$

where k_{chem} is the rate constant for the steps associated with the methyl transfer chemistry (*see* **Notes 8–9**), K_d is the enzyme affinity for the tRNA substrate (K_d (tRNA)), and E_o is the enzyme concentration [31, 32].

2. The procedure can be repeated with the wild-type enzyme to obtain the K_d (AdoMet) or repeated with a mutant enzyme to obtain all three parameters K_d (tRNA), K_d (AdoMet), and k_{chem} to investigate the enzyme structure–function relationship.

4 Notes

1. Calculation of the specific activity of the working stock of AdoMet:

   The working stock consists of 200 μL of the commercial ^{3}H-AdoMet (6.6 μM, 0.55 μCi/μL) with 90 μL of unlabeled AdoMet (1 mM)

   - Total concentration of the commercial ^{3}H-AdoMet in the mixture: 4.55 μM.

   - Total concentration of unlabeled AdoMet in the mixture: 310.34 μM.

- Combined concentration of AdoMet: 314.89 µM.

- Total µCi in the mixture: 0.55 µCi/µL × 200 µL = 110 µCi.

- Specific activity in dpm/pmole: [110 µCi × (2.2 x 10^6 dpm/µCi)]/[314.89 pmole/(µL × 290 µL)] = 2650 dpm/pmole.

2. In the Sequenase reaction, some of the overlapping primers are extended to completion in only 2–3 h while others take longer time. Overnight incubation should assure complete reaction.

3. In the tRNA transcription reaction, GMP (5′-monophosphate guanosine nucleoside) is added in excess of GTP to promote initiation of transcription with GMP, so that the transcript will have a 5′-monophosphate, rather than a 5′-triphosphate. If T7 RNA polymerase is purified from an overproducer strain [27], it should be titrated to the level where 10 µL of the enzyme would give visible precipitation of pyrophosphate that forms aggregates with Mg^{2+} in less than 1 h at 37 °C. The pyrophosphate is released from NTP due to incorporation of NMP during active transcription. The observation of such precipitation is usually a good indication of strong transcription. If no precipitation is observed in more than 1 h, then add more T7 RNA polymerase to the transcription reaction.

4. The 5x for *Mj*Trm5 and *Hs*Trm5 buffer is as follows:

Trm5 component	Volume	5× concentration	1× concentration
1 M Tris–HCl, pH 8.0	500.0 µL	0.5 M	0.1 M
2 M NH$_4$Cl$_2$	250.0 µL	0.5 M	0.1 M
1 M MgCl$_2$	30.0 µL	30.0 mM	6.0 mM
1 M DTT	20.0 µL	20.0 mM	4.0 mM
50 mM EDTA	10.0 µL	0.5 mM	0.1 mM
10 mg/mL BSA	12.0 µL	0.12 mg/mL	0.024 mg/mL
ddH$_2$O	178.0 µL		
Total	1000.0 µL		

5. When measurements are performed on the RQF-3 instruments, the tRNA and AdoMet concentration in syringe #1 should be the 2× of the final, while the enzyme concentration in syringe #2 should be 2× of the final. Both syringe #1 and syringe #2 solutions are made in 1× buffer so that, after mixing at an equal volume of each, the final buffer remains at 1×.

6. For single-turnover assays of methyl transfer, the reaction rate is driven by the enzyme concentration, not by the tRNA concentration [31]. Therefore, a series of reactions are designed,

with increasing concentration of the enzyme. The 6 reactions designed here are to provide an initial evaluation of the data points. More reactions with additional concentrations of the enzyme can be designed to obtain more data points for better curve fitting.

7. Control experiments should be performed to confirm that the mixing order of enzyme, tRNA, and AdoMet does not affect the rate of methyl transfer. This confirmation establishes rapid equilibrium binding.

8. In rapid chemical quench experiments, the k_{obs} is a composite term that includes all of the steps leading to the chemistry of methyl transfer. Under the condition of rapid equilibrium binding, the formation of enzyme complexes ([E-AdoMet], [E-tRNA], and [E-AdoMet-tRNA]) is fast. The k_{chem} in this condition therefore refers to either the pre-chemistry enzyme re-arrangement of the active site or the chemical step of methyl transfer involving the nucleophilic attack of the N^1 atom of G37 on the methyl group of AdoMet.

9. The experiments can be repeated to obtain the K_d of the enzyme affinity for the AdoMet substrate (K_d (AdoMet)). For this purpose, the enzyme concentration should be in excess of the AdoMet substrate, while the tRNA concentration is saturating. Repeat Subheading 3.4, with the modification of **step 5** as follows (*see* **Note 10**).

10. To determine the K_d (AdoMet), prepare the tRNA solution for syringe #1 of RQF-3 as follows:

Syringe 1	Volume	2x concentration	1x concentration
G37-tRNA transcript (100 μM)	60 μL	20.0 μM	10.0 μM
ddH$_2$O	170 μL		
5x buffer	60.0 μL		1x
Working stock of AdoMet (30 μM) (*see* **Note 11**)	10.0 μL	1.0 μM	0.5 μM
Total	300.0 μL		

Follow **steps 6–16** in Subheading 3.4 and proceed through Data analysis in Subheading 3.5. Upon fitting the data to Eq. 2, the obtained K_d is the K_d (AdoMet).

11. The working stock of AdoMet consists of 200 μL of the commercial ^{3}H-AdoMet (6.6 μM, 0.55 μCi/μL) with 6.2 μL of unlabeled AdoMet (1 mM).

   • Total concentration of the commercial ^{3}H-AdoMet in the mixture: 4.55 μM

- Total concentration of unlabeled AdoMet in the mixture: 30 µM

- Combined concentration of AdoMet: 34.55 µM

- Total µCi in the mixture: 0.55 µCi/µL × 200 µL = 110 µCi

- Specific activity in dpm/pmole: $[110 \,\mu Ci \times (2.2 \times 10^6 \,dpm/\mu Ci)]/[34.55 \;\; pmole/(\mu L \times 206.2 \;\; \mu L)] = 3400$ dpm/pmole (*see* **Note 12**).

12. The specific activity of AdoMet (3400 dpm/pmole) for determination of the K_d (AdoMet) is higher relative to the specific activity (2650 dpm/pmole) for determination of the K_d (tRNA). This is advantageous to give a higher signal to compensate for the lower amount of AdoMet used in the reaction.

Acknowledgement

This work was supported by NIH/NIGMS grants R01GM81601 and R01GM108972 to Y.M.H. The author thanks Thomas Christian for discussion and Dr. Isao Masuda and Dr. Megumi Shigematsu for preparation of Fig. 1.

References

1. Wang Y, Li Y, Toth JI, Petroski MD, Zhang Z, Zhao JC (2014) N6-methyladenosine modification destabilizes developmental regulators in embryonic stem cells. Nat Cell Biol 16(2): 191–198

2. Ludwig ML, Matthews RG (1997) Structure-based perspectives on B12-dependent enzymes. Annu Rev Biochem 66:269–313

3. Markham GD, Hafner EW, Tabor CW, Tabor H (1980) S-Adenosylmethionine synthetase from Escherichia coli. J Biol Chem 255(19): 9082–9092

4. Fustin JM, Doi M, Yamaguchi Y, Hida H, Nishimura S, Yoshida M, Isagawa T, Morioka MS, Kakeya H, Manabe I, Okamura H (2013) RNA-methylation-dependent RNA processing controls the speed of the circadian clock. Cell 155(4):793–806

5. Cantoni GL (1975) Biological methylation: selected aspects. Annu Rev Biochem 44:435–451

6. Schubert HL, Blumenthal RM, Cheng X (2003) Many paths to methyltransfer: a chronicle of convergence. Trends Biochem Sci 28(6):329–335

7. Nureki O, Shirouzu M, Hashimoto K, Ishitani R, Terada T, Tamakoshi M, Oshima T, Chijimatsu M, Takio K, Vassylyev DG, Shibata T, Inoue Y, Kuramitsu S, Yokoyama S (2002) An enzyme with a deep trefoil knot for the active-site architecture. Acta Crystallogr D Biol Crystallogr 58(Pt 7):1129–1137

8. Nureki O, Watanabe K, Fukai S, Ishii R, Endo Y, Hori H, Yokoyama S (2004) Deep knot structure for construction of active site and cofactor binding site of tRNA modification enzyme. Structure 12(4):593–602

9. Christian T, Evilia C, Williams S, Hou YM (2004) Distinct origins of tRNA(m1G37) methyltransferase. J Mol Biol 339(4):707–719

10. Ahn HJ, Kim HW, Yoon HJ, Lee BI, Suh SW, Yang JK (2003) Crystal structure of tRNA(m1G37)methyltransferase: insights into tRNA recognition. EMBO J 22(11):2593–2603

11. Elkins PA, Watts JM, Zalacain M, van Thiel A, Vitazka PR, Redlak M, Andraos-Selim C, Rastinejad F, Holmes WM (2003) Insights into catalysis by a knotted TrmD tRNA methyltransferase. J Mol Biol 333(5):931–949

12. Goto-Ito S, Ito T, Ishii R, Muto Y, Bessho Y, Yokoyama S (2008) Crystal structure of archaeal tRNA(m(1)G37)methyltransferase aTrm5. Proteins 72(4):1274–1289

13. Goto-Ito S, Ito T, Kuratani M, Bessho Y, Yokoyama S (2009) Tertiary structure checkpoint at anticodon loop modification in tRNA functional maturation. Nat Struct Mol Biol 16(10):1109–1115

14. Baba T, Ara T, Hasegawa M, Takai Y, Okumura Y, Baba M, Datsenko KA, Tomita M, Wanner BL, Mori H (2006) Construction of Escherichia coli K-12 in-frame, single-gene knockout mutants: the Keio collection. Mol Syst Biol 2(2006):0008

15. O'Dwyer K, Watts JM, Biswas S, Ambrad J, Barber M, Brule H, Petit C, Holmes DJ, Zalacain M, Holmes WM (2004) Characterization of Streptococcus pneumoniae TrmD, a tRNA methyltransferase essential for growth. J Bacteriol 186(8):2346–2354

16. Bjork GR, Wikstrom PM, Bystrom AS (1989) Prevention of translational frameshifting by the modified nucleoside 1-methylguanosine. Science 244(4907):986–989

17. Christian T, Gamper H, Hou YM (2013) Conservation of structure and mechanism by Trm5 enzymes. RNA 19(9):1192–1199

18. Bjork GR, Jacobsson K, Nilsson K, Johansson MJ, Bystrom AS, Persson OP (2001) A primordial tRNA modification required for the evolution of life? EMBO J 20(1-2):231–239

19. Gamper HB, Masuda I, Frenkel-Morgenstern M, Hou YM (2015) Maintenance of protein synthesis reading frame by EF-P and m(1) G37-tRNA. Nat Commun 6:7226

20. Ito T, Masuda I, Yoshida K, Goto-Ito S, Sekine S, Suh SW, Hou YM, Yokoyama S (2015) Structural basis for methyl-donor-dependent and sequence-specific binding to tRNA substrates by knotted methyltransferase TrmD. Proc Natl Acad Sci U S A 112:E4197–E4205

21. Lahoud G, Goto-Ito S, Yoshida K, Ito T, Yokoyama S, Hou YM (2011) Differentiating analogous tRNA methyltransferases by fragments of the methyl donor. RNA 17(7):1236–1246

22. Christian T, Hou YM (2007) Distinct determinants of tRNA recognition by the TrmD and Trm5 methyl transferases. J Mol Biol 373(3):623–632

23. Sakaguchi R, Giessing A, Dai Q, Lahoud G, Liutkeviciute Z, Klimasauskas S, Piccirilli J, Kirpekar F, Hou YM (2012) Recognition of guanosine by dissimilar tRNA methyltransferases. RNA 18(9):1687–1701

24. Christian T, Lahoud G, Liu C, Hou YM (2010) Control of catalytic cycle by a pair of analogous tRNA modification enzymes. J Mol Biol 400(2):204–217

25. Christian T, Evilia C, Hou YM (2006) Catalysis by the second class of tRNA(m1G37) methyl transferase requires a conserved proline. Biochemistry 45(24):7463–7473

26. Swinehart WE, Henderson JC, Jackman JE (2013) Unexpected expansion of tRNA substrate recognition by the yeast m1G9 methyltransferase Trm10. RNA 19(8):1137–1146

27. Tabor S, Huber HE, Richardson CC (1987) Escherichia coli thioredoxin confers processivity on the DNA polymerase activity of the gene 5 protein of bacteriophage T7. J Biol Chem 262(33):16212–16223

28. Grodberg J, Dunn JJ (1988) ompT encodes the Escherichia coli outer membrane protease that cleaves T7 RNA polymerase during purification. J Bacteriol 170(3):1245–1253

29. Redlak M, Andraos-Selim C, Giege R, Florentz C, Holmes WM (1997) Interaction of tRNA with tRNA (guanosine-1)methyltransferase: binding specificity determinants involve the dinucleotide G36pG37 and tertiary structure. Biochemistry 36(29):8699–8709

30. Kao C, Zheng M, Rudisser S (1999) A simple and efficient method to reduce nontemplated nucleotide addition at the 3 terminus of RNAs transcribed by T7 RNA polymerase. RNA 5(9):1268–1272

31. Zhang CM, Perona JJ, Ryu K, Francklyn C, Hou YM (2006) Distinct kinetic mechanisms of the two classes of Aminoacyl-tRNA synthetases. J Mol Biol 361(2):300–311

32. Dupasquier M, Kim S, Halkidis K, Gamper H, Hou YM (2008) tRNA integrity is a prerequisite for rapid CCA addition: implication for quality control. J Mol Biol 379(3):579–588

33. Sakaguchi R, Lahoud G, Christian T, Gamper H, Hou YM (2014) A divalent metal ion-dependent N(1)-methyl transfer to G37-tRNA. Chem Biol 21(10):1351–1360

34. Sprinzl M, Horn C, Brown M, Ioudovitch A, Steinberg S (1998) Compilation of tRNA sequences and sequences of tRNA genes. Nucleic Acids Res 26(1):148–153

Chapter 9

Purification and Functional Reconstitution of Box H/ACA Ribonucleoprotein Particles

Chao Huang, Guowei Wu, and Yi-Tao Yu

Abstract

Pseudouridylation is the most abundant and widespread RNA modification, and it plays an important role in modulating the structure and function of RNA. In eukaryotes and archaea, RNA pseudouridylation is catalyzed largely by box H/ACA ribonucleoproteins (RNPs), a distinct group of RNA–protein complexes each consisting of a unique RNA and four common proteins. The RNA component of the complex serves as a guide that base-pairs with its substrate RNA and specifies the target uridine to be modified. In order to systematically study the function and mechanism of pseudouridylation, it is desirable to have a reconstitution system in which biochemically purified/reconstituted box H/ACA RNPs are capable of introducing pseudouridines into an RNA at any target site. Here, we describe a method for the reconstitution of functional box H/ACA RNPs using designer box H/ACA guide RNAs, which in principle can be adopted to reconstitute other RNA–protein complexes as well.

Key words Pseudouridylation, RNA modification, Immunoprecipitation, Reconstitution, Box H/ACA RNP

1 Introduction

Pseudouridine (Ψ) is the most abundant post-transcriptionally modified nucleotide, and exhibits chemical properties that are distinct from those of all other known nucleotides [1–5]. When compared with uridine, Ψ has an extra hydrogen bond donor and contains, between the base and sugar, a carbon–carbon bond, which is ~10 % longer than the normal nitrogen–carbon glycosidic bond found in all other nucleotides. It has long been known that in all eukaryotic species, tRNAs, rRNAs, and spliceosomal snRNAs are highly enriched with Ψ [6–8]. Most recently, Ψ has also been found in other noncoding RNAs, as well as in mRNAs [9, 10]. It has been reported that Ψs within rRNAs and spliceosomal snRNAs contribute significantly to protein translation and pre-mRNA splicing, respectively [11, 12].

Ren-Jang Lin (ed.), *RNA-Protein Complexes and Interactions: Methods and Protocols*, Methods in Molecular Biology, vol. 1421, DOI 10.1007/978-1-4939-3591-8_9, © Springer Science+Business Media New York 2016

In eukaryotes and archaea, pseudouridylation (U-to-Ψ conversion) is mainly catalyzed by a family of ribonucleoproteins called box H/ACA RNPs, each of which consists of four common core proteins (NAP57/dyskerin/Cbf5, Nhp2, Nop10, and Gar1) and a single unique small RNA (box H/ACA RNA) [13]. In the complex, this RNA component serves as a guide that base-pairs with its substrate RNA, positioning the target uridine at the pseudouridylation pocket (Fig. 1a). The catalytic component (NAP57/dyskerin/Cbf5) of the complex then catalyzes the pseudouridylation reaction, converting the specified (target) uridine into Ψ. Target specificity is thus determined by base-pairing interactions between the guide

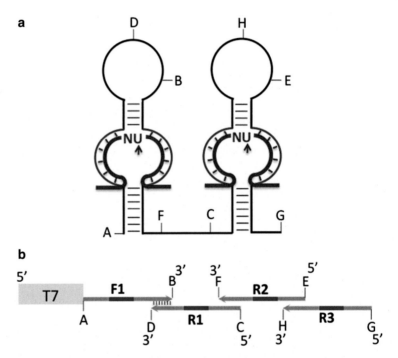

Fig. 1 (**a**) Typical structure of a box H/ACA RNA (bound with its substrate RNA). Box H/ACA RNA is typically ~200-nts long and folds into a hairpin-hinge-hairpin-tail structure (*black and blue lines*). Within each hairpin, there is an internal loop (*blue*) that serves as a guide to base-pair with its substrate RNA (*red*). The arrows indicate the target uridines (U) to be pseudouridylated. N stands for any nucleotide. Here, four short DNA oligos are used to construct a DNA template (by PCR) for the synthesis of an artificial box H/ACA RNA (with desired guide sequences). For illustrative purposes, the box H/ACA RNA is marked at several positions by black letters (A–H), and oligos are designed according to the markers (*see* **b**). (**b**) Design of oligos (one forward sense oligo, F1, and three reverse antisense oligos, R1, R2, and R3) for template DNA amplification. The overlapping complementary region (between B and D) is indicated. The overlapping regions between C and F and between E and H are also shown. The blue segments represent guide sequences that can be altered according to the substrate sequences. The T7 promoter (for in vitro transcription) is also indicated

sequence of the box H/ACA RNA and the target sequence of the substrate RNA [14]. Based on this guide-substrate base-pairing scheme, one can in theory engineer artificial box H/ACA RNAs (with different guide sequences) to introduce pseudouridines into RNA at novel sites. Indeed, by designing an artificial box H/ACA RNA, Karijolich and Yu were able to introduce a Ψ into a premature termination codon (PTC) of an mRNA molecule, promoting nonsense suppression in vivo [15, 16]. Using the same concept, our lab also demonstrated that box H/ACA RNP complexes could be purified (from yeast cells) and functionally reconstituted in vitro [17, 18]. This in vitro reconstitution system has been very helpful for studying the function and mechanism of RNA-guided RNA pseudouridylation [17, 18].

Here, we describe a protocol for reconstitution of functional box H/ACA RNP in vitro. Starting with a yeast strain where one of the four core proteins is TAP (tandem affinity purification)-tagged [19], we purify, using anti-tag immunoprecipitation, the whole family of box H/ACA RNPs from yeast cell lysate. We then treat the purified box H/ACA RNPs with micrococcal nuclease to remove the endogenous box H/ACA RNAs from the complexes. Finally, upon inactivation of micrococcal nuclease, we add in vitro transcribed designer box H/ACA RNAs to the RNA-depleted box H/ACA RNP complexes, thereby allowing reconstitution of functional box H/ACA RNPs capable of directing RNA pseudouridylation at the desired sites. In this protocol, the designer artificial box H/ACA RNA is based on snR81 [17], as illustrated in Fig. 1a. Using this technique, we have been able to introduce pseudouridines into various RNA substrates. We believe that this technique can in principle be used to reconstitute other RNA–protein complexes as well.

2 Materials

2.1 Yeast Cell Culture

1. SD-URA drop out medium: To make powder for 60 L SX-URA drop out medium (X represents the type of sugar that will be added separately), weigh 100.4 g of yeast nitrogen base, 301.6 g of ammonium sulfate, 1.8 g of isoleucine, 9.0 g of valine, 1.2 g of adenine, 1.2 g of arginine, 1.2 g of histidine, 1.8 g of leucine, 1.8 g of lysine, 1.2 g of methionine, 3.0 g of phenylalanine, 1.2 g of tryptophan, and 1.8 g of tyrosine. Add the mixture into a ball mill grinder and grind the mixture overnight. To make SD-URA liquid medium, weigh 3.7 g of SX-URA powder, 10 g of dextrose, then add 500 mL of ddH₂O, autoclave to sterilize it.

2. SR-URA liquid medium: same as SD-URA liquid medium except that raffinose is used instead of dextrose (see **Note 1**).

3. 3×-YPGal: weigh 30 g of yeast extract, 60 g of peptone, 60 g of galactose and add ddH_2O to 1 L. Autoclave to sterilize it.

4. TAP-tagged yeast strain: For reconstitution of Box H/ACA RNP, any of the three core protein-tag (Cbf5-TAP, Gar1-TAP, Nhp2-TAP) in wild type yeast strains (BY4741, BY4742, etc.) can be used. Nop10 cannot be TAP-tagged in our experience [17]. The tag used in this protocol contains successive (6×His-HA-3C-Protein A) tags [20], which is a modified version of the original TAP tag [21]. This tag is used in conjunction with a URA marker (see **Note 2**).

5. UV–VIS Spectrophotometer and plastic cuvette: Measure OD_{600nm} of yeast culture to monitor cell growth.

6. Extraction Buffer: 50 mM Tris–HCl pH 7.5, 1 mM EDTA, 4 mM $MgCl_2$, 10 % glycerol, 1 M NaCl (add β-mercaptoethanol and protease inhibitors immediately before use, see below).

7. Sorvall centrifuge: Equipped with a rotor for 500 mL bottles (e.g., SLA-3000) and a rotor for 50 mL conical tubes (e.g., SH-3000).

2.2 Immuno-precipitation

1. IgG Sepharose 6 Fast Flow Bead: purchased from GE Healthcare Life Sciences.

2. IPP150 buffer: 10 mM Tris–HCl pH 8.0, 150 mM NaCl, 0.1 % NP40.

3. Table top centrifuge, maximum speed is at least 13,000 rpm ($13800 \times g$).

4. β-mercaptoethanol: Before use, add β-mercaptoethanol to Extraction Buffer to 5 mM final concentration (see **Note 3**).

5. Protease inhibitors: PMSF can be dissolved in ethanol at stock concentration of 100 mM and stored at –20 °C. Pepstatin can be dissolved in ethanol at 2.5 mg/mL and used as 1000× stock solution. Leupeptin can be dissolved in ddH_2O at 1 mg/mL and used as 400× stock solution (see **Note 4**).

6. 3C cleavage buffer: 10 mM Tris–HCl pH 8.0, 150 mM NaCl, 0.1 % NP40, 2 mM β-mercaptoethanol.

7. Bleach: Dilute 1 volume of bleach with 9 volume of ddH_2O (This is for decontaminating and discarding used yeast media). Add diluted bleach to the yeast media, and incubate for 20 min before discarding it down the drain.

2.3 Micrococcal Nuclease Treatment

1. Micrococcal nuclease: Use it at a final concentration of ~1 U/μL.

2. 100 mM $CaCl_2$: Weigh 111 g of $CaCl_2$ and dissolve in 1 L of ddH_2O to make 1 M stock solution. Dilute the stock to 100 mM before use.

3. 0.5 M EGTA: dissolve 19.02 g of EGTA in 90 mL of ddH_2O. Use 10 N NaOH to adjust pH to 8.0. Add ddH_2O to 100 mL.

2.4   In Vitro Transcription of RNA

1. DNA oligos (template sequence):

   T7-snR81-F1 template: 5′-*TAATACGACTCACTATA GGG* AATGTCGACTAGGGACTGC NNNNNNN GCGGC GAGG CAGCCCACATCAAGTGGAACTACAC-3′

   snR81-R1 template: 5′-TAGGATTGCTCTTGGGACCG NNNNNNN GCGACAAGGAAGTCTGTGTAGTTCCAC TTGATG-3′

   snR81-R2 template: 5′-TCCGTGGACTGTACAGGTT CAGC GGGGG NNNNNNNNN TTGCTTGTTAGGATTGCT CTTGG-3′

   snR81-R3 template: 5′-AACTGCAGAAGATGTGAAA AAGCG NNNN CCCCCAAATCATATAACTTCTGCACC ATCCGTGGACTG TACAG 3′

   The italic letters represent the T7 promoter sequence, and the "N" letters represent the guide sequence (it varies depending on the substrate sequence).

   Briefly spin the DNA oligo tubes (purchased from IDT) and add ddH$_2$O to dissolve pellets to desired stock concentration of 100 μM. Stock could be further diluted to 10 μM before use.

2. PCR Thermocycler: If the thermocycler does not have heated lid, add a layer of mineral oil on top of the PCR mixture prior to PCR cycling.

3. 1.5 % Agarose Gel: Weigh 0.75 g of agarose and add it to 50 mL of 1× TBE buffer. Microwave the mixture and boil it for 5~10 s. Shake gently in-between to avoid local overheating. Cool the solution down to ~50 °C in water bath and add ethidium bromide to a final concentration of 0.5 μg/mL.

4. 10× transcription buffer: 500 mM Tris–HCl pH 7.5, 100 mM MgCl$_2$, 100 mM DTT, 20 mM spermidine. Aliquot to 1.5 mL tubes and store at −20 °C.

5. DNase I: Typically 1 unit is enough to digest 2 μg of DNA in 20 min.

6. [alpha-P^{32}] GTP, or any other [alpha-P^{32}] NTP (*see* **Note 5**).

7. 6 % denaturing gel: To make 100 mL of gel mix, use 20 mL of 30 % acrylamide mix (acrylamide:bis 19:1), 10 mL of 10× TBE, 48 g of urea, and add ddH$_2$O to 100 mL. Use a stir bar to dissolve and mix the solution. Add 1 mL of 10 % APS, 100 μL of TEMED immediately before casting the gel (*see* **Note 6**).

8. G50 buffer: 20 mM Tris–HCl pH 7.5, 300 mM NaOAc, 2 mM EDTA, 0.2 % SDS. Store at room temperature (*see* **Note 7**).

9. PCA: Phenol/chloroform/isoamyl alcohol (25:24:1) (*see* **Note 8**).

10. 70 % Ethanol: Mix 7 volume of 100 % ethanol with 3 volume of ddH$_2$O.

11. 10× DNA loading dye: Mix 4.9 mL of glycerol, 200 µL 0.5 M EDTA, 0.025 g bromophenol blue, 0.025 g xylene cyanol FF. Bring to a total volume of 10 mL with ddH$_2$O.

12. 2× RNA loading dye: Mix 9.5 mL formamide, 25 µL 10 % (w/v) SDS, 10 µL 0.5 M EDTA, 0.025 g bromophenol blue, 0.025 g xylene cyanol FF, bring to a total volume of 10 mL with ddH$_2$O.

2.5 Functional Reconstitution

1. 5× in vitro pseudouridylation buffer: 1 M Tris–HCl at pH 8.0, 1 M ammonium acetate, 50 mM MgCl$_2$, 2 mM DTT, 0.1 mM EDTA.

2. 2 % Glycogen: Dissolve 0.2 g of glycogen into 10 mL of ddH$_2$O, aliquot and store at −20 °C.

3. Nuclease P1 digestion buffer: 200 mM acetic acid, 10 mM CaCl$_2$, pH 6.0.

4. Nuclease P1: 0.1 µg of nuclease P1 is sufficient to digest 10 pmol of RNA substrate, when incubation is carried out at 37 °C for 1 h.

5. TLC PEI cellulose F plate: Purchased from EMD.

6. TLC developing buffer: isopropanol/HCl/H$_2$O (70:15:15). Make fresh buffer for each experiment. TLC developing chamber should be placed in the chemical hood.

3   Methods

3.1 TAP-Tagged Yeast Cell Culture

1. Thaw glycerol stock (Cbf5-, Gar1-, or Nhp2-TAP-tagged yeast strain) at room temperature (*see* **Note 2**). Immediately transfer 250 µL into a 50 mL conical tube containing 5 mL of SD-URA medium. Grow the cells by shaking at 200 rpm, 30 °C for overnight (*see* **Note 22**).

2. Inoculate 5 mL of overnight culture into 50 mL of SR-URA medium (*see* **Note 9**). Grow by shaking at 200 rpm, 30 °C for 8 h.

3. Measure the OD$_{600nm}$ of the culture. Inoculate the culture into 600 mL of SR-URA medium at a starting OD$_{600nm}$ of 0.02. Grow by shaking at 200 rpm, 30 °C for overnight.

4. When the OD$_{600nm}$ of the overnight culture reached 1.0, add 300 mL of 3×-YPGal to the culture to activate Gal promoter and induce protein expression. Induction is typically carried out for 24 h.

5. Harvest cells by centrifugation at 2000×g for 5 min (SLA-3000 rotor) (*see* **Note 10**). Remove the supernatant (treat with bleach before discard). Keep the cell pellet.

6. Add 80 mL of ddH$_2$O to the cell pellet and resuspend the cells. Transfer the cells to two 50 mL tubes in equal volume (~40 mL each).

7. Centrifuge the 50 mL tubes at 2000×g for 5 min in SH-3000 swing bucket. Discard supernatant and resuspend cell pellets in 30 mL of Extraction Buffer. Spin down the cells at 2000×g and remove the Extraction Buffer. Freeze the cells in liquid nitrogen and store at −80 °C until IgG purification.

3.2 Immuno-precipitation of Box H/ACA RNPs

1. Pre-wash IgG Sepharose beads: Take 200 μL of IgG beads, spin at 5000 rpm (2040 × g) for 1 min to remove supernatant. Add 1 mL of cold IPP150 buffer; invert the tube by hand until all beads are resuspended. Spin at 5000 rpm for 1 min and remove 900 μL of the supernatant (*see* **Note 11**).

2. Repeat **step 1** three times. After wash, keep beads on ice until use (beads plus sample have a volume of ~200 μL).

3. To the Extraction Buffer, add β-mercaptoethanol to a final concentration of 5 mM, PMSF to 1 mM, pepstatin to 2 μg/mL, and leupeptin to 2 μg/mL. Chill on ice, and dissolve the cell pellet in 15 mL of cold Extraction Buffer. Break yeast cells by running through French Press eight times (18,000 PSI). Keep the sample on ice whenever possible (*see* **Note 12**).

4. Spin the cell lysate at 12,000×g for 10 min to remove cell debris.

5. Transfer the supernatant into a cold 50 mL conical tube and measure the volume V. Split the supernatant evenly into two 50 mL conical tubes. To each tube, add 2.83 V volume of IPP150 buffer, 0.005 V volume of 10 % NP40, 100 μL of pre-washed IgG beads, and 0.005 V volume of 100 mM PMSF.

6. Rotate the conical tube from end to end at 4 °C for 2 h.

7. Spin at 400×g for 10 min to precipitate the beads.

8. Carefully remove supernatant without disturbing beads. Resuspend the beads in 1 mL of IPP150 buffer. Transfer the bead suspension from the 50 mL conical tubes to 1.5 mL low binding tubes.

9. Spin at 5000 rpm in Table Top Centrifuge at 4 °C for 2 min, and remove the supernatant.

10. Repeat **steps 8** and **9** at least two times.

11. Wash the beads again with 1 mL of 3C cleavage buffer two times.

12. Resuspend the beads in 150 μL of 3C cleavage buffer. Add 4 μg of 3C protease. Rotate the tube from end to end at 4 °C overnight.

13. Use a flame-heated needle to poke a small hole at the bottom of the tube, and place it on top of another 1.5 mL tube. Spin at 1000 rpm for 2 min to collect the purified box H/ACA RNPs (*see* **Note 13**).

14. Add 100 μL of cold 100 % glycerol to each tube and store the box H/ACA RNPs at –20 °C for short-term use and –80 °C for long-term storage (*see* **Note 14**).

3.3 Depletion of Endogenous Box H/ACA RNAs

1. To each 250 μL of immunoprecipitated RNP (Subheading 3.2, **step 14**), add 5 μL of 100 mM CaCl$_2$ (final 2 mM) and 300 units of micrococcal nuclease. Incubate at 30 °C for 30 min.

2. Add EGTA to a final concentration of 5 mM to inactivate micrococcal nuclease. The micrococcal nuclease-treated complexes are now ready for functional reconstitution (*see* Subheading 3.5 below)

3.4 In Vitro Synthesis of Designer Box H/ACA RNA

1. According to the substrate sequence information, design and purchase DNA oligos (template sequence) (listed in Subheading 2.4, **step 1**, and illustrated in Fig. 1b). Dilute the DNA oligos to 10 μM [22].

2. Set up a standard PCR reaction where the final concentrations of the four oligos are: 1 μM of F1 oligo, 0.1 μM of R1 oligo, 0.1 μM of R2 oligo, 1 μM of R3 oligo. The PCR cycle parameters are: 94 °C for 20 s, 30 cycles of [94 °C for 20 s, 50 °C for 15 s, and 72 °C for 30 s], followed by 72 °C for 3 min.

3. Add 1/10 (v/v) of 10× DNA loading dye to the completed PCR reaction. Load and resolve the PCR product on a freshly prepared 1.5 % agarose gel.

4. Excise the band of expected size (PCR product) and gel-purify the product using the Qiagen gel extraction kit (*see* **Note 15**).

5. Set up a 200-μL in vitro transcription reaction containing 50 mM Tris–HCl pH 7.5, 10 mM MgCl$_2$, 10 mM DTT, 2 mM spermidine, 1 mM of each rNTP, 10 pmol DNA template (purified PCR product) and 5 U/μL of T7 RNA polymerase [23]. Transfer 10 μL of the reaction mixture to a new tube containing ~50,000 cpm (input cpm) [alpha-^{32}P]GTP (less than ~0.5 μL). This ~10-μL reaction is to be used later for locating and quantifying RNA transcript produced (see **Note 23**). Incubate both reactions (190 μL and 10 μL) at 37 °C for 2 h (*see* **Note 16**).

6. Add 5 U of DNase I (RNase free) to the 190-μL reaction and ~1/4 U (1 μL of diluted DNase (1:20 dilution)) to the ~10-μL reaction, and incubate at 37 °C for 20 min.

7. Add G50 buffer to bring both reactions to 300 μL. Add 500 μL of PCA and mix thoroughly. After a brief spin (13,000 rpm, 1–2 min), transfer the aqueous phase to new tubes. Add 750 μL of 100 % ethanol, mix well by vortexing, and place the tubes on dry ice. 20 min later, precipitate the RNA transcript

by centrifugation (bench-top centrifuge, 13,000 rpm, 4 °C, 15 min) (*see* **Note 17**).

8. Wash the RNA pellet once with 70 % ethanol by spinning at top speed for 5 min. Resulting pellet is dried and resuspended in a small volume of water (<10 μL, if possible), add an equal volume of 2× RNA loading dye, and load both samples (from the 190-μL and 10-μL reactions) side-by-side onto a 6 % denaturing polyacrylamide gel (8 M urea).

9. After electrophoresis, excise the bands (containing the RNA transcript) from both lanes, and place them in new tubes. The band in the 10-μL reaction lane (containing ^{32}P) is visualized by autoradiography and used as a marker to pinpoint the position of the RNA transcript (unlabeled) in the neighboring lane (the 190-μL reaction) containing no radioisotope.

10. Add 400 μL of G50 to both tubes, and place the tubes on dry ice for 10 min. Transfer the tubes to room temperature, and elute the RNA (from the gel slices) overnight.

11. Spin down the gel slices, and transfer the eluent to new tubes. Add 400 μL of PCA to each tube containing the RNA eluate, mix well, and spin at top speed (13,000 rpm) in a bench-top centrifuge for 3 min.

12. Carefully transfer the aqueous phase to new tubes, and add 10 μg of glycogen, and then 1 mL of 100 % ethanol. Spin at 13,000 rpm in the bench-top centrifuge at 4 °C for 15 min. Wash the pellet with 250 μL of 70 % ethanol by spinning in the bench-top centrifuge (13,000 rpm) at 4 °C for 5 min. Remove ethanol and air-dry the pellet.

13. Resuspend the pellet in ddH$_2$O, and use scintillation counter to obtain cpm (from the 10-μL reaction). Calculate the percentage of [alpha-^{32}P]GTP-incorporation (incorporated cpm/input cpm; usually ~5 %), and determine the amount of RNA transcript recovered (*see* **Note 23**). Typically, a 190-μL reaction generates ~10–15 μg of RNA transcript. Store the RNA transcript at −80 °C until use.

3.5 Reconstitution of Box H/ACA RNP (Pseudouridylation Assay)

1. Assemble in vitro pseudouridylation reaction: 5 μL of micrococcal nuclease-treated Box H/ACA RNP (*see* Subheading 3.3), 50 ng of in vitro synthesized box H/ACA RNA, 5000 cpm of singly labeled substrate RNA (*see* **Note 18**), 40 μg of yeast tRNA in 100 mM Tris–HCl at pH 8.0, 100 mM ammonium acetate, 5 mM MgCl$_2$, 2 mM DTT, 0.1 mM EDTA (*see* **Note 19**). Add ddH$_2$O to bring to 20 μL, and incubate at 37 °C for 1 h [18].

2. Add 280 μL of G50 buffer and 300 μL of PCA to the reaction, vortex vigorously for 1 min, and spin at 13,000 rpm in the bench-top centrifuge at room temperature for 5 min.

3. Transfer the supernatant to a new tube and add 1 mL of 100 % ethanol, 10 μg of glycogen. Spin at 13,000 rpm in the bench-top centrifuge at 4 °C for 15 min. Wash RNA pellet using 70 % ethanol and air-dry the pellet (*see* **Note 20**).

4. Resuspend the RNA pellet in 10 μL of nuclease P1 digestion buffer. Add 5 U of nuclease P1 and incubate the reaction at 37 °C for 2 h.

5. Spot 1.5 μL of the digestion products onto TLC PEI cellulose plate with 1 cm space between dots. Develop the plate in TLC Developing Buffer for 4–7 h.

6. Dry the plate in chemical hood overnight. Wrap the plate in saran wrap and expose the plate using phosphoimager (*see* **Note 21**). A sample result is shown in Fig. 2.

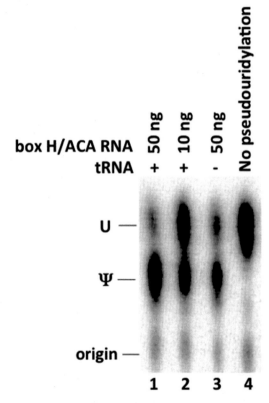

Fig. 2 In vitro pseudouridylation assay. Pseudouridylation is carried out at 37 °C for 1 h. The RNA substrate contains a single ^{32}P at the site 5′ of target uridine. After the reaction, the RNA is recovered, cleaved by RNase P1, and analyzed by TLC. Lane 1, in vitro assay performed as described in the text (Subheading 3.5) where the pseudouridylation reaction contains 50 ng of in vitro transcribed box H/ACA RNA. Lane 2, same as lane 1, except that the reaction contains 10 ng of box H/ACA RNA. Lane 3, same as lane 1, except that tRNA is not included. Lane 4, substrate RNA used directly for RNase P1 digestion (no pseudouridylation)

4 Notes

1. Raffinose is used to bridge the transition from dextrose to galactose, so that all the dextrose will be removed (dextrose represses the expression of a Gal promoter).

2. Yeast TAP-tagged library is available from Open Biosystems of Thermo Scientific [24].

3. Pure β-mercaptoethanol has a concentration of 14.6 M.

4. Protease inhibitors are not stable. PMSF will lose its activity within an hour of use. Prepare protease inhibitors immediately before use and always keep them cold.

5. ^{32}P has a short half-life of 14 days. They should be ordered not long before the experiment.

6. Acrylamide is a neurotoxin. Wear gloves and exercise caution when making the gel. Dissolving urea in water is an endothermic reaction, thus it is desirable to turn on low heat on the stirring device.

7. Keep G50 buffer above 4 °C, otherwise SDS will precipitate.

8. Phenol and chloroform are toxic and harmful in contact with skin. Handle them in the hood and wear safety glasses.

9. Use at least 250 mL flask to grow a 50 mL culture. Adequate air space is necessary for efficient aeration and growth of the culture.

10. Chill the yeast culture, ddH$_2$O, and Extraction Buffer on ice before harvesting cells. Turn on Sorvall centrifuge and keep rotor and centrifugation bottles at 4 °C.

11. To facilitate transfer of beads, cut pipette tips at the fine ends with scissors or razor blades to make a wider opening.

12. To evaluate yeast cell lysis, take 1 μL of the extract and dilute it with 9 μL of ddH$_2$O. Examine 2 μL of diluted samples under a light microscope. (Lysed yeast cells are dark while un-lysed cells appear to be bright and refractile.)

13. The hole should be small enough to prevent beads from going through. If leaky, the leaked beads can be spun down, and the supernatant recovered.

14. Store the Box H/ACA RNP in small aliquots. Avoid repeated freezing and thawing.

15. In our experience, higher molecular weight nonspecific bands might show up in the gel. It is important to cut the right band using DNA ladder as reference to avoid contamination.

16. A radioactive transcript is essential for locating the position of the RNA transcript on the gel and quantifying the amount of RNA recovered.

17. It is necessary to clean up the RNA transcript by phenol/chloroform extraction and concentrate it by ethanol precipitation followed by resuspension into a small volume of water prior to loading onto the gel (a large volume will not fit in the well).

18. The uridine to be pseudouridylated can be site-specifically ^{32}P-labeled. This can be achieved by cleaving the full-length RNA at the site 5′ of the target uridine, followed by phosphorylation of the 3′ fragment with [gamma-^{32}P]ATP and ligation of the labeled 3′ fragment with the 5′ fragment [23, 25].

19. Yeast tRNA is needed to protect substrate RNA from degradation (Fig. 2, compare lane 1 with lane 3).

20. Use the Geiger counter to detect radiolabeled RNA and make sure RNA is precipitated.

21. Make sure the TLC plate is dried completely before being wrapped and exposed to a phosphorimaging screen. The residual developing buffer could otherwise damage and reduce the sensitivity of the phosphorimaging screen.

22. All cell culture procedures should be performed under sterile conditions (close to a flame, and on a clean bench top). The operation should be done in a short period of time to avoid possible contamination, especially for operation involving nutrient-rich 3×-YPGal medium.

23. If the percentage of [alpha-^{32}P]GTP-incorporation [dividing the incorporated cpm (assume ~2500) by the input cpm (~50,000)] is ~5 %, the amount of RNA generated from the 10 μL reaction can be calculated as follows: 1 mM (GTP concentration) × 10 μL (reaction volume) × 5 % (percentage of [alpha-^{32}P]GTP-incorporation) × 4 (assuming the 4 nucleotides, A, U, C, and G, are equally distributed in the RNA molecule) x 330 (average nucleotide molecular weight) = 660 ng of RNA. Thus, a 190 μL reaction will generate 660 ng × 19 = 12,540 ng (or 12.54 μg) of RNA.

Acknowledgement

We thank the members of the Yu laboratory for inspiring discussions. This work was supported by grants GM104077 and AG039559 (to Y.-T. Y.) from the National Institutes of Health.

References

1. Davis DR (1995) Stabilization of RNA stacking by pseudouridine. Nucleic Acids Res 23: 5020–5026

2. Davis FF, Allen FW (1957) Ribonucleic acids from yeast which contain a fifth nucleotide. J Biol Chem 227:907–915

3. Charette M, Gray MW (2000) Pseudouridine in RNA: what, where, how, and why. IUBMB Life 49:341–351

4. Newby MI, Greenbaum NL (2001) A conserved pseudouridine modification in eukaryotic U2 snRNA induces a change in branch-site architecture. RNA 7:833–845

5. Kierzek E, Malgowska M, Lisowiec J, Turner DH, Gdaniec Z, Kierzek R (2014) The contribution of pseudouridine to stabilities and structure of RNAs. Nucleic Acids Res 42:3492–3501

6. Reddy R, Busch H (1988) Small nuclear RNAs: RNA sequences, structure, and modifications. In: Birnsteil ML (ed) Structure and function of major and minor small nuclear ribonucleoprotein particles. Springer-Verlag Press, Heidelberg, pp 1–37

7. Ofengand J, Fournier MJ (1998) The pseudouridine residues of rRNA: number, location, biosynthesis and function. In: Grosjean H, Benne R (eds) Modification and Editing of RNA: the Alteration of RNA structure and Function. ASM press, Washington, D.C., pp 229–253

8. Grosjean H, Sprinzl M, Steinberg S (1995) Posttranscriptionally modified nucleosides in transfer RNA: their locations and frequencies. Biochimie 77:139–141

9. Carlile TM, Rojas-Duran MF, Zinshteyn B, Shin H, Bartoli KM, Gilbert WV (2014) Pseudouridine profiling reveals regulated mRNA pseudouridylation in yeast and human cells. Nature 515:143–146

10. Schwartz S, Bernstein DA, Mumbach MR, Jovanovic M, Herbst RH, Leon-Ricardo BX, Engreitz JM, Guttman M, Satija R, Lander ES, Fink G, Regev A (2014) Transcriptome-wide mapping reveals widespread dynamic-regulated pseudouridylation of ncRNA and mRNA. Cell 159:148–162

11. King TH, Liu B, McCully RR, Fournier MJ (2003) Ribosome structure and activity are altered in cells lacking snoRNPs that form pseudouridines in the peptidyl transferase center. Mol Cell 11:425–435

12. Yu AT, Ge J, Yu YT (2011) Pseudouridines in spliceosomal snRNAs. Protein Cell 2:712–725

13. Karijolich J, Yu YT (2008) Insight into the protein components of the box H/ACA RNP. Curr Proteomics 5:129–137

14. Ge J, Yu YT (2013) RNA pseudouridylation: new insights into an old modification. Trends Biochem Sci 38:210–218

15. Karijolich J, Yu YT (2011) Converting nonsense codons into sense codons by targeted pseudouridylation. Nature 474:395–398

16. Fernandez IS, Ng CL, Kelley AC, Wu G, Yu YT, Ramakrishnan V (2013) Unusual base pairing during the decoding of a stop codon by the ribosome. Nature 500:107–110

17. Ma X, Yang C, Alexandrov A, Grayhack EJ, Behm-Ansmant I, Yu YT (2005) Pseudouridylation of yeast U2 snRNA is catalyzed by either an RNA-guided or RNA-independent mechanism. EMBO J 24:2403–2413

18. Xiao M, Yang C, Schattner P, Yu YT (2009) Functionality and substrate specificity of human box H/ACA guide RNAs. RNA 15:176–186

19. Karijolich J, Stephenson D, Yu YT (2007) Biochemical purification of box H/ACA RNPs involved in pseudouridylation. Methods Enzymol 425:241–262

20. Alexandrov A, Grayhack EJ, Phizicky EM (2005) tRNA m7G methyltransferase Trm8p/Trm82p: evidence linking activity to a growth phenotype and implicating Trm82p in maintaining levels of active Trm8p. RNA 11:821–830

21. Rigaut G, Shevchenko A, Rutz B, Wilm M, Mann M, Seraphin B (1999) A generic protein purification method for protein complex characterization and proteome exploration. Nat Biotechnol 17:1030–1032

22. Huang C, Wu G, Yu YT (2012) Inducing nonsense suppression by targeted pseudouridylation. Nat Protoc 7:789–800

23. Huang C, Yu YT (2013) Synthesis and labeling of RNA in vitro. Curr Protoc Mol Biol. Chapter 4: Unit4 15

24. Gelperin DM, White MA, Wilkinson ML, Kon Y, Kung LA, Wise KJ, Lopez-Hoyo N, Jiang L, Piccirillo S, Yu H, Gerstein M, Dumont ME, Phizicky EM, Snyder M, Grayhack EJ (2005) Biochemical and genetic analysis of the yeast proteome with a movable ORF collection. Genes Dev 19:2816–2826

25. Zhao X, Yu YT (2004) Detection and quantitation of RNA base modifications. RNA 10:996–1002

Chapter 10

Northwestern Blot Analysis: Detecting RNA–Protein Interaction After Gel Separation of Protein Mixture

Shangbing Zang and Ren-Jang Lin

Abstract

Northwestern assays detect a direct binding of a given RNA molecule to a protein immobilized on a nitrocellulose membrane. Here, we describe protocols to prepare ^{32}P-labeled RNA probes and to use them to assay for RNA–protein interactions after partially purified protein preparations are resolved on denaturing SDS-polyacrylamide gels. The method can unambiguously determine whether the protein of interest can directly and independently bind RNA even in the presence of contaminating bacterial proteins or degradation products that at times may hinder interpretation of results obtained from gel mobility shift or RNP immunoprecipitation assays.

Key words ^{32}P-labeling of RNA, SP6 or T7 RNA polymerase, In vitro transcription, Cobalt affinity purification of histidine-tagged protein, Protein renaturation on membrane, RNA-binding protein, Direct RNA binding

1 Introduction

Northwestern detection of RNA–protein interaction is used when one wants to assay the binding of a known RNA to a known or an unknown protein. The method was first described by Bowen et al. when they developed a "protein blotting" method to detect DNA–protein and RNA–protein interactions [1]. They separated a nuclear protein lysate on an SDS-polyacrylamide gel and blotted the proteins onto nitrocellulose filters, which were then probed with ^{32}P-labeled DNA, ^{32}P-labeled RNA, or ^{125}I-labeled histone. Upon washing away nonspecific binding, filter-anchored proteins that can bind to the specific DNA sequence, RNA sequence, or histone are detected by autoradiography. The assay allows one to identify in a protein mixture the number of proteins—and their sizes—that can bind to, for example, a specific RNA sequence. It is worth noting that even the protein was denatured in SDS and then immobilized onto the filter, the protein was able to refold and bind its RNA target.

Ren-Jang Lin (ed.), *RNA-Protein Complexes and Interactions: Methods and Protocols*, Methods in Molecular Biology, vol. 1421,
DOI 10.1007/978-1-4939-3591-8_10, © Springer Science+Business Media New York 2016

This "Northwestern" method was further modified by Jeffrey Wilusz and his colleagues to screen cDNA expression libraries for proteins that can bind RNA with a specific sequence [2, 3]. A bacteriophage expression library was grown on agar plates and then blotted onto a nitrocellulose membrane. The proteins on the membrane were renatured and probed with radioactive RNA, and the positive clones were identified by aligning the film, the filter, and the plate. A Northwestern method using biotinylated, nonradioactive RNA probes has also been developed [4].

Gel mobility shift assay (GMSA) is widely used for measuring RNA–protein interactions and the method can be quantitative [5]. However, GMSA requires highly purified proteins and the assay can be ineffective or even misleading when just a small amount of nucleases or other RNA-binding proteins are present in the protein preparation. The Northwestern method, on the other hand, can overcome these shortcomings because contaminating proteins can be separated from the protein of interest on a SDS gel. Moreover, Northwestern can also be used to confirm that the "putative" RNA-binding protein indeed binds directly, without the aid of other protein subunits for example, to the RNA target. Some recent examples of how Northwestern is used for the investigation of RNA–protein interaction are provided here [6–11].

Here, we describe a Northwestern protocol to detect RNA–protein interaction using partially purified recombinant human proteins from *E. coli*. RNA probes are made by transcribing linear DNA template in the presence of ^{32}P-UTP and gel purified. The protocol is straight forward and can yield semi-quantitative information if the protein amounts can be applied accurately. Even in the presence of contaminating bacterial proteins or degradation products that can bind ^{32}P-labeled RNA under the conditions used, Northwestern assays can still readily discern whether the protein of interest can directly and independently bind RNA or not.

2 Materials

Prepare all buffers and solutions with autoclaved, deionized, or glass double-distilled water. Common reagents from reliable suppliers should work fine with these protocols. Vendor information is provided for some items for convenience to users unless specified otherwise.

2.1 Preparation of RNA Probes

2.1.1 DNA Templates

1. Plasmid DNAs containing sequence of interest.

2. Restriction enzymes, 10× reaction buffers, and bovine serum albumin (BSA).

3. TE buffer: 10 mM Tris–HCl, pH 8.0, 1 mM EDTA.

4. TE-saturated phenol, pH 6.6 (Fisher Scientific). Phenol is toxic and is irritating to the eyes, skin, and respiratory system; wear gloves and work in a fume hood when handle it.

5. CIA mix: chloroform:isoamyl alcohol = 24:1 (v/v).

6. 3 M ammonium acetate.

2.1.2 In Vitro Transcription

1. 100 mM dithiothreitol (DTT).

2. A solution contains 25 mM each of ATP, CTP, GTP, and UTP (for example, 10× NTP mix, pH 7.5 by New England Biolabs).

3. α-[^{32}P] uridine triphosphate (UTP) (800 Ci/mmol, 10 µCi/µL; MP Biomedical).

4. RNasin (Promega).

5. T7 RNA polymerase (New England Biolabs).

6. SP6 RNA polymerase (New England Biolabs).

7. 10× RNA polymerase reaction buffer (*see* **Note 1**).

2.1.3 Gel Purification of RNA

1. 10× TBE buffer. To make 1 L, dissolve in water 108 g Tris-base, 55 g boric acid, and 9.3 g Na_2EDTA (or 40 mL of EDTA-NaOH, pH 8.0). Autoclave and store at room temperature.

2. 8 M urea/1× TBE. To make 100 mL, dissolve in water 48 g of ultrapure urea with 10 mL of 10× TBE buffer. Store at 4 °C.

3. 40 % acrylamide (19:1). To make 100 mL, dissolve in water 38 g of acrylamide and 2 g of bis-acrylamide. Store at 4 °C.

4. 20 % acrylamide (19:1)/8 M urea/1× TBE. To make 50 mL, combine 24 g of ultrapure urea with 5 mL of 10× TBE and 25 mL of 40 % acrylamide (19:1). Filter the solution through a Whatman filter paper in a Büchner funnel and store at 4 °C (*see* **Note 2**).

5. Formamide sample buffer: 1× TBE, 90 % (v/v) deionized formamide, trace amounts of bromophenol blue and xylene cyanol (~0.2 % each). Store in small aliquots at –20 °C.

6. 10 % ammonium persulfate. To make 10 mL, dissolve in water 1 g of ammonium persulfate. Store in small aliquots at –20 °C.

7. 10 % SDS. To make 100 mL, dissolve in water 10 g of sodium dodecyl sulfate. Store at room temperature. In solid powder form, SDS is irritating to the eyes, skin, and respiratory system; wear gloves and face mask, and handle it in a fume hood.

8. RNA elution buffer: 0.5 M ammonium acetate, 1 mM EDTA, and 0.1 % SDS.

9. 20 mg/mL glycogen (*see* **Note 3**). Dissolve 1 g of glycogen in 1 mL of water and extract once with one volume of buffer-saturated phenol. Take a new tube and measure its

weight; transfer the supernatant to that new tube, add 1.5 mL of 100 % ethanol, mix and let stand at –20 °C overnight. Collect the precipitate by centrifugation and air-dry the pellet. Measure the weight of the recovered glycogen, add autoclaved water to 20 mg/mL, aliquot and store at –20 °C.

10. KODAK X-OMAT LS Film.

11. Phosphorescent tape.

2.2   Preparation of Protein Blots

2.2.1   Protein Isolation

1. 1 M IPTG (Isopropyl β-D-1-thiogalactopyranoside).

2. Phosphate buffer: 50 mM Na_2HPO_4, pH 8.0, 300 mM NaCl.

3. Lysis Buffer: Phosphate buffer plus 0.1 % Triton-X-100 and 100 μg/mL lysozyme.

4. Cobalt-charged beads: Chelating Sepharose Fast Flow beads charged with cobalt, store in 20 % ethanol at 4 °C.

5. 1 M Imidazole.

6. Dialysis buffer: 20 mM Hepes-KOH, pH 7.9, 50 mM KCl, 2 mM $MgCl_2$.

7. Chromatography Columns.

2.2.2   SDS-PAGE

1. Resolving gel buffer: 1.5 M Tris–HCl, pH 8.8. To make 1 L, dissolve in about 850 mL of water 181.7 g of Tris-base, adjust pH with HCl to 8.8. Store at room temperature or 4 °C.

2. Stacking gel buffer: 0.5 M Tris–HCl, pH 6.8. To make 1 L, dissolve 60.6 g Tris-base and adjust to pH 6.8 with HCl.

3. 5× SDS-PAGE running buffer. To make 500 mL, dissolve in about 450 mL of water 7.55 g of Tris-base and 47 g of glycine, and add 25 mL of 10 %SDS.

4. 30 % acrylamide/bis-acrylamide (37.5:1) (Sigma-Aldrich). Store at 4 °C.

5. 2× SDS sample buffer: 0.5 % 2-mercaptoethanol, 3 % SDS, 0.5× resolving gel buffer, 25 % glycerol, a trace amount of bromophenol blue.

2.2.3   Protein Blotting

1. 5× Transfer buffer. To make 500 mL, dissolve in about 450 mL of water 7.55 g of Tris-base and 36 g of glycine, and add 25 mL of 10 % SDS.

2. Methanol buffer: 1× Transfer buffer, 20 % (v/v) methanol. Combining 3 volumes of water, 1 volume of 5× Transfer buffer, and 1 volume of 100 % methanol.

3. 0.1 % Ponceau S staining solution: To make 1 L, dissolve in water 1 g of Ponceau S with 50 mL of acetic acid. Store at 4 °C.

4. PVDF (polyvinyl difluoride) membranes (Fisher Scientific).

5. Trans-Blot Semi-Dry Transfer Cell (Bio-Rad).

**2.3 Binding of RNA
to Immobilized
Protein—
Northwestern**

1. Northwestern (NW) buffer: 10 mM Tris–HCl, pH 7.5, 50 mM NaCl, 1 mM EDTA, 0.02 % (w/v) Ficoll 400, and 0.02 % (w/v) polyvinylpyrrolidone-40 (PVP-40).

2. Renaturing buffer: The NW buffer plus 1 % BSA (w/v).

3. Hybridization Tubes (Fisher Scientific).

4. Hybridization oven (Fisher Scientific).

3 Methods

**3.1 Preparation
of RNA Probes**

**3.1.1 Prepare Linearized
DNA Templates (see
Note 4)**

In order to examine sequence-specific binding, RNA with different sequences will need to be prepared. In this illustration, we use three plasmid DNAs to make 3 RNAs: (1) pRG1 is a derivative of plasmid pMINX [12] containing an SP6 promoter and a β-globin sequence with two exons and an intron [13], (2) pCR2.1-T7.U2 contains a T7 promoter and a U2 snRNA sequence, and (3) pCR2.1-T7.U4 contains a T7 promoter and a U4 snRNA sequence [6]. These plasmids need to be cleaved at a restriction site downstream of the β-globin or snRNA sequence for use in the in vitro transcription to generate run-off transcripts (*see* **Note 5**).

1. Mix 10 μg plasmid DNA, 5 μL 10× reaction buffer, 2 μL restriction enzyme, add water to 50 μL (*see* **Note 6**).

2. Incubate the mixture for 4 h at 37 °C or an appropriate temperature.

3. Check 1 μL of the digested material on an agarose gel to verify the completion of digestion.

4. When digestion is complete, add 150 μL of TE buffer and 200 μL of buffer-saturated phenol.

5. Vortex for 2 min and then centrifuge in a microcentrifuge at the maximum speed (commonly at $17,000 \times g$) for 5 min.

6. Transfer the supernatant to a new tube and add an equal volume of CIA mix and repeat extraction (*see* **step 5**).

7. Transfer the supernatant to a new tube and add 1/10 volume of 3 M ammonium acetate and 2.5 volumes of 100 % ethanol. Keep at –80 °C for 1 h.

8. Centrifuge at the maximum speed for 30 min. Discard the supernatant. Add 1 mL of cold 70 % ethanol to the pellet; briefly mix, and then centrifuge at the maximum speed for 5 min.

9. Remove the supernatant. Let the pellet air-dry with the lid open or dry the DNA pellet under vacuum in a SpeedVac.

10. Resuspend the DNA in 20 μL of TE buffer. Measure the DNA concentration (for example, using a NanoDrop). The DNA solution can be stored at –20 °C.

<table>
<tr><td>

3.1.2 *Generate*
^{32}P-Labeled RNA by In Vitro
Transcription

</td><td>

^{32}P-labeled RNA is prepared by transcribing the linear DNA template by SP6 or T7 RNA polymerase. To assemble a 25 μL reaction, the following are added in order to a prechilled microcentrifuge tube:

</td></tr>
</table>

Water (25 μL minus the total volume of all the other reagents).

2.5 μL of 10× RNA polymerase reaction buffer.

2.5 μL of 10× NTP mix.

2.5 μL of 100 mM DTT.

1 μL of 10 U/μL RNasin.

3 μL of α-[^{32}P] UTP.

1.5 μg of the linearized DNA template.

1 μL of T7 (50 units) or SP6 (20 units) RNA polymerase.

Incubate at 37 °C for 60 min (*see* **Note 7**).

3.1.3 *Purify ^{32}P-Labeled RNA from Gel*

The RNA generated by in vitro transcription can be purified by denaturing polyacrylamide gel electrophoresis. The ^{32}P-labeled RNA band can be visualized through the aid of an X-ray film and can be cut out using a razor blade. The RNA can then be eluted with an overnight incubation in RNA elution buffer.

1. In vitro transcription reactions are terminated by adding an equal volume of the formamide sample buffer, heated at 95 °C for 1 min, and then chilled on ice.

2. Cast a 0.4 mm thick, 6 % polyacrylamide, 8 M urea denaturing gel (*see* **Note 8**).

3. Pre-run the gel at 500 V for 30 min; just before loading the samples, use a Pasteur pipette or micropipet to flush urea out of the wells (*see* **Note 9**). The samples are electrophoresed at 500 V for 1–2 h in 1× TBE buffer until the xylene cyanol tracking dye is about 3 cm from the gel bottom.

4. After gel electrophoresis, the gel is transferred to a used X-ray film. The ^{32}P-labeled RNAs are visualized by exposing to a new X-ray film (Fig. 1). The full-length RNA gel piece is excised (*see* **Note 10**).

5. Put the gel slice containing the RNA into a microcentrifuge tube, and add 400 μL of RNA elution buffer. The RNA is eluted at room temperature overnight (or at 37 °C for 4 h) in a shaking incubator.

6. Transfer 350 μL of the eluate to a new tube (use caution to avoid transferring any small gel piece). Add 0.5 μL of 20 mg/mL glycogen and precipitate the RNA by adding 350 μL of isopropanol. Incubate at –80 °C for at least 30 min (*see* **Note 11**).

7. Recover the precipitated RNA by microcentrifugation at the maximum speed (17,000×g) at 4 °C for 30 min. Wash the

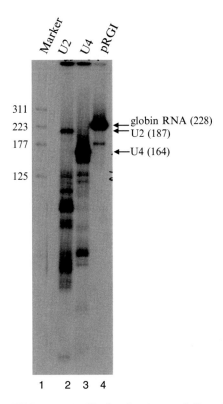

Fig. 1 **[32]P-labeled RNA generated by in vitro transcription of linearized DNA templates**. Restriction enzyme linearized plasmid DNAs (T7.U2, T7.U4, and pRG1) were transcribed by SP6 or T7 RNA polymerase in the presence of α-^{32}P-UTP. The reaction mixture was electrophoresed on a 6 % polyacrylamide urea gel and the ^{32}P-labeled RNA was detected by an X-ray film. Size markers to the left are labeled with number of nucleotides. Full-length transcripts are labeled on the right with the sizes labeled in parenthesis. Full-length transcripts were isolated from the gel and used for subsequent Northwestern binding assays

pellet with ice-cold 70 % ethanol, dried, and resuspended in 30 μL of RNase-free water. Count the radioactivity by Cerenkov counting; RNA can be stored at –20 °C (*see* **Note 12**).

3.2 Preparation of Protein Blots

3.2.1 Preparation of Protein

Proteins of interest can be extracted from cells and used as unpurified or partially purified. In this illustration, we describe the isolation from *E. coli* of several human recombinant proteins: hnRNP A1, GPKOW, GPKOW with point substitutions, and GPKOW with deletions (Fig. 2). HnRNP A1 is a heterogeneous nuclear ribonucleoprotein with a high affinity for many RNA sequences [14]. GPKOW contains G-patch and KOW domains and is recently shown to interact with RNA [6, 15]. All the recombinant proteins contain a histidine-tag and can be purified by a

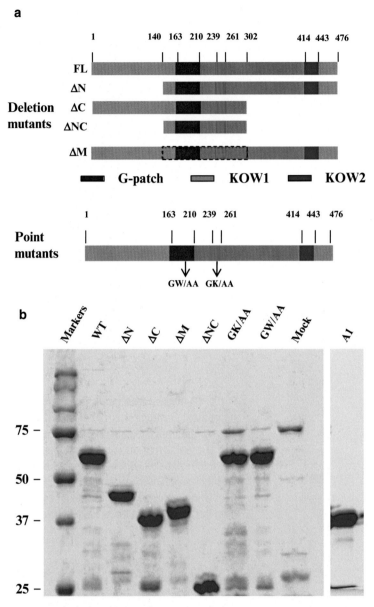

Fig. 2 Recombinant human proteins expressed and partially purified from *E. coli*. (**a**) A diagram depicting the domain arrangement in GPKOW: *red*, G-patch; *green*, KOW1; *blue*, KOW2. The numbers above the diagram are the amino acid positions counted from the amino-terminus. Deletion mutants (GPKOW-ΔN, ΔC, ΔM, and ΔNC) and double-point mutants (GPKOW-GK/AA and GW/AA) are indicated. (**b**) His-tagged recombinant proteins (GPKOW-WT, ΔN, ΔC, ΔNC, ΔM, GK/AA, and GW/AA, and hnRNP A1) were purified by cobalt-affinity chromatography, separated on SDS-PAGE gels, and stained with Coomassie blue. Mock: proteins purified from cells with the "empty" plasmid vector

nickel or cobalt affinity chromatography. Each protein is expressed from an IPTG-inducible promoter-carrying plasmid in *E. coli*.

1. Grow 50 mL of bacterial culture carrying the plasmid expressing the protein of interest to OD_{600} between 0.1 and 0.2.

2. Save and chill 1 mL of the uninduced culture as a negative control.

3. Add IPTG to the culture to a final concentration of 1 mM and continue culturing for 5 h for induction.

4. Save and chill 1 mL of the induced culture as a positive control.

5. Transfer the main culture to a test tube and harvest the cells by centrifugation at 4 °C. The cell pellets can be kept in a −80 °C freezer for later purification steps.

6. To prepare total protein lysates from control samples, resuspend the cell pellet in 100 μL of Phosphate buffer, add 100 μL of 2× SDS-PAGE loading buffer, pass through a 25-gauge needle several times to reduce the viscosity, and heat at 95 °C for 5 min (*see* **Note 13**).

7. To purify the recombinant protein from the main culture, resuspend the cell pellet in 5 mL of cold Lysis buffer and incubate on ice for 30 min.

8. Sonicate the cell lysate for 20 s at 30 % power/amplitude, rest 30 s on ice, and repeat this sonication cycle 5 more times (a total of 2 min of sonication).

9. Transfer the lysate to several microcentrifuge tubes and centrifuge at maximum speed in a microcentrifuge (about $17,000 \times g$) at 4 °C for 15 min and collect the supernatant for subsequent protein purification.

10. Pack a chromatography column with 500 μL of the Cobalt-charged beads, wash the column with 1 mL of water and then equilibrate with 1 mL of Phosphate buffer.

11. Add the protein lysate from **step 9** to the column, cap the top and the bottom of the column, and gently rotate at 4 °C for 30 min.

12. Set the column up and let the solution flow through by gravity.

13. Wash the column with 30 Column Volume (CV) of Phosphate buffer containing 0.1 % Triton-X100 and 10 mM imidazole.

14. Elute the protein with 5 CV of Phosphate buffer containing 200 mM imidazole and collect fractions of 250 μL each.

15. Measure the protein quantity by common colorimetric methods and check the protein purify by SDS-PAGE (*see* **Note 13** and Fig. 2). Pool the desired protein fractions and dialyzed in

Dialysis buffer for 2 h, aliquot, freeze in liquid nitrogen, and store at −80 °C.

3.2.2 Separation of Proteins by SDS-PAGE

1. Assemble gel plates. Wash the glass plates with detergent really well, rinse, and wipe clean with 70 % ethanol.

2. Mix 2.5 mL of resolving gel buffer, 3.33 mL of 30 % acrylamide (29:1), and 4 mL of water. Add 100 μL of 10 % SDS, 80 μL of 10 % ammonium persulfate, 10 μL of TEMED, and cast within a 7.25 cm × 10 cm × 1.5 mm plate assembly. Leave a space for the stacking gel. Gently overlay with water.

3. After the gel polymerizes, remove water. Prepare the stacking gel by mixing 1.25 mL of stacking gel buffer, 0.67 mL of 30 % acrylamide, and 3 mL water. Add 100 μL of 10 % SDS, 40 μL of 10 % ammonium persulfate, 5 μL of TEMED, and cast over the resolving gel. Insert the gel comb without introducing air bubbles.

4. Prepare protein samples in SDS sample buffer: 1.5 μg of each protein preparation, 0.3 μg of hn-RNPA1 as positive control, and 1.5 μg of BSA as negative control (*see* **Note 14**).

5. Load protein samples in the middle lanes of the gel. Load protein standards in a flanking lane and SDS sample buffer to the unused wells. Electrophorese at 100 V till the dye front reaches the bottom of the gel.

6. Separate the gel plates with the help of a spatula. The gel should remain on one plate. Remove the stacking gel. Use water to gently remove the residual SDS-PAGE running buffer from the gel.

7. Carefully transfer the gel to a container with Methanol buffer.

3.2.3 Protein Transfer and Ponseau S Staining

1. Cut a PVDF membrane to the size of the gel. Immerse the membrane in methanol for 20 min, rinse twice in water, and leave in Methanol buffer.

2. Cut 6 pieces of Whatman no. 3 filter paper to the size of the gel. Submerge and soak the filters in Methanol buffer for at least 5 min.

3. Gently rinse the inside of the Trans-Blot Semi-Dry Transfer Cell. Lay 3 pieces of the soaked filters on the bottom electrode of the Transfer Cell. Squeeze out any air bubbles by rolling a pipet over the filters.

4. Place the PVDF membrane on top of the filters paper; remove any air bubbles.

5. Carefully pick the gel up and place directly onto the membrane.

6. Finally top with three more soaked filters; remove any air bubbles.

7. Connect the top electrode. Apply 5 V for 60 min.

8. Disassemble the set up, but never let the membrane get dry. Cut off a corner of the membrane for orientation and keep the protein side of the membrane up throughout all subsequent steps.

9. Rinse the membrane with water 2 times. Stain the membrane with 0.1 % Ponseau S staining solution for 3 min at room temperature (*see* **Note 15**). The proteins on the membrane are stained red/pink, which can be photographed (Fig. 3).

10. Rinse the membrane with water 3 times, 5 min for each time. The membrane is ready for RNA binding assays.

3.3   Binding of RNA to Membrane-Bound, Renatured Protein— Northwestern

1. Submerge the membrane in 10 mL of Renaturing buffer for 60 min at room temperature with gentle shaking. This is a key step in renaturing or refolding the proteins on the membrane.

2. Wash 3 times in 10 mL of NW buffer, 5 min for each time.

3. Insert the membrane into a tube designed for nucleic acid hybridization (100 mL in volume). Add 5 mL of NW buffer plus $2-5 \times 10^5$ cpm/mL of ^{32}P-labeled RNA and 50 µg/mL of yeast tRNA into the tube (*see* **Note 16**).

4. Place the hybridization tube on the rotation scaffold in a hybridization oven. Incubate with gentle rotation for 2 h at room temperature to allow binding of RNA to the membrane-bound protein.

5. Remove the solution from the tube. Add 10 mL of NW buffer and incubate at room temperature with gentle rotation for 10 min. Repeat this washing procedure for 4 more times (*see* **Note 17**).

6.   Wrap the membrane in Saran Wrap or let the membrane air dry before exposing it to an X-ray film (Fig. 3). Measure the binding efficiency by quantifying the ^{32}P signal and the amount of protein on the membrane.

As shown in Fig. 3b, all three RNAs (β-globin RNA, U2 snRNA, and U4 snRNA) bind to hnRNP A1 and the derivatives of GPKOW, but do not bind to BSA at all. By estimating the ^{32}P intensity (Fig. 3b) versus the amount of protein (Figs. 2 and 3a), it appears that GPKOW-ΔM has a higher RNA affinity than other GPKOW mutants, but none of them show a strong preference for RNA sequence. Nonetheless, the Northwestern assay can provide clear evidence that these proteins bind directly to the RNA activity even when some contaminating RNA-binding protein may exist in the partially purified protein samples (*see* **Note 18**).

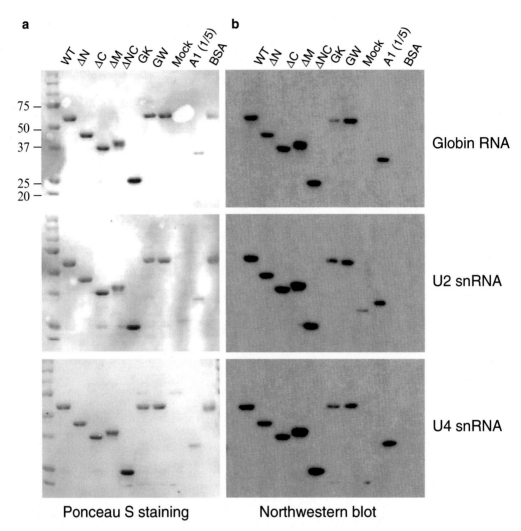

Fig. 3 Northwestern blots showing direct binding of RNA to protein renatured on the membrane. Recombinant His6-tagged GPKOW (WT), GPKOW deletion mutants (ΔN, ΔC, ΔM, and ΔNC) and double-point mutants (GK/ AA and GW/AA), and hnRNP A1 proteins (*see* Fig. 2) were electrophoresed on SDS-polyacrylamide gels, transferred to PVDF membranes, and stained with Ponceau S (Panel **a**). The membranes were incubated with ^{32}P-labeled RNA as indicated and autoradiographed (Panel **b**). An equal amount of protein was used in all samples except that 1/5 of that amount was used in His6-tagged hnRNP A1. Mock: empty vector mimics the protein purification. The same amount of BSA was loaded as negative control (panel **b**)

4   Notes

1. 10× buffer for RNA polymerase reaction is usually supplied by the vendors with the RNA polymerase. The generic SP6 10× buffer contains 400 mM Tris–HCl, pH 7.5, 60 mM MgCl$_2$, and 40 mM spermidine. The generic T7 10× buffer contains 400 mM Tris–HCl, pH 7.9, 60 mM MgCl$_2$, and 20 mM spermidine.

2. The bulk of urea takes up a lot of volume and usually no additional water is needed. Clearing up the solution by filtering produces sharper RNA bands during gel electrophoresis.

3. Glycogen-like yeast tRNA is often used as a carrier for nucleic acid precipitation. Glycogen is used when the procedure calls for an inert carrier that does not participate in or interfere with the subsequent reaction or manipulation.

4. DNA template for in vitro transcription can also be generated by using PCR with a promoter-sequence-containing forward primer and a sequence-of-interest reverse primer [16].

5. To generate linear DNA template by restriction digestion, the restriction enzyme shall not cut between the promoter and the sequence of interest.

6. Water is added to the tube first. Use the buffer recommended by the supplier for the restriction enzyme digestion. Buffer conditions can also be found in the New England Biolabs catalog or website. Bovine serum albumin (BSA) may be required.

7. When using SP6 RNA polymerase, incubation at 39 °C may produce more RNA transcripts than incubation at 37 °C. Longer incubation times, like 2 h, may also increase yield.

8. To make the gel, mix 6 mL of 20 % polyacrylamide (19:1)/8 M urea/1× TBE, 14 mL of 8 M urea/1× TBE, 45 μL of 10 % ammonium persulfate, and 24 μL of TEMED (N,N,N',N'-Tetramethylethylenediamine). This is sufficient for casting a $195 \times 160 \times 0.4$ mm gel. The thickness of 0.4 mm allows the RNA to be eluted from the gel slice without being diced or crushed.

9. The pre-run of denaturing gel removes ammonium persulfate, free acrylic acid, and other free ions that may interact with RNA or protein. The pre-run also heats up the gel to help denaturing the RNA. Urea that diffuses into the wells of the gel may interfere with sample loading as well as cause uneven RNA bands.

10. After gel electrophoresis, separate the glass plates and let the gel set on one of the glass plates. Gently put a used X-ray film on the gel and careful lift the film to detach the gel from the glass plate; wrap the film and the gel with plastic wrap. Cut two small pieces of phosphorescent tape into irregular shapes and affix them onto the sides of the wrapped gel. In a dark room, place the plastic-wrapped gel face down on a new X-ray film and exposed for the desired length of time (e.g., 1 min). Develop the film and align it to the gel using the marks from the phosphorescent tape. The gel band corresponding to the full-length RNA product is cut out with a sharp, ethanol-cleaned razor blade.

11. To avoid denaturing the RNA, the eluted sample is not extracted with phenol or chloroform but simply purified by isopropanol or ethanol precipitation.

12. Cerenkov counting is a measurement technique for samples that deliver high-energy β-particles, such as ^{32}P, and has the advantage of not requiring the addition of scintillation cocktail. The tube containing the entire prep or an aliquot transferred to another tube is placed in a vial and counted with the full window of detection in a scintillation counter. As ^{32}P has a half-life of 2 weeks, the sensitivity of detection will drop if the RNA is stored for an extensive period.

13. It is important to verify by analyzing the negative and positive protein lysates by SDS-PAGE with Coomassie blue staining that the induction of recombinant protein expression is efficient. Purified protein samples are analyzed on SDS-PAGE gels and stained with Coomassie blue. This simple procedure can yield information regarding the complexity and purification of the protein samples (Fig. 2).

14. Protein samples in SDS sample buffer can be loaded to the gel without heating to 95 °C.

15. Ponseau S stains protein transiently and allows a quick estimate of the transfer efficiency and constancy. The stain is easily removed in the subsequent steps.

16. The amount of ^{32}P-labeled RNA used in the experiment depends on the RNA–protein affinity. In our experiment, 2×10^5 cpm/mL of ^{32}P-labeled RNA is sufficient to detect RNA–protein binding. Adding 50 μg/mL of yeast tRNA is to eliminate nonspecific binding of the ^{32}P-labeled RNA to the protein.

17. We usually collect the washing solution after each washing step and measure the radioactivity in it. This will help to determine if additional washing is needed.

18. Some contaminating bacterial proteins or degradation products are found to bind ^{32}P-labeled RNA (Fig. 3). Even with these drawbacks, direct binding of protein of interest to the RNA can be readily discerned on the Northwestern blot.

Acknowledgment

This work was supported by grants from the Beckman Research Institute to R-J L.

References

1. Bowen B, Steinberg J, Laemmli UK, Weintraub H (1980) The detection of DNA-binding proteins by protein blotting. Nucleic Acids Res 8(1):1–20

2. Qian Z, Wilusz J (1993) Cloning of a cDNA encoding an RNA binding protein by screening expression libraries using a northwestern strategy. Anal Biochem 212(2):547–554

3. Bagga PS, Wilusz J (1999) Northwestern screening of expression libraries. Methods Mol Biol 118:245–256. doi:10.1385/1-59259-676-2:245

4. Rodriguez PL, Carrasco L (1994) Non-radioactive northwestern analysis using biotinylated riboprobes. Biotechniques 17(4):702, 704, 706–707

5. Ryder SP, Recht MI, Williamson JR (2008) Quantitative analysis of protein-RNA interactions by gel mobility shift. Methods Mol Biol 488:99–115. doi:10.1007/978-1-60327-475-3_7

6. Zang S, Lin TY, Chen X, Gencheva M, Newo AN, Yang L, Rossi D, Hu J, Lin SB, Huang A, Lin RJ (2014) GPKOW is essential for pre-mRNA splicing in vitro and suppresses splicing defect caused by dominant-negative DHX16 mutation in vivo. Biosci Rep 34(6), e00163. doi:10.1042/BSR20140142

7. Zhao S, Xue Y, Hao J, Liang C (2015) The RNA-binding properties and domain of Rice stripe virus nucleocapsid protein. Virus Genes 51(2):276–282. doi:10.1007/s11262-015-1235-4

8. Islam A, Schulz S, Afroz S, Babiuk LA, van Drunen Littel-van den Hurk S (2015) Interaction of VP8 with mRNAs of bovine herpesvirus-1. Virus Res 197:116–126. doi:10.1016/j.virusres.2014.12.017

9. Salinas T, El Farouk-Ameqrane S, Ubrig E, Sauter C, Duchene AM, Marechal-Drouard L (2014) Molecular basis for the differential interaction of plant mitochondrial VDAC proteins with tRNAs. Nucleic Acids Res 42(15):9937–9948. doi:10.1093/nar/gku728

10. Calla-Choque JS, Figueroa-Angulo EE, Avila-Gonzalez L, Arroyo R (2014) Alpha-Actinin TvACTN3 of Trichomonas vaginalis is an RNA-binding protein that could participate in its posttranscriptional iron regulatory mechanism. Biomed Res Int. doi:10.1155/2014/424767

11. Doroshenk KA, Tian L, Crofts AJ, Kumamaru T, Okita TW (2014) Characterization of RNA binding protein RBP-P reveals a possible role in rice glutelin gene expression and RNA localization. Plant Mol Biol 85(4–5):381–394. doi:10.1007/s11103-014-0191-z

12. Zillmann M, Zapp ML, Berget SM (1988) Gel electrophoretic isolation of splicing complexes containing U1 small nuclear ribonucleoprotein particles. Mol Cell Biol 8(2):814–821

13. Gaur RK, McLaughlin LW, Green MR (1997) Functional group substitutions of the branch-point adenosine in a nuclear pre-mRNA and a group II intron. RNA 3(8):861–869

14. Burd CG, Dreyfuss G (1994) RNA binding specificity of hnRNP A1: significance of hnRNP A1 high-affinity binding sites in pre-mRNA splicing. EMBO J 13(5):1197–1204

15. Aksaas AK, Larsen AC, Rogne M, Rosendal K, Kvissel AK, Skalhegg BS (2011) G-patch domain and KOW motifs-containing protein, GPKOW; a nuclear RNA-binding protein regulated by protein kinase A. J Mol Signal 6:10. doi:10.1186/1750-2187-6-10

16. Dery KJ, Yean SL, Lin RJ (2008) Assembly and glycerol gradient isolation of yeast spliceosomes containing transcribed or synthetic U6 snRNA. Methods Mol Biol 488:41–63. doi:10.1007/978-1-60327-475-3_4

Chapter 11

Measuring mRNA Translation by Polysome Profiling

Marek Kudla and Fedor V. Karginov

Abstract

Determination of mRNA translation rates is essential to understanding the regulatory pathways governing eukaryotic gene expression. In this chapter, we present a transcriptome-wide method to assess translation by association of mRNAs with polysomes on sucrose density gradients. After sedimentation, the fractions are spiked with a control RNA mixture and the RNA content is measured by high-throughput sequencing. Normalization to the spike-ins provides a global quantitative view on the translational status of cellular mRNAs, with the ability to measure changes and identify active and silent subpopulations of each.

Key words Translation, mRNA, Sucrose gradient, Polysome

1 Introduction

Translational control of mRNA expression is important in many normal and pathological cellular processes. Accordingly, the need to assess the translational status of mRNAs is crucial to understanding the regulatory pathways involved. Regulation of translation is achieved by multiple mechanisms [1], and involves scores of RNA-binding proteins and microRNAs. Early in vitro methods of measuring translation rates in reconstituted systems from reticulocyte, wheat germ, and E. coli lysates established much of the mechanics and principles of translation, but often failed to recapitulate the more complex regulatory events that govern eukaryotic gene expression [2]. In vivo methods that do not suffer from these drawbacks include metabolic pulse labeling with ^{35}S-methionine and/or cysteine, followed by immunoprecipitation and SDS-PAGE analysis and radiometric quantitation. However, such approaches are more labor-intensive, focus on individual proteins, and require quality antibodies for the protein under study.

For decades, researchers also used a common and versatile semi-quantitative method of assessing translation that relied on the detection of mRNAs in association with ribosomes/polysomes. Here, the number of ribosomes bound to a particular transcript

Ren-Jang Lin (ed.), *RNA-Protein Complexes and Interactions: Methods and Protocols*, Methods in Molecular Biology, vol. 1421, DOI 10.1007/978-1-4939-3591-8_11, © Springer Science+Business Media New York 2016

provides a measure of its translation rate. In such polysome profiling methods, the polysomes are arrested on mRNAs by cycloheximide treatment of cells, and lysates are subjected to ultracentrifugation in sucrose buffer gradients of increasing density. Individual ribosomal subunits, monosomes and polysomes show increasing sedimentation velocities according to their size. Classically, fractions are collected and probed for the presence and abundance of specific mRNAs by northern blot. Thus, appearance of an mRNA in the heavier fractions indicates a substantial extent of translation. Furthermore, shifts of the measured profile upon changes in cellular state demonstrate an altered translation capacity. While the ability to hybridize any mRNA of sufficient abundance ensured the wide applicability of the method, the need for individual testing of mRNAs limited its utility.

Recent technological advances have substantially improved our ability to measure levels of biomolecules on a genome-wide scale. Here, we present a method to profile the translational status of cellular mRNAs by combining sucrose gradient polysome profiling with high-throughput mRNA sequencing [3]. Similar analyses have also been carried out by microarrays [4]. A related, recently developed method of ribosome footprint profiling [5] relies on quantification of mRNA fragments associated with and protected by ribosomes/polysomes. A clear advantage of this technique is the spatial resolution in observing the ribosomes' locations along the transcript, allowing for detailed analysis of initiation, elongation, pausing, frameshifting and termination events and their regulation. In comparison, advantages of the polysome profiling method presented here lie in the relative simplicity of the protocol, and the ability to identify and quantify the heterogeneity of each mRNA's pool with respect to its translational status, manifesting itself as single or multiple peaks in the profile.

Briefly, cellular lysates are separated by traditional sucrose gradient ultracentrifugation and fractionation. Importantly for the quantitation, the fractions are spiked with a fixed amount of a control RNA mixture. The spike-in RNAs undergo all of the subsequent steps along with the sample, allowing relative abundances across samples to be established. Following RNA isolation, processing, library construction and sequencing, reads uniquely mapping to each mRNA species are counted and normalized by fraction-specific scaling factors computed from the abundances of spike-in reads. The result is a quantitative profile of each mRNA's distribution across the gradient, which serves as an indirect measure of its translational status. Finally, changes in translational efficiency can be quantified by comparing the distributions between two conditions.

2 Materials

2.1 Equipment and Supplies

1. Beckman polyallomer ultracentrifuge tubes, catalog # 331372.
2. Beckman SW41 rotor and suitable ultracentrifuge.
3. Gradient maker (optional), Hoefer catalog # SG15.
4. Brandel BR-188 Density Gradient Fractionation System or similar.
5. Measurement Computing USB-1208FS Data Acquisition Device (optional).
6. Eppendorf tubes.

2.2 Reagents and Buffers

1. 10 % sucrose buffer: 20 mM Hepes pH 7.5, 100 mM KCl, 5 mM MgCl$_2$, 1 mM DTT, 10 % sucrose (w/v).
2. 50 % sucrose buffer: 20 mM Hepes pH 7.5, 100 mM KCl, 5 mM MgCl$_2$, 1 mM DTT, 50 % sucrose (w/v).
3. Chase solution: 60 % sucrose (w/v), bromophenol blue.
4. Puromycin.
5. Cycloheximide.
6. Phosphate buffered saline (PBS).
7. Hypotonic lysis buffer: 5 mM Tris pH 7.5, 2.5 mM MgCl$_2$, 1.5 mM KCl.
8. RNasin Plus (Promega).
9. Dithiothreitol (DTT).
10. 10 % Triton X-100.
11. 10 % sodium deoxycholate.
12. ERCC RNA spike-in mix 1 (Ambion catalog # 4456740).
13. Trizol LS.
14. Acid phenol-chloroform.
15. Chloroform.

3 Methods

3.1 Gradient Preparation

We prepare the sucrose gradients in ultracentrifuge tubes (*see* **Note 1**) using a Hoefer gradient maker. The device consists of two vertical chambers connected at the bottom by a valve, with a closable outflow tube from the bottom of the second chamber.

1. Pour 5.5 mL of the 10 % and 50 % sucrose buffers in the first and second chamber, respectively.
2. Place the gradient maker on a magnetic stir plate with a mini stir bar in the second chamber, and direct the outflow tube to

the top of an ultracentrifuge tube positioned slightly below the stir plate to assist the flow by gravity (allow the gradient to drip down the side of the tube to avoid excessive mixing).

3. With rigorous stirring, open both the connection and the out-flow valve. As the heavier sucrose buffer flows out to the tube, it will be progressively replaced and mixed with the lighter buffer (*see* **Note 2**).

4. Using this method, the gradients are prepared with cold buffers on the day of centrifugation and kept at 4 °C. Avoid disturbing the gradient by shaking.

An alternative preparation method involves freeze/thawing of a stepwise gradient from a series of 10, 20, 30, 40, and 50 % sucrose buffers [6].

1. Pour 2.2 mL of the 50 % sucrose into ultracentrifuge tubes, and freeze upright at –80 °C.

2. Repeat the procedure with each successive lighter buffer.

3. Gradients can remain frozen for months, and need to be thawed at 4 °C overnight before use. A more continuous gradient is established by diffusion during and after thawing.

3.2 Cell Lysate Preparation and Centrifugation

Lysates from up to a 15 cm plate of attached confluent cells can be reasonably resolved in a single 11 mL sucrose gradient. The protocol below is suited for one 15 cm plate of HEK293S cells at 70 % confluency. In parallel, cells may be subjected to puromycin treatment (200 μM for 20 min prior collection) to verify the translation status of mRNAs: actively translating ribosomes are disrupted, and the polysome fraction, along with the mRNAs in it, is shifted toward lighter fractions (Fig. 1). Note that the heavier polysomes are preferentially disrupted, while fractions containing 2–5 ribosomes are more resilient to dissociation.

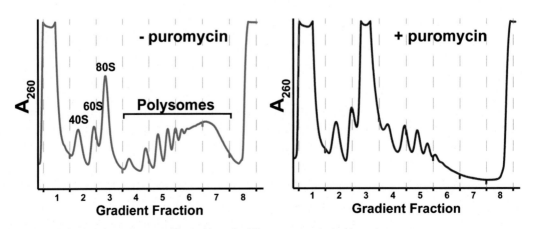

Fig. 1 Ribosome and polysome profiles with and without puromycin treatment

1. Following your desired experimental treatment, wash cells twice with cold PBS with 100 µg/mL cycloheximide.

2. Harvest the cells by scraping in the same buffer and centrifuge. All further steps should be performed at or near 4 °C.

3. Resuspend the cell pellet in 850 µL of hypotonic lysis buffer in an Eppendorf tube.

4. Add 12 µL of RNasin Plus. Add cycloheximide and DTT to a final concentration of 100 µg/mL and 2 mM respectively. Vortex.

5. Add 25 µL of 10 % Triton X-100 and 25 µL of 10 % sodium deoxycholate. Vortex.

6. Centrifuge for 2 min at 14,000 rpm ($16,000 \times g$) in a tabletop centrifuge and collect the cleared lysate supernatant.

7. Carefully load the cleared lysate to the top of the sucrose gradient tubes. Balance the weight of the tubes by carefully adding lysis buffer to the top of the tube.

8. Centrifuge at 38,000 rpm ($178,000 \times g$ at r_{av}) for 2 h at 4 °C in an SW41 rotor. During centrifuge spin down, do not apply brakes below 800 rpm to prevent disturbing the gradient.

3.3 Gradient Fractionation

Reading out and fractionation of the centrifuged sucrose gradient can be performed using a wide range of systems. Here, we describe the typical operation using a Brandel/Teledyne-ISCO system. The gradient fractionator consists of several modules: (1) pump unit with flow controls, (2) tube piercer and UV detector, (3) main unit with electronics and paper tape plotter, (4) fractionator unit, and (5) an optional third-party digitizing unit connected to the main unit for recording traces into a PC. The principle of operation is straightforward: the tube is loaded vertically into the piercer, punctured from the bottom and the pump is engaged to start filling the tube from the bottom with high-density solution. Gradient liquid that is displaced from the tube goes upward through the UV detector to the fractionator that dispenses into eppendorf tubes (*see* **Note 3**). It is best if the gradient fractionation is set up in a cold room or deli case at 4 °C.

1. Power up the main unit, pump, fractionator device, and digitizer unit. Switch the main unit from Standby mode to Operate. Wait 20–30 min for lamp warm-up (*see* **Notes 4** and **5**).

2. Fill the pump syringe with chase solution by detaching the tubing from the needle assembly and placing it into the solution (*see* **Note 6**). Run the pump in reverse to fill the syringe.

3. With the tubing detached, run the pump forward slowly to remove any air bubbles from the pump and tube path, tapping if required.

4. The piercer needle can trap air that will severely disturb the gradient during operation. To remove the air, attach the tubing to the needle. Advance the needle through the polymer seal in order to expose the outlet hole on the side of the needle. Run the pump forward until all air bubbles exit and several drops run out. Wipe the spilled solution with absorbent paper without reintroducing air into the needle. Lower the needle back into the polymer seal.

5. Clean the downstream flowpath with water. This step is required prior to run if the system is not dry, but was left in solution or in-between runs, since droplets of chasing solution from the previous run that remain in the flowpath can affect the reading. Mount a tube of water into the piercer and pierce the bottom with the needle. Run the pump forward with a flow rate of 1.5 mL/min. Ensure the water can be seen flowing out of the fractionator waste tubing. Dismount the tube.

6. Clean the tube mounting collar and o-ring at the top of piercer with lint-free absorbent tissue. This step ensures no remaining droplets of the dense chasing liquid can drop back into the next sample, disturbing the gradient.

7. Mount the tube in the piercer: lower the needle and loosen the mounting collar. Place the tube about 1.2 cm (0.5 in.) into the mounting collar and tighten carefully (this is a common leakage point).

8. Pierce the tube with the needle: with the needle lowered into the bottom seal, raise the needle assembly so that the tube is well supported by the bottom seal and remains vertical. Advance the needle until the side outlet hole is visible in the gradient.

9. Start the pump with a 1.5 mL/min flow (*see* **Note 7**), start the data collection application, and start the fractionator. It takes approximately 1.5 min for the top of the gradient to reach the detector. Typically, we collect eight 1.5 mL fractions (*see* **Note 8**).

10. After all gradients have been run, clean the system with water as above. For prolonged periods, flush with air and store dry.

3.4 RNA-seq Library Preparation

Messenger RNA abundance in the fractions can be quantified by any of the available commercial or custom high-throughput RNA sequencing methods. Importantly, to compare abundance across the fractions, a fixed amount of spike-in RNA must be added to each fraction before library preparation. We have used the ERCC spike-in mix, consisting of 92 polyadenylated RNAs of varying length, mimicking eukaryotic mRNAs [3]. Alternatively, any RNA with little sequence similarity to your organism (e.g., yeast total

RNA for mammalian samples) can be used [7]. If the RNA-sequencing method uses oligo-dT selection, ensure that the spike-in RNAs are polyadenylated.

1. Spike 300–1000 μL from each fraction with 1 μL of a 1:10 dilution of ERCC RNA spike-ins.

2. Extract total RNA with Trizol LS with the following modifications: after separation and collection of the aqueous layer, perform an additional acid phenol-chloroform extraction, followed by a chloroform extraction, to exclude DNA contamination. Note that the last few fractions with higher sucrose concentrations will be heavier than the phenol-chloroform phase and will be at the bottom.

3. Prepare RNA-seq libraries from the isolated RNA. We have previously used the not-so-random primer method [8].

4. Sequence the prepared libraries.

3.5   Data Analysis

Many variations of high-throughput sequencing analysis pipelines exist based on the sequencing technology platform and choice of downstream alignment and processing software. Thus, we will outline a general workflow that is not specific to a particular technology.

1. For each sequenced fraction, process the reads as necessary for your sequencing platform (trim off adaptor sequences, reverse-complement, etc.).

2. Align the processed reads to a database constructed of the ERCC spike-in sequences. A commonly used alignment suite is bowtie2.

3. Count the number of uniquely mapping reads for each ERCC sequence. Exclude ERCC sequences receiving too few counts (e.g., less than 100) from further analysis. Divide the counts by the length of the ERCC sequence.

4. Compute the scaling factors for the fractions: use one of the fractions as a reference. For each of the other fractions, perform a linear regression of its length-normalized ERCC counts ($\mathbf{y}$) against the reference counts ($\mathbf{x}$) using the formula $\mathbf{y} = a\mathbf{x}$. The resulting parameter $\mathbf{a}$ is the scaling factor.

5. Align the reads that did not map to the ERCC database to the genome/transcriptome of study, accepting uniquely mapping reads only. Annotate the reads that map to mRNAs based on the genomic alignment if necessary.

6. Count the per-gene or per-transcript exonic reads.

7. Normalize the per-gene counts by the above scaling factors across fractions.

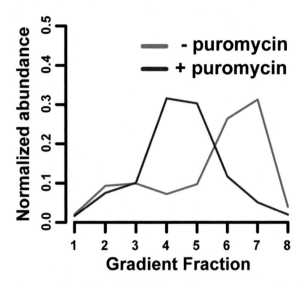

Fig. 2 Distribution of the PTBP1 mRNA in the polysome gradient with and without puromycin treatment. Note the shift of PTBP1 mRNA toward less ribosome occupancy upon treatment

8. To compute the proportional distribution of each mRNA across the gradient fractions, divide the normalized counts by their sum across all fractions for the mRNA (Fig. 2).

9. A measure of translational activity can be computed as the center of mass of the mRNA in the gradient. Arithmetically, this is the average of the fraction numbers weighted by the corresponding proportion of the mRNA in the fractions.

4   Notes

1. Use only polyallomer tubes, since other materials may break the needle of the gradient fractionator.

2. If the sucrose buffers in the two chambers are not mixing during gradient preparation due to air bubbles in the connecting tube, move the bubble out by closing and pressing down on the top of the chamber with your finger.

3. In a test run, time how long it takes for the liquid to travel from the detector to the outlet of the fractionator, and program this delay into the fractionator. This will ensure that the fraction markings on the recorded trace will correspond with the actual collected fractions.

4. The warm-up period is required for any spectrophotometer with a mercury-containing UV lamp. Skipping the warm-up may cause fluctuation in the baseline of UV radiation and cause unpredictable shifts in the readout.

5. Ensure that the tubing above the UV lamp is firmly attached to the flow cell. Fastening should be tight. This is a point where minor leakage can occur.

6. The blue color of the chasing solution is easily spotted in the transparent tubes of the system, while also giving strong absorption signal in the UV spectrum (it delineates the end of the gradient in the UV trace). This is helpful in tracking progress of fractionation, as well as for troubleshooting. Use of degassed solutions is advisable, although it does not appear to be a strict requirement. A commonly recommended chasing liquid is fluorinert, although we found it to be expensive and hard to use, since it has to be collected for disposal.

7. Do not exceed 3 mL/min with the tube in the machine, since it will start leaking. The slower the flow rate, the better the resolution of the gradient, however the gradient will take more time to collect.

8. The fractions can be collected by time (where the flow rate will determine the volume collected) or by the number of drops per eppendorf tube. We have found that the drop volume changes with the sucrose percentage, making the second option unreliable. The switch between the tubes is signaled to the main unit and appears both on the paper and on the digitized version of the run as a tick mark in the trace.

References

1. Sonenberg N, Hinnebusch AG (2009) Regulation of translation initiation in eukaryotes: mechanisms and biological targets. Cell 136(4):731–745. doi:10.1016/j.cell.2009.01.042

2. Merrick WC (1994) Eukaryotic protein synthesis: an in vitro analysis. Biochimie 76(9): 822–830

3. Karginov FV, Hannon GJ (2013) Remodeling of Ago2-mRNA interactions upon cellular stress reflects miRNA complementarity and correlates with altered translation rates. Genes Dev 27(14):1624–1632. doi:10.1101/gad. 215939.113

4. Arava Y, Wang Y, Storey JD, Liu CL, Brown PO, Herschlag D (2003) Genome-wide analysis of mRNA translation profiles in *Saccharomyces cerevisiae*. Proc Natl Acad Sci U S A 100(7):3889–3894. doi:10.1073/pnas. 0635171100

5. Ingolia NT, Ghaemmaghami S, Newman JR, Weissman JS (2009) Genome-wide analysis in vivo of translation with nucleotide resolution using ribosome profiling. Science 324(5924):218–223. doi:10.1126/science.1168978

6. Luthe DS (1983) A simple technique for the preparation and storage of sucrose gradients. Anal Biochem 135(1):230–232

7. Spies N, Burge CB, Bartel DP (2013) 3' UTR-isoform choice has limited influence on the stability and translational efficiency of most mRNAs in mouse fibroblasts. Genome Res 23(12):2078–2090. doi:10.1101/gr.156919.113

8. Armour CD, Castle JC, Chen R, Babak T, Loerch P, Jackson S, Shah JK, Dey J, Rohl CA, Johnson JM, Raymond CK (2009) Digital transcriptome profiling using selective hexamer priming for cDNA synthesis. Nat Methods 6(9):647–649. doi:10.1038/nmeth.1360

Chapter 12

Genome-Wide Profiling of RNA–Protein Interactions Using CLIP-Seq

Cheryl Stork and Sika Zheng

Abstract

UV crosslinking immunoprecipitation (CLIP) is an increasingly popular technique to study protein–RNA interactions in tissues and cells. Whole cells or tissues are ultraviolet irradiated to generate a covalent bond between RNA and proteins that are in close contact. After partial RNase digestion, antibodies specific to an RNA binding protein (RBP) or a protein-epitope tag is then used to immunoprecipitate the protein–RNA complexes. After stringent washing and gel separation the RBP–RNA complex is excised. The RBP is protease digested to allow purification of the bound RNA. Reverse transcription of the RNA followed by high-throughput sequencing of the cDNA library is now often used to identify protein bound RNA on a genome-wide scale. UV irradiation can result in cDNA truncations and/or mutations at the crosslink sites, which complicates the alignment of the sequencing library to the reference genome and the identification of the crosslinking sites. Meanwhile, one or more amino acids of a crosslinked RBP can remain attached to its bound RNA due to incomplete digestion of the protein. As a result, reverse transcriptase may not read through the crosslink sites, and produce cDNA ending at the crosslinked nucleotide. This is harnessed by one variant of CLIP methods to identify crosslinking sites at a nucleotide resolution. This method, individual nucleotide resolution CLIP (iCLIP) circularizes cDNA to capture the truncated cDNA and also increases the efficiency of ligating sequencing adapters to the library. Here, we describe the detailed procedure of iCLIP.

Key words UV crosslinking immunoprecipitation, RNA binding proteins, Immunoprecipitation, RBP–RNA complex, iCLIP

1 Introduction

RNA–protein interaction plays an important role in post-transcriptional regulation [1]. Currently, different methods study RNA–RBP interactions with varying resolutions. Electrophoretic mobility shift assay (EMSA) is a simple and sensitive technique to identify RNA–RBP interaction based on the observation that RNA–RBP complexes migrate slower in a native gel than the RNA alone. EMSA has many limitations including (1) possible complex disassembly and mis-assembly during electrophoresis, (2) poor sequence resolution to determine interacting nucleotides,

Ren-Jang Lin (ed.), *RNA-Protein Complexes and Interactions: Methods and Protocols*, Methods in Molecular Biology, vol. 1421, DOI 10.1007/978-1-4939-3591-8_12, © Springer Science+Business Media New York 2016

and (3) loss of in vivo cellular context [2]. Traditional RNA binding protein immunoprecipitation (RIP) allows for quantifiable identification of RNA targets bound to an RBP in cellular extracts. Immunoprecipitation of endogenous RNA–protein complexes using antibodies targeting the RBP is followed by purification and profiling of the bound RNA by microarray or high-throughput sequencing. When additional RBPs are co-immunoprecipitated with the assayed RBP (e.g. as part of a larger RNP complex), RIP is unable to distinguish indirectly bound RNA [3]. Promiscuous protein–RNA binding in cell lysates is also a significant source of false positives.

To circumvent the limitations of the previously developed methods, crosslinking immunonoprecipitation (CLIP) was developed. Crosslinking RNA–protein complexes using UV irradiation combined with immunoprecipitation has allowed us to map the RNA sequence bound to a RNA binding protein in vivo with high resolution and specificity [4–9]. RNA–RBP complexes are crosslinked using UV irradiation, partially digested with RNase, and purified by immunoprecipitation; the RNA is then extracted after protease digestion, ligated to an adapter, reverse-transcribed to cDNA, and sequenced. Kirk Jensen and Jenej Ule first applied CLIP to the study of RNA binding protein Nova in brain in 2003 [4]. They were able to identify 340 new RNA tags that were crosslinked to Nova1, 18 of which flanked alternative exons suggesting a role of Nova in alternative splicing [4]. Various methods of CLIP have since been developed to increase the resolution of RNA–protein binding to the near nucleotide level allowing for the identification of not only specific binding sites but specific motifs or patterns of RNA–protein binding [6].

CLIP experiments often have limited yield partly due to inefficient crosslinking using UV 254 nm. To improve RNA recovery, Thomas Tuschl's group developed photoactivatable-ribonucleoside-enhanced crosslinking and immunoprecipitation (PAR-CLIP). PAR-CLIP labels living cells with photoactivatable nucleosides such as 4-thiouridine (4-SU) and 6-thioguanosine (6-SG) that are incorporated into nascent RNA prior to irradiation with UV at 365 nm [8, 10]. When crosslinked, incorporation of 4-SU or 6-SG results in U to C and G to A mutations respectively, allowing researchers to use mutational analysis to identify crosslinked sites and filter out noise. An increase in crosslinking efficiency and resolution are the advantages of PAR-CLIP. However, PAR-CLIP relies on metabolic labeling which has inherent biases as well as difficulties for in vivo application.

A third variant of CLIP or iCLIP, developed by Ule's group, allows identification of crosslinking sites at the near nucleotide level (Fig. 1) [11]. Unlike other CLIP methods that ligate adapters to RNA before reverse transcription, iCLIP introduces a circularization step after reverse transcription to ligate the cleavable 5' end

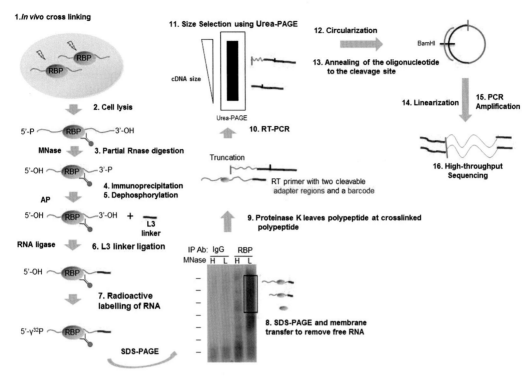

Fig. 1 Schematic representation of iCLIP. RNA binding protein (RBP) and RNA are covalently bound in vivo using UV radiation (step 1). Potential RBP–RNA complexes are then purified together (steps 2–5). L3 linker adapter ligation to the 3' end allows for sequence-specific priming of reverse transcription and the 5' is radioactively labeled (steps 6 and 7). RBP–RNA complexes are purified from free RNA using SDS-PAGE and wet membrane transfer (step 8). Complexes are recovered from the membrane using Proteinase K (step 9). Reverse transcription truncates at the remaining polypeptide and introduces two cleavable adapter regions and a barcode (step 10). Size selection using Urea-PAGE removes the RT primer prior to circularization of the cDNA (steps 11–13). Linearization generates templates for PCR amplification (steps 14 and 15). High-throughput sequencing generates reads where the barcode sequences are immediately before the last nucleotide of the input cDNA (step 16). This nucleotide is one position upstream of the crosslinked nucleotide, allowing for identification of the RBP–RNA binding site with high resolution

adapter to the 3' end of the cDNA [12]. Because reverse transcriptase often stops at the crosslink sites and does not necessarily extend to the 3' adapter that is used by other CLIP methods, circularization allows iCLIP to capture these truncated cDNA. The cleavable 5' end adapter includes random barcodes to eliminate PCR artifacts. iCLIP was successfully used to globally identify RNA that were bound to HNRPNPC, TDP-43, FUS, and TIA1/TIAL1 [12–15].

Despite the power of CLIP, cautions should be taken in analyzing any CLIP result in particular due to the sequence-dependent UV crosslinking biases. RBPs are multifunctional. To understand their complete functionality, other genomic methods are needed in conjunction with CLIP to globally identify their functionally relevant RNA targets.

2 Materials

Prepare all reagents in ultrapure, sterile water at room temperature unless stated otherwise. Store stock solutions and buffers at room temperature unless otherwise specified.

2.1 Buffer Components

1. TE Buffer: 1 M Tris–HCl pH 8.0, 0.5 M ethylenediaminetetraacetic acid (EDTA) pH 8.0, H_2O.

2. Lysis Buffer: 50 mM Tris–HCl pH 7.4, 100 mM NaCl, 1 % Igepal CA-630, 0.1 % SDS, 0.5 % sodium deoxycholate, on the day add 1/100 volume of Protease Inhibitor Cocktail Set III. For tissues use 1/1000 of ANTI-RNase (*see* **Note 1**).

3. High Salt Wash: 50 mM Tris–HCl pH 7.4, 1 M NaCl, 1 mM EDTA, 1 % Igepal CA-630, 0.1 % SDS, 0.5 % sodium deoxycholate.

4. PNK Buffer: 20 mM Tris–HCl, pH 7.4, 10 mM $MgCl_2$, 0.2 % Tween-20.

5. 5× PNK pH 6.5 Buffer: 350 mM Tris–HCl pH 6.5, 50 mM $MgCl_2$, 5 mM dithiothreitol (DTT). Freeze aliquots of the buffer. Avoid freeze–thaw (*see* **Note 2**).

6. 6 4× Ligation Buffer: 200 mM Tris–HCl pH 7.8, 40 mM $MgCl_2$, 4 mM dithiothreitol, Freeze aliquots of the buffer. Avoid freeze–thaw.

7. PK Buffer: 100 mM Tris–HCl, pH 7.4, 50 mM NaCl, 10 mM EDTA.

8. PK buffer + 7 M Urea: 10 mM Tris–HCl pH 7.4, 50 mM NaCl, 10 mM EDTA, 7 M Urea.

2.2 UV Crosslinking and Immuno-precipitation

1. Stratalinker UV Crosslinker (Stratagene, La Jolla, CA).

2. 4-Thiouridine (Sigma, St. Louis, MO) (*see* **Note 3**).

3. Protein G Dynabeads (Life Technologies, Carlsbad, CA).

4. Protein A Dynabeads (Life Technologies).

5. Protease Inhibitor Cocktail Set III (Calbiochem/Merck).

6. ANTI-RNase (Life Technologies).

7. RNase I (Life Technologies).

8. Turbo DNase (Life Technologies).

9. T4 PNK plus 10× PNK Buffer (NEB, Ipswich, MA).

10. RNasin (Promega, Madison, WI).

11. Proteus Clarification Mini Spin Columns (Generon, Houston, TX).

12. T4 RNA Ligase I (NEB).

13. Pre-adenylated adapter L3-App (IDT, rAppAGATCGGAA-CAGCGGTTCAG/ddC/).

14. ATP [γ-³²P] (Perkin Elmer, Waltham, MA).

2.3 Immunoblotting Components

1. 4–12 % NuPAGE gels (Life Technologies).
2. Electrophoresis chamber (Life Technologies).
3. Novex wet transfer apparatus (Life Technologies).
4. LDS-4× Sample Buffers (Life Technologies).
5. Pre-stained protein marker (Fermentas, Pittsburg, PA).
6. Nitrocellulose membrane (VWR, Radnor, PA).
7. Protran BA85.
8. Sponge pads for XCell II blotting (Life Technologies).
9. 20× transfer buffer (Life Technologies).
10. 20× MOPS-SDS Running Buffer (Life Technologies).
11. Whatman filter paper (GE Healthcare, Pittsburg, PA).
12. Film.

2.4 RNA Isolation Components

1. Proteinase K (Fisher Scientific, Waltham, MA).
2. 19G syringe needles.
3. Phenol/Chloroform pH 6.7 (Sigma) (*see* **Note 4**).
4. Phase lock gel heavy tube (VWR).
5. Glycoblue (Ambion).
6. 3 M sodium acetate pH 5.5 (Life Technologies).
7. Thermomixer (Eppendorf).

2.5 Reverse Transcription

1. PCR tubes.
2. dNTPs (Promega).
3. SuperScript III (Life Technologies).
4. 5× First Strand Buffer (Life Technologies).
5. DTT (Life Technologies).
6. 1 M HEPES pH 7.3 (Thermo Scientific).
7. TE buffer.
8. R#Clip Sequences (N denotes any nucleotide).
   Rt1clip: /5Phos/NNAACCNNNAGATCGGAAG
   AGCGTCGTGgatcCTGAACCGC
   Rt2clip: /5Phos/NNACAANNNAGATCGGAAG
   AGCGTCGTGgatcCTGAACCGC
   Rt3clip: /5Phos/NNATTGNNNAGATCGGAAG
   AGCGTCGTGgatcCTGAACCGC

Rt4clip: /5Phos/NNAGGTNNNAGATCGGAAG
AGCGTCGTGgatcCTGAACCGC

Rt6clip: /5Phos/NNCCGGNNNAGATCGGAAG
AGCGTCGTGgatcCTGAACCGC

Rt7clip: /5Phos/NNCTAANNNAGATCGGAAG
AGCGTCGTGgatcCTGAACCGC

Rt8clip: /5Phos/NNCATTNNNAGATCGGAAG
AGCGTCGTGgatcCTGAACCGC

Rt9clip: /5Phos/NNGCCANNNAGATCGGAAG
AGCGTCGTGgatcCTGAACCGC

Rt11clip: /5Phos/NNGGTTNNNAGATCGGAAG
AGCGTCGTGgatcCTGAACCGC

Rt12clip: /5Phos/NNGTGGNNNAGATCGGAAG
AGCGTCGTGgatcCTGAACCGC

Rt13clip: /5Phos/NNTCCGNNNAGATCGGAAG
AGCGTCGTGgatcCTGAACCGC

Rt14clip: /5Phos/NNTGCCNNNAGATCGGAAG
AGCGTCGTGgatcCTGAACCGC

Rt15clip: /5Phos/NNTATTNNNAGATCGGAAG
AGCGTCGTGgatcCTGAACCGC

Rt16clip: /5Phos/NNTTAANNNAGATCGGAAG
AGCGTCGTGgatc CTGAACCGC

2.6 cDNA Isolation

1. 2× TBE-urea loading buffer (Life Technologies).
2. 6 % TBE-urea (pre-cast gels) (Life Technologies).
3. Low molecular weight marker (NEB).
4. TBE Running buffer (Life Technologies).
5. SYBR Green II (Life Technologies).
6. Glass Pre-filters (Whatman).
7. Phase lock gel heavy (VWR).
8. Costar SpinX Column (Corning Incorporated, Corning, NY).

2.7 5' End Ligation

1. 10× CircLigase Buffer (Cambio, Cambridge, UK).
2. CircLigase II (Cambio).
3. Cut_oligo: GTTCAGGATCCACGACGCTCTTCaaaa [8].
4. MnCL$_2$ (Cambio).
5. Bam HI (Fermentas, Pittsburg, PA).
6. Fast digest Buffer (Fermentas, Pittsburg, PA).

2.8 PCR
Amplification

1. Accruprime Supermix I (Life Technologies).

2. SYBR green I (Life Technologies).

3. P5Solexa: AATGATACGGCGACCACCGAGATCTACACTC TTTCCCTACACGACGCTCTTCCGATCT [4, 11].

4. P3Solexa: CAAGCAGAAGACGGVATACGAGATCGGTCTC GGCATTCCTGCTGAACCGCTCTTCCGATCT [4, 11].

2.9 qPCR
Quantification

1. Accuprime Supermix I (Life Technologies).

2. PhiX Control Library (RNAseq/ChIPseq) (Illumina, San Diego, CA).

3. Platinum Taq Mastermix (Life Technologies, Carlsbad, CA).

4. DLP
   [6FAM]
   CCCTACACGACGCTCTTCCGATCT
   [TAMRA] [11].

5. Primer 1
   AATGATACGGCGACCACCGAGATC [11].

6. Primer 2
CAAGCAGAAGACGGCATACGAGATC [11].

2.10 Buffers
for Stringent-Urea
iCLIP Protocol

1. Urea Cracking buffer: 50 mM Tris–HCl pH 7.4, 6 M Urea, 1 % SDS, 25 % PBS.

2. T-20 Buffer: 50 mM Tris–HCl pH 7.4, 150 mM NaCl, 0.5 % Tween 20, and 0.1 mM EDTA.

2.11 Stringent-
Urea iCLIP

1. Dynabeads Antibody Coupling Kit (Life Technologies).

2. SUPERase in RNase inhibitor (Life Technologies).

3. NuPAGE sample reducing agent (Life Technologies).

4. NuPAGE antioxidant (Life Technologies).

5. Linear acrylamide (Life Technologies).

3 Methods

3.1 UV-C
Crosslinking

1. After culturing adherent cells in 10 cm plate, remove medium, add 6 mL ice-cold PBS and place on ice.

2. Remove lid and irradiate cells with 150 mJ/cm² (needs to be optimized for different proteins) at 254 nm in a Stratalinker 2400 (*see* **Note 5**).

3. (**Optional**) Have non-irradiated cells as a negative control of the crosslinking.

4. Scrape off the cells with cell lifters and transfer the cell suspension into three 2 mL microtubes. Centrifuge at $800 \times g$ at 4 °C for 1 min to pellet cells, and then remove supernatant.

5. Snap-freeze the cell pellets on dry ice and *store at –80°C until further use.*

3.2 4-Thiouridine Labeling and UV-A Crosslinking (Alternative to UV-C)

1. Add 50 µL of 4SU (stock concentration 100 mM 4SU) to a 10 cm plate of cells culture in 10 mL DMEM supplemented with FBS, PenStrep to get a final concentration of 500 µM 4SU.

2. Alternatively add 10 µL of 4-thiouridine (stock concentration: 100 mM) to get a final concentration of 100 µM 4SU (*see* **Note 6**).

3. Incubate the cells with 4-thiouridine for 60 min at 500 µM or 8 h at 100 µM: afterward check the viability of your cells. Have cells that are not incubated with 4SU as a negative control.

4. Aspirate the medium and add 6 mL of ice-cold PBS to cells growing in a 10 cm plate. Remove lid and place on ice.

5. Irradiate twice at once with 2×400 mJ/cm^2 in a Stratalinker 2400 with 365 nm bulbs.

6. Scrape off the cells with cell lifters.

7. Transfer the cell suspension into three 2 mL microtubes. Centrifuge at $800 \times g$ at 4 °C for 1 min to pellet cells, and then remove the supernatant.

8. Snap-freeze the cell pellets on dry ice and *store at –80°C until further use.*

3.3 Bead Preparation for Immunoprecipitation

1. Add 100 µL of protein G dynabeads per experiment to a fresh microtube.

2. Wash beads twice with 900 µL lysis buffer (without protease inhibitor).

3. Resuspend the beads in 100 µL lysis buffer with 2–10 µg antibody per experiment (*see* **Note 7**).

4. Rotate tubes for 30–60 min at room temperature.

5. Wash 3 times with 900 µL lysis buffer and leave in the last wash until ready to proceed with immunoprecipitation.

3.4 Cell Lysis and Partial RNA Digestion

1. Resuspend the cell pellet in 1 mL lysis buffer (with 1:100 Protease Inhibitor Cocktail Set III) and transfer to a 1.5 mL microtube. Prepare enough lysate for two experimental samples and appropriate controls (*see* **Notes 8** and **9**).

2. Sonication of samples (Note: Optional).

3. *Option 1*: Sonicate the sample on ice using a probe sonicator. The probe should be approximately 0.5 cm from the bottom of

the tube and not touching the tube sides in order to avoid foaming. Sonicate twice with 10 s bursts at 5 dB. Clean the probe by sonicating H_2O before and after sample treatment.

4. *Option 2*: Use Bioruptor plus for five cycles with alternating 30 s on/30 s off at low intensity.

5. Prepare two dilutions of RNase I in PBS. 1:50 for High RNase I and 1:250 for low RNAseI. This will need to be carefully optimized (*see* **Note 10**).

6. Add 10 μL of low or high RNAse I dilution and 2 μL Turbo DNase to the cell lysate and immediately place the samples at 37 °C for 3 min shaking at 1100 rpm. Immediately afterward transfer to ice for > 3 min.

7. Centrifuge for 10 min at 22,000×g and at 4 °C to clear the lysate. Carefully collect the supernatant.

8. *Optional*: load 500 μL of the lysate onto a Proteus Clarification Mini Spin Column. Centrifuge for 1 min at 22,000×g at 4 °C. Transfer flow-through to a new tube and place on ice. Repeat for the rest of the lysate and combine the fractions.

3.5 Immunoprecipitation

1. Add the lysate to the beads and rotate for 1 h or overnight at 4 °C.

2. Place on magnet and discard the supernatant and wash three times with high-salt wash (rotate the third wash for at least 1 min at 4 °C) (*see* **Note 11**).

3. Wash twice with PNK Buffer and then leave in 1 mL PNK Buffer and proceed to 3' end dephosphorylation step.

3.6 RNA 3' End Dephosphorylation

1. Discard the supernatant and resuspend the beads in 20 μL of the following mix: 4 μL 5× PNK pH 6.5 Buffer, 0.5 μL PNK, 0.5 μL RNasin, and 15 μL H_2O.

2. Incubate for 20 min at 37 °C in a thermomixer at 1100 rpm.

3. Wash with PNK Buffer.

4. Wash with high-salt wash (rotate wash in the cold room for at least 1 min).

5. Wash twice with PNK Buffer.

3.7 L3 Adapter Ligation

1. Carefully remove the supernatant and resuspend the beads in 20 μL of the following mix: 8 μL H_2O, 5 μL 4× ligation buffer, 1 μL RNA ligase, 0.5 μL RNasin, 1.5 μL pre-adenylated adapter L3-App (20 μM), and 4 μL PEG400.

2. Incubate overnight at 16 °C in thermomixer at 1100 rpm.

3. Add 500 μL PNK Buffer.

4. Wash twice with high-salt buffer, rotating each wash in the cold room for 5 min, and twice with PNK Buffer.

3.8 RNA 5′ End Labeling

1. Remove supernatant and resuspend beads in 4 μL Hot PNK mix: 0.2 μL PNK, 0.4 μL γ-[^{32}P]-ATP, 0.4 μL 10× PNK Buffer, and 3 μL H_2O.

2. Incubate for 5 min at 37 °C in a thermomixer at 1100 rpm. Discard the supernatant.

3. Remove the supernatant as radioactive waste and add 20 μL 1× NuPAGE loading buffer (*see* **Note 12**).

4. Incubate on a thermomixer at 70 °C for 10 min.

5. Place on the magnet to precipitate the beads, collect the supernatant and load it on a gel.

3.9 SDS-PAGE and Nitrocellulose Transfer

1. Load the samples on a 4–12 % NuPAGE Bis-Tris gel. Also load 5 μL of a pre-stained protein size marker. Use 500 mL 1× MOPS running buffer.

2. Run the gel at 180 V for 50 min.

3. Cut off the dye front and discard it as solid radioactive waste.

4. Transfer the protein–RNA complexes from the gel to a Nitrocellulose membrane using a Novex wet transfer apparatus.

5. Transfer at 30 V for 1 h and overnight.

6. After transfer, rinse the membrane in PBS buffer, then wrap it in saran wrap and expose it to film at –80 °C. Expose for 30 min, 1 h, and overnight.

3.10 RNA Isolation

1. A smear will be seen in the low RNase condition and not the high RNase condition. Isolate the protein–RNA complexes from the low RNase condition using the autoradiograph as a mask for cutting the respective region (usually 20–60 kDa above the expected molecular weight) of the membrane. Cut into many pieces and place into a 1.5 mL microtube (*see* **Note 13**).

2. Add 10 μL Proteinase K in 200 μL PK buffer to the pieces making sure all pieces are submerged. Incubate pieces for 20 min shaking at 1100 rpm at 37 °C.

3. Add 200 μL PK Buffer + 7 M Urea and incubate for 20 min at 1100 rpm at 37 °C.

4. Collect the solution and add it together with 400 μL phenol/chloroform pH 6.7 to a 2 mL Phase Lock Gel Heavy Tube (*see* **Note 14**).

5. Incubate for 5 min at 30 °C shaking at 1100 rpm. Separate the phases by spinning for 5 min at full speed and room temperature.

6. Transfer the aqueous layer into a new tube. Be careful not to touch the gel matrix with the pipette. *Optional*: spin again for 1 min and transfer into a new tube.

7. Precipitate by adding 0.75 µL glycoblue and 40 µL 3 M sodium acetate pH 5.5. Then mix and add 1 mL 100 % ethanol, mix again and place at –20 °C overnight.

8. Centrifuge at 15,000 rpm at 4 °C for 20 min. Remove the supernatant and wash the pellet with 0.9 mL 80 % ethanol and spin for 5 min (*see* **Note 15**).

9. Resuspend pellet in 5 µL H$_2$O and transfer to a PCR tube.

3.11 Reverse Transcription

1. Add 1 µL primer Rt#clip (0.5 pmol/µL) and 1 µL dNTP mix (10 mM) to the resuspended pellet. For each experiment or replicate, use a different Rclip primer containing individual barcode sequences (*see* **Note 16**).

2. Incubate at 70 °C for 5 min. Cool to 25 °C. Hold at 25 °C until RT mix is added.

3. Add RT mix to the resuspended pellet: 7 µL H$_2$O, 4 µL 5× first strand buffer, 1 µL 0.1 M DTT, 0.5 µL RNasin, and 0.5 µL Superscript III.

4. Run RT Program: 25 °C for 5 min, 43 °C for 20 min, 50 °C for 40 min, 80 °C for 5 min, 4 °C hold.

5. Add 1.65 µL 1 M NaOH and incubate at 98 °C for 20 min. Then add 20 µL 1 M HEPES-NaOH pH 7.3 to eliminate radioactivity from strongly labeled samples and to prevent RNA from interfering with subsequent reactions.

6. Add 350 µL TE buffer, 0.75 µL glycoblue, and 40 µL 3 M sodium acetate pH 5.5. Mix and then add 1 mL 100 % ethanol. Mix again and precipitate at –20 °C overnight.

3.12 Gel Purification of cDNA

1. Centrifuge for 15 min at 15,000 rpm at 4 °C. Remove the supernatant and wash the pellet with 500 µL 80 % EtOH. Centrifuge again, remove supernatant, and resuspend the pellet in 6 µL H$_2$O.

2. Add 6 µL 2× TBE-urea loading buffer to the cDNA. *Recommended*: add 6 µL loading buffer to DNA size marker.

3. Heat samples to 80 °C for 5 min immediately before loading. *Optional*: leave one lane free between each sample to avoid cross-contamination.

4. Prepare 800 mL of 1× TBE running buffer and fill the upper chamber with 200 mL and the lower chamber with 600 mL. Use a P1000 pipette to flush remaining urea out of the wells before loading 12 µL of each sample. Load the marker into the last lane.

5. Run 6 % TBE-urea gel for 40 min at 180 V until the lower dye (dark blue) is close to the bottom.

6. Cut off the last line containing the size marker and incubate it gently shaking for 10 min in 20 mL TBE buffer with 2 µL SYBR green II. Wash once with TBE and visualize by UV

transillumination with 100 % scaling and use as a mask to guide cDNA excision from the rest of the gel.

7. Together the full L3-App and primer sequence accounts for 52 nt of the cDNA. Cut three bands: at 70–80 nt, 80–100 nt, and 100–150 nt.

8. Add 400 µL TE to each gel piece and crush gel piece with a 1 mL syringe plunger.

9. Incubate shaking for 1 h at 1100 rpm at 37 °C, place on dry ice for 2 min, and place back for 1 h at 1100 rpm at 37 °C. Transfer the liquid portion to Costar SpinX column, into which you placed two 1 cm glass pre-filters.

10. Centrifuge at full speed for 1 min. collect the solution and add it together with 400 µL DNA phenol/chloroform to a 2 mL Phase Lock Gel Heavy Tube.

11. Incubate for 5 min shaking at 1100 rpm at 30 °C. Separate the phases by spinning 5 min at full speed at RT.

12. Transfer the aqueous layer into a new tube. Centrifuge again for 1 min and transfer into a new tube.

13. Add 1 µL glycoblue and 40 µL 3 M sodium acetate, pH 5.5, mix. Then add 1 mL 100 % ethanol. Mix again and precipitate overnight at –20 °C.

3.13 Ligation of Primer to 5' End of cDNA

1. Centrifuge the precipitate to pellet the cDNA. Wash pellet with 500 µL 80 % ethanol. Resuspend pellet in 8 µL ligation mix (6.5 µL H_2O, 0.8 µL 10× CircLigase buffer II, 0.4 µL 50 mM $MnCl_2$, 0.3 µL CircLigase II) in PCR tubes.

2. Transfer PCR tube and incubate at 60 °C for 1 h.

3. Add 30 µL annealing mix: 6 µL H_2O, 3 µL Fast digest buffer, 1 µL 10 µM Cut_oligo. Anneal the oligonucleotide using program 95 °C for 2 min, successive 20 s cycles starting at 95 °C and decreasing 1 °C down to 25 °C, hold at 25 °C.

4. Add 2 µLl BamHI and incubate for 30 min at 37 °C, then incubate for 5 min at 80 °C.

5. Add 350 µL TE, 0.75 µL glycoblue, and 40 µL 3 M sodium acetate, pH 5.5, and mix. Then add 1 mL 100 % EtOH. Mix again and precipitate at –20 °C overnight.

3.14 PCR Amplification

1. Centrifuge precipitate and wash with 500 µL 80 % EtOH. Resuspend in 21 µL of H_2O.

2. Prepare PCR mix: 1 µL cDNA, 0.25 µL primer, 5 µL Accuprime Supermix I, and 3.75 µL H_2O.

3. Run PCR reaction: 94 °C for 2 min, [94 °C 15 s, 65 °C 30 s, 68 °C 30 s] for 25–35 cycles, 68 °C for 3 min, 25 °C hold.

4. Mix 8 µL PCR product with 2 µL 5× TBE Loading Buffer and load on a 6 % TBE gel and stain with SYBR green I (*see* **Note 16**).

3.15 qPCR Quantitation

1. Set up serial dilutions (1:10 to 1:10,000) in a 96-well plate. Barcodes in Rclip primers allow for multiplex.

2. Prepare master mix: 10 μL Invitrogen Platinum qPCR Supermix w/ ROX, 0.5 μL DLP oligo, 0.6 μL Primer 1 (10μM), 0.6 μL Primer 2 (10 μM), and 6.3 μL H₂O. Add 18 μL per well in qPCR plate.

3. Add 2 μL template from serial dilutions and seal the plate with an adhesive film.

4. Run the following qPCR program (normal setting, detector FAM-TAMRA, passive reference ROX): 54 °C 2 min, 94 °C 10 min, [94 °C 15 s, 62 °C 1 min, 72 °C 30 s] for 40 cycles.

5. Plot concentrations vs. Ct value for the standard on a log scale and fit a line to the graph. Use standard line to obtain concentrations of the unknown samples using their Ct values.

6. Dilute the iCLIP library to 10 nM, submit 10 μL for sequencing, and store the rest.

3.16 High-Throughput Sequencing

1. Libraries can be sequenced using standard Illumina protocols. 50-nt single end runs are recommended.

4   Notes

1. Lysis Buffer without protease inhibitors can be made in advance. However, Protease Inhibitor Cocktail III or ANTI-RNase should be added on the day of use.

2. High concentrations of DTT can lower IP efficiency, so lower DTT was used to increase recovery of protein complexes.

3. Crosslinking with 4-thiouridine allows crosslinking with UV-A light as seen in photoactivatable-ribonucleoside-enhanced CLIP (PAR-CLIP). Addition is optional and used to enhance crosslinking of certain proteins. 4-Thiouridine is light-sensitive and cells should be placed back in the incubator immediately after addition.

4. Phenol/chloroform is used at a pH of 6.7 to reduce DNA–RNA hybrids in the phenol phase.

5. Crosslinking depends on the protein that you are working with and method used and should be optimized. However, increasing crosslinking dose could distort library preparation by damaging RNA.

6. Photoreactive nucleosides are toxic to cells. You should determine optimal duration of incubation prior to use.

7. The amount of antibody to add depends on quality and purity and will need to be optimized prior to use.

8. Measure protein concentration using Nanodrop 2000c or a Bradford assay. Normalize samples to lowest concentration. 2 mg/mL protein is recommended.

9. Recommended controls include absence of RBP in cells, absence of crosslinking, or absence of antibody during IP.

10. RNase I has been shown to deactivate after prolonged incubation with 0.1 % SDS in lysis buffer.

11. Save 15 μL from supernatant to monitor depletion efficiency.

12. SDS-PAGE gels changes it's pH during its run which can result in alkaline hydrolysis of RNA. The NuPAGE system keeps the pH at around 7 when used with the appropriate buffer.

13. To determine the specificity of the protein–RNA you will need to check the high RNase I condition. The radioactive band should be ~5 kDa above the MW. The negative controls should have no bands. The band in low RNase I should be diffuse. RNA is usually ~20 kDa. Cutting 15–80 kDa above the expected MW of the protein is recommended.

14. Try to not disturb the pellet. If you disturb the pellet, repeat spin. Leave on bench for 3 min with cap open to dry.

15. Use distinct Rtclip primers (Rt1clip–Rt16clip) on the controls and experimental samples. Each Rtclip contains a 4 nt barcode sequences that differs by two nucleotides to ensure that mutations cannot convert one barcode to another. Barcodes enable the user to remove PCR artifacts and control for cross-contamination during multiplexing.

16. It is recommended that cDNA work be done on a designated bench. Avoid taking cDNA from PCR to an area where iCLIP RNA was used.

Acknowledgements

This work was supported by the NIH grant R00 MH 096807.

References

1. Modic M, Ule J, Sibley CR (2013) CLIPing the brain: studies of protein–RNA interactions important for neurodegenerative disorders. Mol Cell Neurosci 56:429–435

2. Hellman LM, Fried MG (2007) Electrophoretic Mobility Shift Assay (EMSA) for detecting protein-nucleic acid interactions. Nat Protoc 2:1849–1861

3. Keene JD, Komisarow JM, Friedersdorf MB (2006) RIP-Chip: the isolation and identification of mRNAs, microRNAs, and protein components of ribonucleoprotein complexes from cell extracts. Nat Protoc 1:302–307

4. Ule J, Jensen KB, Ruggiu M et al (2003) Clip identifies nova-regulated RNA networks in the brain. Science 302:1212–1215

5. Darnell RB (2010) *HITS-CLIP*: panoramic views of protein–RNA regulation in living cells. Wiley Interdiscip Rev RNA 1: 266–286

6. Licatalosi DD, Aldo M, Fak JJ et al (2008) HITS-CLIP yields genome-wide insights into brain alternative RNA processing. Nature 456:464–470

7. Milek M, Wyler E, Landthaler M (2012) Transcriptome-wide analysis of protein–RNA interactions using high-throughput sequencing. Semin Cell Dev Biol 23:206–212

8. Ascano M, Hafner M, Cekan P et al (2012) Identification of RNA–protein networks using PAR-CLIP. Wiley Interdiscip Rev RNA 3:159–177

9. Huppertz I, Attig J, D'Ambrogio A et al (2014) iCLIP: protein–RNA interactions at nucleotide resolution. Methods 65:274–287

10. Hafner M, Lanthaler M, Burger L et al (2010) Transcriptome-wide identification of RNA-binding protein and microRNA target sites by PAR-CLIP. Cell 141:129–141

11. Konig J, Zarnack K, Rot G et al (2011) iCLIP-transcriptome-wide mapping of protein–RNA interactions with individual nucleotide resolution. J Vis Exp. doi:10.3891/2638

12. Konig J, Rot G, Curk T et al (2010) iCLIP reveals the function of hnRNP particles in splicing at individual nucleotide resolution. Nat Struct Mol Biol 17:909–915

13. Tollervey JR, Curk T, Rogelj B et al (2011) Characterizing the RNA targets and position-dependent splicing regulation by TDP-43. Nat Neurosci 14:452–458

14. Rogelj B, Easton LE, Gireesh KB et al (2012) Widespread binding of FUS along nascent RNA regulates alternative splicing in the brain. Sci Rep. doi:10.1038/srep00603

15. Wang Z, Kayikci M, Briese M et al (2010) iCLIP predicts the dual splicing effects of TIA-RNA interactions. PLoS Biol. doi:10.1371/journal.pbio.1000530

Chapter 13

Identification of Endogenous mRNA-Binding Proteins in Yeast Using Crosslinking and PolyA Enrichment

Sarah F. Mitchell and Roy Parker

Abstract

The maturation, localization, stability, and translation of messenger RNAs (mRNAs) are regulated by a wide variety of mRNA-binding proteins. Identification of the complete set of mRNA-binding proteins is a key step in understanding the regulation of gene expression. Herein, we describe a method for identifying yeast mRNA-binding proteins in a systematic manner using UV crosslinking, purification of polyA(+) mRNAs under denaturing conditions, and mass spectrometry to identify covalently bound proteins.

Key words mRNA, mRNA-binding proteins, Yeast, PolyA, Mass spectrometry

1   Introduction

Identification of mRNA-binding proteins, whether individually or globally, has long been an important goal for biologists interested in the regulation of gene expression. mRNA-binding proteins have often been identified from the study of RNA processing, translation, degradation, as well as the factors that bind to cis-acting elements that regulate these events. One useful approach has been purification of mRNPs (mRNA–protein complexes) under denaturing conditions after crosslinking in vivo, followed by identification of the crosslinked proteins. Early attempts at identifying mRNA-binding proteins on a global scale utilized protein crosslinking to mRNA followed by antibody development to the enriched mRNA-binding proteins. This approach led to the identification of the hnRNP proteins in mammalian cells [1].

Genome-wide approaches have also been used to characterize mRNP components globally. For example, individual mRNPs have been characterized through protein pull-down and identification of associated mRNA and proteins [2, 3]. In addition, in vitro methodologies have identified a variety of potential novel mRNA-binding proteins, leading to the conclusion that many

Ren-Jang Lin (ed.), *RNA-Protein Complexes and Interactions: Methods and Protocols*, Methods in Molecular Biology, vol. 1421, DOI 10.1007/978-1-4939-3591-8_13, © Springer Science+Business Media New York 2016

mRNA-binding proteins do not have a canonical mRNA-binding motif [4, 5]. One limitation of the in vitro approaches, or purification of non-covalently bound complexes, is that in vitro conditions could alter the nature of interactions between proteins and RNA, and proteins can exchange on mRNAs in cell lysates [6]. Given these limitations, developing robust methods for the identification of proteins bound to mRNAs in vivo was needed.

One approach that has been highly fruitful has been to couple the crosslinking of proteins to mRNAs in vivo to modern mass spectroscopy methods to identify the bound proteins. This approach has been used to identify hundreds of potential new mRNA-binding proteins in both yeast and mammals [7–10]. Herein, we describe the methods of performing this process on yeast cells.

This method begins with the growth and harvesting of yeast cells. In the experiments we have performed to date, we subjected the yeast to glucose deprivation stress. We utilized stress because the process of UV crosslinking causes stress as measured by the induction of mRNP granules such as P-bodies [11]. By depriving the cells of glucose we are able to uniformly stress the cells in a manner that has been well characterized. In addition, stress alters the post-transcriptional regulation of gene expression and may lead to interesting changes in the mRNA-binding proteome [12]. However, in principle this method could be applied to unstressed cells keeping in mind the caveat that UV crosslinking triggers some stress response.

In any experiment, one half of the cells are exposed to UV light, crosslinking the mRNAs to those proteins that are in direct contact with them (Fig. 1). The remaining cells are used as a

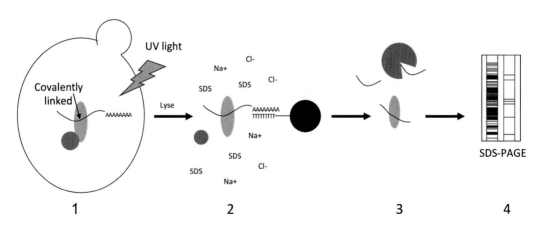

Fig. 1 A flow chart of the procedure. After cells are harvested and, if desired, stressed by glucose deprivation, one half of the cells are crosslinked using UV light (**step 1**). Cells are then lysed and the lysate is bound to oligo(dT) cellulose, washed with denaturing buffers and eluted with low salt (**step 2**). The eluent is concentrated and nuclease-treated (**step 3**). The resultant proteins are then run on an SDS-PAGE gel, proteins are trypsin digested before being extracted from the gel and then analyzed by LC-MS/MS (**step 4**)

control (−UV control) to identify proteins that are pulled-down non-specifically, or which tightly associate with the mRNA despite the denaturing conditions of the pull-down. Cells are then lysed using a ball mill grinder, the lysate is clarified and bound to prepared oligo(dT) cellulose. The matrix is washed with buffer containing SDS to denature and wash away proteins that are not crosslinked to the RNA, and with low salt to remove loosely associated RNA without a polyA tail. After elution in salt-free buffer, the RNA is digested and the protein is separated by SDS-PAGE. Samples are then analyzed by LC-MS/MS. The method described here identified a total of 120 mRNA-binding proteins in yeast, including 66 novel mRNA-binding proteins [8]. These proteins include ribosome processing factors, tRNA modification enzymes and kinases.

2   Materials

(*see* **Note 1**).

2.1   Cell Growth and Crosslinking

1. YEPD + Ade Media: 10 g yeast extract, 20 g bacto peptone, 55 mg adenine, 2 % dextrose per liter, autoclaved.

2. GSA (or equivalent) rotor (Thermo Fisher Scientific, Waltham, MA, USA).

3. 1× Phosphate-buffered saline (PBS): 137 mM NaCl, 2.7 mM KCl, 10 mM Na_2HPO_4, 1.8 mM KH_2PO_4.

4. Petri dishes, 10 cm diameter.

5. 10 % Nonidet P-40 (sold as IGEPAL CA-630 by Sigma Aldrich, St. Louis, MO, USA).

6. UV crosslinker (such as the CL-1000 Short Range UV 254 nm crosslinker, UVP, Upland, CA, USA).

7. Lysis Buffer: 10 mM Tris–HCl pH 7.4, 600 mM NaCl, 0.2 % SDS, 10 mM EDTA, 2 mM DTT, 10 mM ribonucleoside vanadyl complex (New England BioLabs, Ipswich, MA, USA).

8. Liquid nitrogen.

2.2   Lysis and Pull-Down

1. Planetary ball mill (Retsch, Haan, Germany).

2. Oligo(dT) cellulose type 7 (GE Healthcare Life Sciences, Pittsburgh, PA, USA).

3. 0.1 M NaOH.

4. 200 mM Ribonucleoside-vanadyl complex (New England BioLabs, Ipswich, MA, USA).

5. Mini complete EDTA-free protease inhibitor tablets (Roche, Indianapolis, IN, USA).

6. Low Salt, High SDS Buffer: 10 mM Tris–HCl pH 7.4, 150 mM NaCl, 1 % SDS, 10 mM EDTA, 2 mM DTT.

7. Low Salt, No SDS Buffer: 10 mM Tris–HCl pH 7.4, 150 mM NaCl, 10 mM EDTA, 2 mM DTT.

8. TE: 10 mM Tris–HCl pH 8.0, 1 mM EDTA.

2.3 Sample Preparation and Mass Spectrometry

1. 100 mM $CaCl_2$.

2. Micrococcal nuclease (2,000,000 Gel U/mL) and 10× buffer (New England BioLabs, Ipswich, MA, USA).

3. 500 mM EGTA.

4. 10 % SDS.

5. 4 M NaCl.

6. Amicon Ultra 0.5 mL Centrifugal Filters, 10,000 NMWL (Millipore, Billerica, MA, USA).

7. 5× SDS-PAGE Loading Dye: 250 mM Tris–HCl pH 6.8, 10 % SDS, 50 % glycerol, 62.5 mM EDTA, 0.1 % bromophenol blue, 250 mM DTT.

8. Precision Plus Protein Unstained Standards (Bio-Rad, Hercules, CA, USA).

9. NuPAGE Novex 4–12 % Bis-Tris 1.0 mm, 10-well protein gel (Life Technologies, Grand Island, NY, USA).

10. 10 % methanol, 7 % acetic acid.

11. SYPRO Ruby Protein Gel Stain (Life Technologies, Grand Island, NY, USA).

12. Methanol.

13. Typhoon Scanner (GE Healthcare Life Sciences, Pittsburgh, PA, USA).

14. UV light box.

15. Razor blades.

16. Thermo Protein Discoverer 1.2 software (Thermo Fisher Scientific, Waltham, MA, USA).

3 Methods

3.1 Cell Growth and Crosslinking

1. Set up overnight culture.

2. Inoculate large cultures at 0.1 OD^{600}. Grow at 30 °C with shaking until OD^{600} reaches 0.4–0.6 (*see* **Note 2**).

3. Spin down cultures in a GSA (or equivalent) rotor at 10,000 × *g* for 10 min.

4. To stress cells if desired (if not stressing cells *see* **Note 3**): resuspend cell pellets in 1× PBS to rinse, transfer to smaller tubes

(such as 50 mL conical tubes) and pellet again. Resuspend cell pellets in 25 mL 1× PBS per liter of culture. Incubate at room temperature for 30 min if crosslinking cells, 1 h if cells are not going to be crosslinked.

5. If preparing cells for −UV control continue on to **step** 7 of Subheading 3.1. It is recommended to do a control that is not crosslinked to identify those proteins that are pulled-down non-specifically. Otherwise, remove 10 mL of cell suspension and divide equally between three petri dishes (10 cm diameter). Add 6 μL of Nonidet P-40 to each petri dish (*see* **Note 4**).

6. Crosslink cells in a UV crosslinker, using 3 cycles of 4000×100 μJ/cm^2 of power. Between cycles cool the cells on ice for 2 min, frequently swirling the dishes to speed cooling and mix cells.

7. Pellet cells in 50 mL conicals. Resuspend cell pellet in Lysis Buffer. Use a volume of Lysis Buffer 1/3 the mass of the cell pellet (*see* **Note 5**).

8. Fill a clean (or foil lined), pre-cooled container with liquid nitrogen. Using an automatic pipette set on slow, take the cell slurry up in a 5 mL pipet. Drop the slurry into the liquid nitrogen slowly, forming individual balls of frozen cells (*see* **Note 6**).

9. Move the pellets to a dry, pre-cooled container and cover loosely. Store at −80 °C until ready to lyse cells (*see* **Note 7**).

10. Use a planetary ball mill grinder to lyse cells. In a 125 mL cup, cells from 12 L of culture can be lysed at one time. Pre-cool the cup, lid and 20 mm diameter ball bearings in liquid nitrogen. Add frozen cell slurry pellets and ball bearings to cups. Lyse using ten 3-min rounds at 400 rpm. Cool the cups in liquid nitrogen between rounds. Transfer lysed powder to a pre-cooled container and cover loosely. Store at −80 °C until ready to use (*see* **Note 8**).

3.2 Pull-Down

1. Prepare the oligo(dT) column, or if preparing + UV and − UV samples in parallel, two columns. Resuspend 2 g of oligo(dT) cellulose per column in ddH$_2$O and pour into a 1–2 cm diameter column. Rinse the column with 20 mL of 0.1 M NaOH. Equilibrate the column with Lysis Buffer until the pH of the flow through is between pH 7 and 8. Column may be stored tightly sealed at 4 °C overnight.

2. Thaw lysates. Add Lysis Buffer to bring total volume to 100 mL for each sample (assuming a starting amount of 24 L of culture for each sample). Add 500 μL of 200 mM ribonucleoside–vanadyl complex and two mini complete EDTA-free protease inhibitor tablets that have been dissolved in a small amount of Lysis Buffer.

3. Clarify lysate by spinning at $2300 \times g$ for 5 min (*see* **Note 9**).

4. Transfer supernatant to a fresh tube. The pellet will be soft and somewhat "spongy" because of the soft spin and concentrated resuspension. To obtain more clarified lysate from the lysed cells, replace the removed volume with fresh Lysis Buffer, mix the tubes by inverting several times and repeat the spin. Add the resulting supernatant to the previously obtained supernatant to create the complete clarified lysate.

5. Remove oligo(dT) cellulose from columns and divide evenly among tubes containing clarified lysate (usually this is two 50-mL conical tubes per sample). Rotate the tubes at room temperature for 1 h to bind the mRNA to the column.

6. Spin tubes at $200 \times g$ for 1 min. Remove and set aside supernatant. Resuspend beads in 10 mL of Lysis Buffer and pour into the column used to prepare the beads.

7. Rinse columns with 20 mL of Lysis Buffer (*see* **Note 10**).

8. Rinse columns with 30 mL of Low Salt, High SDS Buffer.

9. Rinse columns with 30 mL of Low Salt, No SDS Buffer.

10. To elute the column, heat 20 mL of TE to 65 °C. Elute the column with 4 1-mL fractions followed by 16 250-μL fractions.

11. Find the concentration of RNA in these fractions by measuring the OD^{260}. Pool fractions containing RNA. Eluted RNA–protein complexes may be stored at –20 °C until needed (*see* **Note 11**).

3.3 Sample Preparation and Mass Spectrometry

1. Using Amicon Ultra 0.5 mL Centrifugal Filters 10,000 NMWL, concentrate the eluted mRNA sample to ~3 mL.

2. To digest RNA: add $CaCl_2$ from a 100 mM stock to a final concentration of 1 mM. Add 10× Micrococcal Nuclease Buffer to 1× and 21 μL of Micrococcal Nuclease (2,000,000 Gel U/mL). Mix well. Incubate at 37 °C for 15 min. Quench reaction by adding EGTA to a final concentration of 12.5 mM from a 500 mM stock solution. Mix well and incubate for 5 min on ice.

3. To prevent aggregation, bring sample to 0.2 % SDS and 300 mM NaCl. Concentrate to 20–30 μL using Amicon Ultra 0.5 mL Centrifugal Filters 10,000 NMWL (*see* **Note 12**).

4. Add 5× SDS-PAGE Loading Dye to sample. Sample can be stored at –20 °C overnight before running gel if desired.

5. Prepare a NuPAGE Novex 4–12 % Bis-Tris 1.0 mm, 10-well protein gel in MOPS buffer. Heat the samples for 5 min at 100 °C and immediately load into three wells each (more or less depending upon protein concentration). Also prepare and load an unstained protein ladder such as Precision Plus Protein

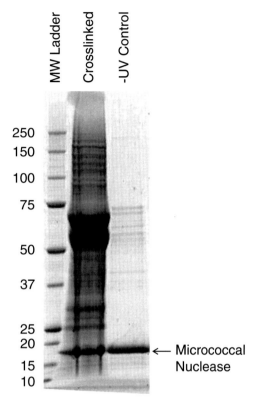

Fig. 2 A gel of the proteins pulled-down using this technique. The second lane contains proteins purified from one-third of a crosslinked sample, the third lane contains one-third of the –UV control. Micrococcal nuclease is the band indicated that runs between 15 and 20 kD. The Precision Plus Protein Unstained Ladder is the molecular weight ladder shown here (Bio-Rad, Hercules, CA, USA). The gel was stained with Sypro Ruby dye and imaged using a Typhoon phosphorimager

Unstained Standards. Run gel at 150 V until dye front is near the bottom of the gel (approximately 1.5 h).

6. Disassemble gel and stain with Sypro Ruby Protein Gel Stain overnight according to product directions.

7. Rinse gel for 30 min in 10 % methanol and 7 % acetic acid on a rocking platform. Rinse 3 times in ddH$_2$O for 5 min each on a rocking platform.

8. Image gel on a Typhoon Scanner using the 620BP30 emission filter and 532 nm (green) excitation laser. A representative gel is shown in Fig. 2 (*see* **Note 13**).

9. Excise gel fragments: In a hood, cover a UV light box with fresh plastic wrap. Place the gel on the wrap. Using fresh razor blades for each piece, slice each lane into five pieces with approximately equal protein amounts and place each slice into

a clean, labeled 2 mL tube. Excise and discard the band of micrococcal nuclease that runs around 17 kD (*see* Fig. 2) (*see* **Note 14**).

10. The sample is now ready to be sent to a mass spectrometry facility for in-gel trypsin digestion, and LC-MS/MS analysis (*see* **Note 15**).

11. Data from the MS/MS spectra were searched against a UniProtKB *S. cerevisiae* protein database using Thermo Protein Discoverer 1.2. Proteins identified by two or more unique peptides with greater than 99 % confidence for each peptide were included in further analysis. Proteins with twofold or greater enrichment by peptide count in the crosslinked sample over the control were considered to be enriched (*see* **Note 16**).

4   Notes

1. Vendor information is provided only for convenience. It is not essential to use products from the vendors mentioned here.

2. A large amount of culture must be grown to have sufficient material for identification of proteins by mass spectrometry. In the past a total of 24 L of crosslinked and 24 L of –UV control cells have been grown. Unless you are in possession of a small army of incubating shakers, rotors, centrifuges, strata-linkers, and skilled undergraduates, it is not possible to grow and process this much culture in one day. In a convenient working day 8 L can be grown, harvested, stressed, crosslinked, and frozen, requiring 6 days total for this process. To make the process more efficient, consider growing some of the cultures overnight so that they can be processed in the morning while additional cultures are growing. It is helpful to stagger the times or densities at which the cultures are started such that each set of 2 L will be done growing approximately 30–45 min apart, or ½ a doubling time for yeast in rich media at 30 °C. This allows sufficient time to crosslink one batch of cells while the next is still growing preventing overgrowth during this time.

3. If you opt not to stress the cells using glucose deprivation, keep the cells in media containing glucose throughout the crosslinking procedure. Even a brief washing in buffer or media without glucose will cause a stress response. One way to make such a procedure efficient may be to utilize a tube-shaped UV crosslinker through which culture may be pumped (such as the biologic ultraviolet water purifier, ultraviolet.com) to process large volumes so that the cells do not have to be pelleted and concentrated, a process which may cause stress. In order to facilitate crosslinking in the media, grow cells in minimal media

that will absorb less UV light than darker colored nutrient-rich media. We were not able to optimize this procedure to make it as efficient as using a UV crosslinking box, but with some effort it should prove adequate and would save time either with or without stressing the culture.

4. Top layers of cells will block lower layers from the UV light. The addition of Nonidet P-40 causes the cell suspension to spread evenly across the petri dish in a thin layer, exposing a greater number of cells to the light. Be sure to remove the covers from the petri dishes before crosslinking as those will also block UV light. If there is sufficient room in the UV crosslinking box, or multiple UV crosslinking boxes are available, more than 10 mL of cell suspension may be crosslinked at one time.

5. While the mass of the cell pellets varies depending upon the precise density at which the cells were harvested, the volume of Lysis Buffer used to resuspend the cells is usually ~1.5 mL per 2 L of culture.

6. It is important to drop the slurry slowly enough that the cells do not form one large clump. As the cells cool completely the droplets harden so that they no longer stick together and fall to the bottom of the container. If groups of only a few droplets stick together this does not cause a problem. Alternatively, if the pipet squirts suddenly and a long "noodle" of cells forms this will not be problematic either. However, a large, solid mass of cells will inhibit the grinding process.

7. Do not tightly cover the cells as they may still contain some liquid nitrogen that could evaporate causing pressure to build up.

8. A planetary ball mill grinder proved to be the most efficient method of cell lysis for this large volume of culture in our hands. Other methods of lysis that are efficient may also be used here. As the pellets break down the powder can form a hard layer on the bottom of the cup. This reduces lysis efficiency. Scraping this layer away from the cup periodically using a cold spatula will help.

9. Clarification is performed at a slow speed to prevent large mRNP granules such as P-bodies and stress granules from pelleting out.

10. The viscosity of the lysate remaining in the column will frequently clog the columns so that dripping is extremely slow. It has sometimes been necessary to mix the media and buffer in the column so that any particulate matter that has settled and is clogging the column can be dislodged.

11. A nanodrop (Thermo Fisher Scientific, Waltham, MA, USA) is useful for quickly finding RNA concentration without wasting

a large amount of material. Total RNA eluted when starting from 24 L of culture is usually about 2 mg.

12. The high concentration of protein in the sample can cause a gel state to form, this will prevent the sample from running into the SDS-PAGE gel efficiently. The addition of salt and SDS will help to solubilize the protein preventing the gel from forming.

13. Other procedures for imaging the gel using UV light are possible. See the instructions included with the Sypro Ruby stain for details.

14. It is important to take care to reduce keratin contamination throughout the entire procedure, but particularly after the gel has been run as keratin will no longer be isolated to one sample. To reduce contamination, always wear a lab coat, and gloves when handling the gel. Clean all surfaces with methanol. Use newly opened containers of tubes and razor blades. When using the UV light box protect your eyes with appropriate goggles or a face shield.

15. Our lab is not in possession of mass spectrometry equipment and outsourced the analysis of the protein content to a core facility. A C18 ZipTip (Millipore) was used for sample clean up. Then the sample was run over a C18 pre-column (100 μm inner diameter×2 cm, Thermo Fisher Scientific, Waltham, MA, USA). This was followed by a C18 analytical column (75 μm inner diameter×10 cm, Thermo Fisher Scientific, Waltham, MA, USA) run at a flow rate of 400 nl/min with a 5–20 % gradient of solvent B (0.1 % formic acid, acetonitrile) over 65 min, and then a 20–35 % gradient of solvent A (0.1 % formic acid in water) over 25 min. The sample was then analyzed on an LTQ Orbitrap Velos mass spectrometer (Thermo Fisher Scientific, Waltham, MA, USA) with a Advion Nanomate ESI source (Advion, Ithaca, NY, USA).

16. There are many ways to assess the enrichment of proteins in the crosslinked sample. The protocol user should feel free to use the statistical method best for his or her data, taking into account either or both peptide count and peak area. An important check to perform is to be sure that those proteins identified do not correlate with protein abundance in the cell.

Acknowledgements

Members of the Parker lab provided essential assistance and feedback in developing the method described here. Saumya Jain provided helpful comments on this manuscript. The Arizona Proteomics Consortium at the University of Arizona performed related mass spectrometry and assisted in analysis. Development of the method described was supported by funding from the US National Institutes of Health grant 7R37 GM045443 (R.P.), a Howard Hughes Medical Institute grant (R.P.), and a Leukemia and Lymphoma Society fellowship 5687-13 (S.F.M.).

References

1. Dreyfuss G, Choi YD, Adam SA (1984) Characterization of heterogeneous nuclear RNA-protein complexes in vivo with monoclonal antibodies. Mol Cell Biol 4:1104–1114

2. Batisse J, Batisse C, Budd A et al (2009) Purification of nuclear poly(A)-binding protein Nab2 reveals association with the yeast transcriptome and a messenger ribonucleoprotein core structure. J Biol Chem 284:34911–34917

3. Klass DM, Scheibe M, Butter F et al (2013) Quantitative proteomic analysis reveals concurrent RNA-protein interactions and identifies new RNA-binding proteins in Saccharomyces cerevisiae. Genome Res 23:1028–1038

4. Tsvetanova NG, Klass DM, Salzman J et al (2010) Proteome-wide search reveals unexpected RNA-binding proteins in saccharomyces cerevisiae. PLoS One 5:e12671

5. Scherrer T, Mittal N, Janga SC et al (2010) A screen for RNA-binding proteins in yeast indicates dual functions for many enzymes. PLoS One 5:e15499

6. Riley KJ, Steitz JA (2013) The "observer effect" in genome-wide surveys of protein-RNA interactions. Mol Cell 49:601–604

7. Kwon SC, Yi H, Eichelbaum K et al (2013) The RNA-binding protein repertoire of embryonic stem cells. Nat Struct Mol Biol 20(9):1122–1130

8. Mitchell SF, Jain S, She M et al (2013) Global analysis of yeast mRNPs. Nat Struct Mol Biol 20:127–133

9. Baltz AG, Munschauer M, Schwanhäusser B et al (2012) The mRNA-bound proteome and its global occupancy profile on protein-coding transcripts. Mol Cell 46:674–690

10. Castello A, Fischer B, Eichelbaum K et al (2012) Insights into RNA biology from an atlas of mammalian mRNA-binding proteins. Cell 149:1393–1406

11. Teixeira D (2005) Processing bodies require RNA for assembly and contain nontranslating mRNAs. RNA 11:371–382

12. Holcik M, Sonenberg N (2005) Translational control in stress and apoptosis. Nat Rev Mol Cell Biol 6:318–327

Chapter 14

Ribo-Proteomics Approach to Profile RNA–Protein and Protein–Protein Interaction Networks

Hsin-Sung Yeh, Jae-Woong Chang, and Jeongsik Yong

Abstract

Characterizing protein–protein and protein–RNA interaction networks is a fundamental step to understanding the function of an RNA-binding protein. In many cases, these interactions are transient and highly dynamic. Therefore, capturing stable as well as transient interactions in living cells for the identification of protein-binding partners and the mapping of RNA-binding sequences is key to a successful establishment of the molecular interaction network. In this chapter, we will describe a method for capturing the molecular interactions in living cells using formaldehyde as a crosslinker and enriching a specific RNA–protein complex from cell extracts followed by mass spectrometry and Next-Gen sequencing analyses.

Key words Formaldehyde crosslinking, Crosslinking immunoprecipitation, Mass spectrometry, Next-Gen sequencing, RNA-binding proteins

1 Introduction

RNA-binding proteins (RBPs) play an important role in modulating various events in RNA metabolism, including alternative splicing, alternative polyadenylation, localization, translation, stability, etc. [1]. Nevertheless, not many RBPs have been studied systematically. To characterize the function of a particular RBP, it is crucial to know what *cis*-acting elements and other *trans*-acting factors that the RBP interacts with in different cellular contexts. Typically, co-immunoprecipitation of target RNAs that bind to RBPs using specific antibodies can be performed to study the function of RBPs in RNA metabolism. Alternatively, epitope-tagged RBPs can be used for the same approach. However, it has been indicated that RNA–protein complexes are dynamic and can change their compositions rapidly in response to physiological context changes [2]. In addition, RBPs could re-associate with their target RNAs during co-immunoprecipitation after cell lysis [3]. Due to the nature of these interactions, crosslinking of these interacting partners in situ is essential prior to their isolation using antibodies specific to the protein of interest.

Ren-Jang Lin (ed.), *RNA-Protein Complexes and Interactions: Methods and Protocols*, Methods in Molecular Biology, vol. 1421, DOI 10.1007/978-1-4939-3591-8_14, © Springer Science+Business Media New York 2016

UV-crosslinking at 254 nm is a simple and common way to crosslink RNA–protein interactions in vivo. It induces covalent bonding between nucleotide bases and aromatic side chains of RBPs [4–6]. This allows single-nucleotide-resolution mapping of the RNA binding by an RBP. However, different RBPs have varying UV-crosslinkability due to differences in their availability and spatial orientation of the aromatic side chains near the RNA-binding site, which inherently renders UV-mediated crosslinking biased among RBPs. Moreover, this method also suffers from low crosslinking efficiency in general [7].

As an improvement, a photoactivable ribonucleoside-enhanced (PAR) UV-crosslinking technique is developed [8]. In this method, cells are treated with photoreactive ribonucleoside analogs to be incorporated into nascent RNA transcripts. Irradiation with UV at 365 nm can efficiently crosslink RNA transcripts containing photoreactive nucleosides with the bound RBPs. While the crosslinking efficiency is increased with PAR, it still suffers from bias caused by varying base compositions in the consensus binding sequences of RBPs. For example, if 4-thiouridine is used as a photoreactive ribonucleoside analog, RBPs binding to U-rich sequences will be more enriched than RBPs whose binding sequences are not U-rich.

Apart from UV-mediated crosslinking, formaldehyde is perhaps the most popular crosslinker in biology. Three macromolecular interactions in cells (protein–protein, protein–DNA, and protein–RNA) can be efficiently crosslinked by formaldehyde. Formaldehyde-mediated crosslinking has been a powerful method in capturing transient or dynamic associations between DNA and protein [9, 10]. This chromatin immunoprecipitation (ChIP) assay becomes more powerful when modern profiling technologies like microarray and next-generation sequencing technologies are used for outcome analysis. The Garcia-Blanco laboratory applied the workflow of ChIP assay to study RNA–protein interactions in living cells (the RNP immunoprecipitation (RIP) assay) [11]. The Moore group also adopted formaldehyde crosslinking in their approach for footprinting RNA–protein complexes (the RIPiT-seq); in their study, they showed the superiority of formaldehyde crosslinking over UV crosslinking with increased crosslinking efficiency and reduced bias [7].

With formaldehyde, nucleic acid (DNA or RNA) and protein interactions can be crosslinked through Schiff base formation (Fig. 1a) [12, 13]; this crosslinking can easily be reversed by treatment of high salt and high temperature with a reducing reagent. In contrast, crosslinking of protein–protein interaction is more complicated and may not be reversible. Treatment of formaldehyde forms three representative reaction groups among amino acids (methylol group—reversible, Schiff base—reversible, and methylene bridge—irreversible (Fig. 1b)) that will react to an adjacent amino acid, resulting in covalent bonds (crosslinking). Many primary amino or thiol group-containing amino acids including arginine, asparagine,

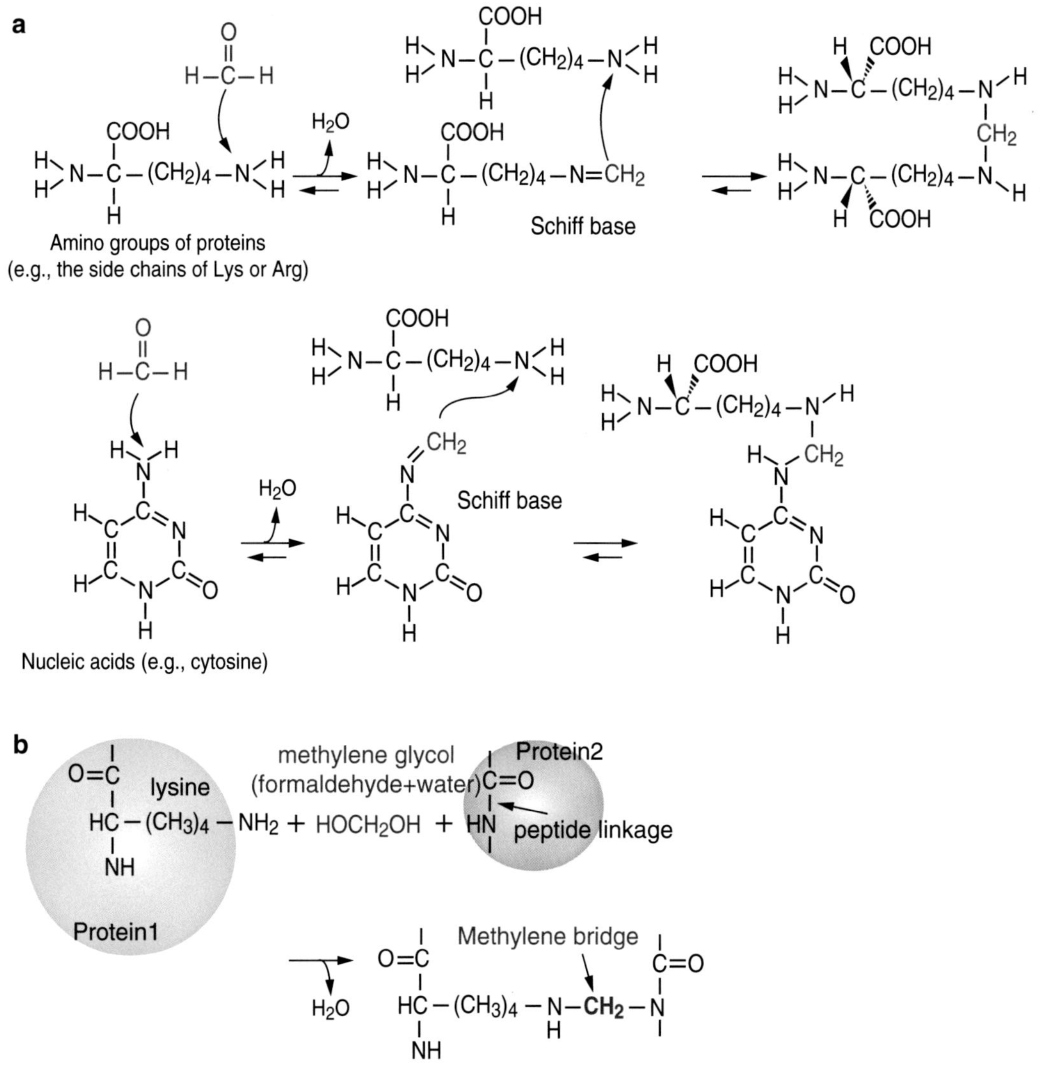

Fig. 1 Nucleic acid and protein interactions can be crosslinked by formaldehyde through Schiff-base or methylene glycol formations. (**a**) Formaldehyde can react with amino groups on the bases of nucleic acids or protein side chains and yield methylol adducts. When partially dehydrated, labile Schiff-bases are formed, which can then crosslink with other amines covalently. (**b**) When formaldehyde is dissolved in water, methylene glycol is formed. It can then crosslink a primary amine to a secondary amine of certain residues through methylene bridges

glutamine, histidine, tryptophan, and tyrosine will react with formaldehyde to form methylene bridges and render the protein–protein crosslinking irreversible [13]. For this reason, it is critical to optimize the crosslinking condition particularly for protein–protein interactions since too much irreversible crosslinking (caused by long reaction time and high reagent concentration) will result in increased background (i.e., lack of specificity) and hinder the enzymatic digestion of the protein complexes for downstream analysis.

In our method, we use formaldehyde to perform in vivo crosslinking. Following crosslinking and immunoprecipitation, we adopt a ribo-proteomic approach to profile the enriched molecular complexes (Fig. 2) [14]: mass spectrometry is used to identify the protein-binding partners of the target protein, and the RNA fragment in the complex is sequenced with Next-Gen sequencing technologies [15].

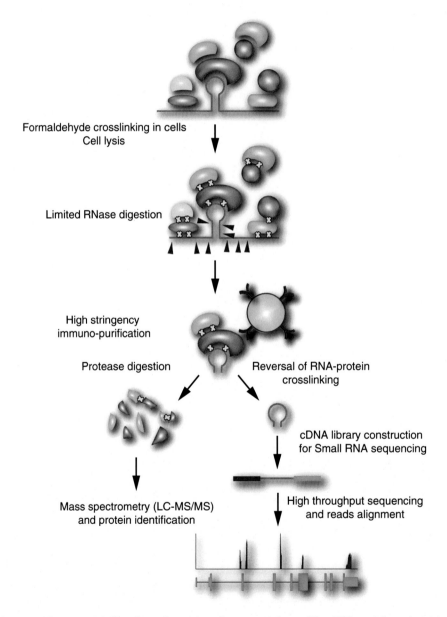

Formaldehyde crosslinking in cells
Cell lysis

Limited RNase digestion

High stringency
immuno-purification

Protease digestion

Reversal of RNA-protein
crosslinking

cDNA library construction
for Small RNA sequencing

Mass spectrometry (LC-MS/MS)
and protein identification

High throughput sequencing
and reads alignment

Fig. 2 Diagrammatic representation of our ribo-proteomic approach for profiling RNA–protein and protein–protein interaction networks. Following in vivo crosslinking with formaldehyde, cells are lysed with sonication. A specific antibody against the RBP of interest is used to immunoprecipitate the crosslinked RNP complexes. Limited nuclease digestion is performed to eliminate proteins that are crosslinked with the RBP of interest through the RNA molecule, rather than through direct protein–protein interactions. Mass spectrometry is used to identify the proteins that are pulled-down along with the RBP of interest; RNA fragments are purified and sequenced with Next-Gen sequencing technologies to map the bindings of the RBP of interest in the transcriptome

2 Materials

2.1 In Vivo Crosslinking

1. 1× PBS: 137 mM NaCl, 2.7 mM KCl, 10 mM Na$_2$HPO$_4$, 1.8 mM KH$_2$PO$_4$ (stored at room temperature).
2. Formaldehyde solution (37 %).
3. 1.5 M Glycine-NaOH, pH 7.5.

2.2 Immunoprecipitation

1. Empigen buffer: 20 mM Tris–HCl, pH 7.5, 500 mM NaCl, 2.5 mM MgCl$_2$, 1 % Empigen, 0.5 % Triton X-100, 2 mg/ml Heparin sulfate.
2. Protein G sepharose bead slurry.
3. 1× PBS with 0.05 % Triton X-100.
4. RSB-100: 20 mM Tris–HCl, pH 7.5–8.0, 100 mM NaCl, 2.5 mM MgCl$_2$.
5. Triton X-100.
6. RNase T1.
7. Reverse crosslinking buffer: 20 mM Tris–HCl, pH 8.0, 500 mM NaCl, 1 mM EDTA, 10 mM beta-mercaptoethanol.

2.3 cDNA Library Preparation

1. 10 % (w/v) SDS.
2. Proteinase K (20 mg/ml).
3. Tris-saturated Phenol.
4. Chloroform.
5. 100 % EtOH.
6. Glycogen (20 mg/ml).
7. 75 % EtOH (room temperature).
8. TruSeq Small RNA Sample Preparation Kit (Illumina).

2.4 Sample Preparation for Mass Spectrometry

1. Elution Buffer: RSB-100 with 0.25 % of SDS.
2. Organic precipitation solution: 50 % ethanol, 25 % methanol, 25 % acetone (store at −20 °C).
3. Resuspension Buffer: 8 M urea, 0.5 M NH$_4$HCO$_3$–NH$_4$OH, pH 8.0, and 4 mM DTT.
4. 20 mM Iodoacetamide.
5. Trypsin, mass spectrometry grade.
6. Trifluoroacetic acid.
7. Suitable solid phase extraction material. For example, Pierce C18 spin columns (product No. 89870).

3 Methods

3.1 Formaldehyde Crosslinking of RNA–Protein Complexes

1. Harvest cells and wash twice with ice-cold PBS (this is usually done in 15 ml conical tube). For abundant RNA–protein complexes or stable interactions, 10,100 mm plates of HeLa or 293 T cells (75–80 % confluent) will suffice. If the complex is less abundant or the interaction is transient, starting with 20 plates is recommended.

2. After the final washing, add 9 ml of PBS containing 0.2 % formaldehyde to the cell pellet and incubate at room temperature for 10 min with gentle agitation (*see* **Note 1**).

3. Add 1 ml of 1.5 M glycine, pH 7.5 (final concentration 0.15 M) and incubate for 5 min to quench the crosslinking by formaldehyde (*see* **Note 2**).

4. Pellet the cells and wash with PBS twice (*see* **Note 3**). At this point, you can use the crosslinked cells immediately to make the extracts for immunoprecipitation or save them at –80 °C for future use.

3.2 Immunoprecipitation of Crosslinked RNPs and Construction of cDNA Library for Illumina Genome Analyzer (see Notes 4 and 5)

1. Resuspend the crosslinked cell pellet in the high stringent empigen buffer (*see* **Notes 6** and **7**). The volume for an IP reaction is 300–500 µl (depending on the availability of crosslinked cells). Use 1.5 ml of the buffer for ten plates of cells (about 3 IPs).

2. Sonicate the cell suspension at the power output of 4 W for 4 times, 10 s each; with 5 s intervals (*see* **Note 8**).

3. While preparing the cell lysates, immobilize antibody to protein G sepharose beads. For each IP, take 15–20 µl of bead slurry and add to 500 µl of 1× PBS with 0.05 % Triton X-100. Do a quick spin, discard supernatant, and add fresh 500 µl of 1× PBS with 0.05 % Triton X-100. Add 2–5 µg of antibodies and incubate at 4 °C for at least an hour on an end-over-end rotator. Spin down the beads and discard the supernatant (*see* **Note 9**).

4. For each IP, add 500 µl of extracts to the antibody-bound protein G sepharose beads prepared in the previous step and incubate for 2 h (*see* **Notes 10** and **11**).

5. Wash 5 times with the empigen buffer. For the first two washes, add 500 µl of the buffer and rotate at 4 °C for 10 min. Use 1 ml of the buffer for the last three quick washes (no 10 min incubation at 4 °C) (*see* **Note 12**).

6. Add to the beads 200 µl of RSB-100 with 0.01 %Triton X-100 and RNase T1 to 0.1 unit/µl and incubate for 5–10 min at room temperature (*see* **Note 13**).

5. Wash with the empigen buffer 2 times: add 800 μl of the buffer directly to the RNase T1 digestion and consider this as one wash. Do the second quick wash with 1 ml of the empigen buffer.

6. Immediately add 250 μl reverse crosslinking buffer to the bead and incubate at 70 °C overnight (*see* **Note 14**).

7. The next day, add SDS directly to the reverse crosslinking reaction to the final concentration of 2 % (*see* **Note 15**).

8. Add 20 μl proteinase K (20 mg/ml) and incubate at 30 °C for 20 min.

9. Perform phenol/chloroform extraction (*see* **Note 16**).

10. Precipitate the RNA fragments with 20 μg of Glycogen, 0.3 M NaOAc, and 2.5 volumes of 100 % ethanol. Wash the pellet with 75 % EtOH (room temperature).

11. Now your RNA is ready for the cDNA library construction using TruSeq Small RNA Sample Preparation Kit. Please follow the manufacturer's instructions (*see* **Notes 17–22**).

3.3 Immunoprecipitation of Crosslinked RNPs for Proteomics Analysis

1. Follow the IP protocol provided above (**steps 1–5** in Subheading 3.2) (*see* **Note 23**).

2. Elute the IP-ed RNPs with 200 μl elution buffer (RSB-100 + 0.25 % of SDS). Elution should be done at room temperature for 1 h.

3. Take supernatant and add 1.2 ml of the organic precipitation solution.

4. Incubate the mixture at −20 °C for at least 2 h and precipitate the proteins by centrifugation in a standard microcentrifuge at the maximum speed at 4 °C for 20 min.

5. Wash the pellet (usually not visible) with 70 % ethanol very gently and centrifuge at maximum speed at room temperature for 10 min.

6. Remove the supernatant and proceed with pellet.

7. Resuspend pellet in 25 μl of the Resuspension Buffer. Incubate at 37 °C for 45 min.

8. Cool down the solution to room temperature for 5 min then add 25 μl of 20 mM iodoacetamide. Incubate at room temperature in the dark for 30 min.

9. Add 50 μl ultrapure water to reduce urea concentration to 2 M.

10. Digest with 20 ng of trypsin at 37 °C overnight. After the digestion, lower the pH of the solution to pH < 3.0 with trifluoroacetic acid (measure pH using pH paper).

11. Purify the sample with Pierce C18 spin columns or any suitable solid phase extraction materials and submit the purified peptides for mass spectrometry analysis.

4   Notes

1. Be sure to bring the PBS to the room temperature before starting this step. Add formaldehyde to the PBS just before the use to make 0.2 % solution. The crosslinking is to be performed at the room temperature. Usually, 0.2 % formaldehyde for 10 min is enough to crosslink most of the RNA–protein interactions. You may want to find an optimal condition for your proteins and RNAs of interest. However, higher formaldehyde concentration and longer incubation time are not recommended.

2. During the crosslinking and quenching, the clumping of cells will be visible and this is a good indication that the crosslinking occurs. But at the same time, 30–40 % of the cells may be lost after the crosslinking.

3. Most of the time, you will notice some clump of cells floating around at this time. It is normal and usually these are removed during the washing step.

4. You can use any IP protocol you feel comfortable with. Two things need to be remembered. First, antibodies you are using have to be of high quality (ChIP grade). Second, the IP efficiency will be decreased compared to the extracts from non-crosslinked cells.

5. Since the abundance of RNA and protein of your interest vary, you will want to estimate how much material (amount of extracts) you will need for IP and Solexa sequencing. The optimal amount of RNA for the construction of cDNA library for Solexa sequencing is about 250 ng–500 ng. So you will have to calculate how many IPs you need to reach this amount of RNA in your IPs (for example, if you get 100 ng of RNA from one IP, 3–5 IPs is recommended).

6. When making extracts, please use extremely stringent condition so non-crosslinked interactions dissociate. Buffers with high detergent and high salt compositions are recommended (such as RIPA buffer). Nevertheless, the empigen buffer described here is highly recommended (for proteomics purpose).

7. Empigen can be purchased from Sigma (37 % stock). Heparin sulfate can be prepared as 100 mg/ml stock solution and diluted to 2 mg/ml as working concentration. Heparin sulfate can significantly decrease the ribosomal RNA and protein contamination.

8. Since high concentration of detergent is used, excessive sonication is unnecessary. The condition described here should be more than enough.

9. The amount of antibodies should be determined by the end user. The end user should consider the efficiency of antibodies in IP and the abundance of the target protein.

10. The incubation time for IP varies based on the efficiency of the antibodies. Overnight incubation is not recommended (2 h is maximum). Keep the volume between 300 and 500 μl as the IP efficiency drops a lot if the volume exceeds 500 μl. If more material is needed for mass spectrometry or deep sequencing, always multiply the number of tubes rather than increasing the volumes of the IPs. Also, do not over-saturate the extracts by using a small volume of extraction buffer for a large cell pellet. This will significantly increase the background.

11. Use about 15–20 μl of bead slurry. Using more beads will increase the background significantly.

12. During the last three quick washes, try to pool the beads together (the rule of thumb is 3 tubes into 1 tube).

13. Activity of RNase T1 at low concentration is sensitive to the secondary structure of RNAs. It is also dependent on the base composition of RNA. As an alternative, micrococcal nuclease (MNase) can be used. In the case of MNase digestion, the buffer should contain 3 mM of $CaCl_2$ as a cofactor and MNase should be deactivated by adding EDTA (15 mM final concentration) to chelate $CaCl_2$. The concentration of RNase T1 and the incubation time should be calibrated for each RNP.

14. Without reverse crosslinking, the efficiency of adapter ligation drops significantly. This is because proteinase K digestion leaves a couple of residual amino acids at the crosslinked site. These residual amino acids are steric hindrances for RNA ligase reaction.

15. This step gets rid of proteins reverse crosslinked from RNAs. Do not do proteinase K digestion before the reverse crosslinking because any trace amount of RNAse contamination will degrade immunoprecipitated RNAs.

16. We do not recommend using Trizol. At this point, RNAs to be purified exist as smaller RNA fragments. Ethanol precipitation with glycogen is more efficient in precipitating small pieces of RNAs.

17. When constructing cDNA library, perform the 3′ adapter ligation first.

18. 5 pmole of adapter is good for 250–500 ng of RNA ligation (in 10 μl reaction).

19. Adding 1 μl of 50 % PEG8000 (or 6000) will help the efficiency of ligation.

20. Ligation overnight at 4 °C is better than incubation at 20 °C and move to 4 °C (as suggested by Illumina protocol).

21. After final PCR, use agarose gel electrophoresis and gel purification kit to recover the PCR product. Purchase a special agarose to make 3–5 % agarose gel (for example, NuSieve 3:1 Agarose cat. No 50090 from Lonza).

22. If nuclease or RNase T1 is used, T4 kinase reaction to phosphorylate your RNA fragment is required.

23. IP can be done in the presence or absence of RNase cocktail (2 μl/IP). The treatment of RNase cocktail will eliminate the protein–protein interactions mediated through RNAs.

References

1. Glisovic T, Bachorik JL, Yong J et al (2008) RNA-binding proteins and post-transcriptional gene regulation. FEBS Lett 582(14):1977–1986. doi:10.1016/j.febslet.2008.03.004
2. Lunde BM, Moore C, Varani G (2007) RNA-binding proteins: modular design for efficient function. Nat Rev Mol Cell Biol 8(6):479–490
3. Mili S, Steitz JA (2004) Evidence for reassociation of RNA-binding proteins after cell lysis: implications for the interpretation of immunoprecipitation analyses. RNA 10(11):1692–1694. doi:10.1261/rna.7151404
4. Dreyfuss G, Choi YD, Adam SA (1984) Characterization of heterogeneous nuclear RNA–protein complexes in vivo with monoclonal antibodies. Mol Cell Biol 4(6):1104–1114
5. Huang Y, Steitz JA (2001) Splicing factors SRp20 and 9G8 promote the nucleocytoplasmic export of mRNA. Mol Cell 7(4):899–905. doi:10.1016/S1097-2765(01)00233-7
6. Mayrand S, Setyono B, Greenberg JR et al (1981) Structure of nuclear ribonucleoprotein: identification of proteins in contact with poly(A)+heterogeneous nuclear RNA in living HeLa cells. J Cell Biol 90(2):380–384. doi:10.1083/jcb.90.2.380
7. Singh G, Ricci EP, Moore MJ (2014) RIPiT-seq: a high-throughput approach for footprinting RNA:protein complexes. Methods 65(3):320–332. doi:10.1016/j.ymeth.2013.09.013
8. Hafner M, Landthaler M, Burger L et al (2010) Transcriptome-wide identification of RNA-binding protein and microRNA target sites by PAR-CLIP. Cell 141(1):129–141. doi:10.1016/j.cell.2010.03.009
9. Jackson V (1978) Studies on histone organization in the nucleosome using formaldehyde as a reversible cross-linking agent. Cell 15(3):945–954. doi:10.1016/0092-8674(78)90278-7
10. Orlando V, Strutt H, Paro R (1997) Analysis of chromatin structure by in vivo formaldehyde cross-linking. Methods 11(2):205–214. doi:10.1006/meth.1996.0407
11. Niranjanakumari S, Lasda E, Brazas R et al (2002) Reversible cross-linking combined with immunoprecipitation to study RNA–protein interactions in vivo. Methods 26(2):182–190. doi:10.1016/S1046-2023(02)00021-X
12. Fraenkel-Conrat H, Olcott HS (1948) Reaction of formaldehyde with proteins: VI. Cross-linking of amino groups with phenol, imidazole, or indole groups. J Biol Chem 174(3):827–843
13. Metz B, Kersten GFA, Hoogerhout P et al (2004) Identification of formaldehyde-induced modifications in proteins: reactions with model peptides. J Biol Chem 279(8):6235–6243. doi:10.1074/jbc.M310752200
14. Yong J, Kasim M, Bachorik JL et al (2010) Gemin5 delivers snRNA precursors to the SMN complex for snRNP biogenesis. Mol Cell 38(4):551–562. doi:10.1016/j.molcel.2010.03.014
15. Morozova O, Marra MA (2008) Applications of next-generation sequencing technologies in functional genomics. Genomics 92(5):255–264. doi:10.1016/j.ygeno.2008.07.001

Chapter 15

Detection of Protein–Protein Interaction Within an RNA–Protein Complex Via Unnatural-Amino-Acid-Mediated Photochemical Crosslinking

Fu-Lung Yeh, Luh Tung, and Tien-Hsien Chang

Abstract

Although DExD/H-box proteins are known to unwind RNA duplexes and modulate RNA structures in vitro, it is highly plausible that, in vivo, some may function to remodel RNA–protein complexes. Precisely how the latter is achieved remains a mystery. We investigated this critical issue by using yeast Prp28p, an evolutionarily conserved DExD/H-box splicing factor, as a model system. To probe how Prp28p interacts with spliceosome, we strategically placed *p*-benzoyl-phenylalanine (BPA), a photoactivatable unnatural amino acid, along the body of Prp28p in vivo. Extracts prepared from these engineered strains were then used to assemble in vitro splicing reactions for BPA-mediated protein–protein crosslinkings. This enabled us, for the first time, to "capture" Prp28p in action. This approach may be applicable to studying the roles of other DExD/H-box proteins functioning in diverse RNA-related pathways, as well as to investigating protein–protein contacts within an RNA–protein complex.

Key words DExD/H-box protein, RNA helicase, RNPase, RNA-dependent ATPase, Pre-mRNA splicing, Prp28p, Prp8p, *p*-benzoyl-phenylalanine, BPA, UV crosslinking

1 Introduction

DExD/H-box proteins are ubiquitous enzymes known to participate in practically all RNA-related pathways, such as pre-mRNA splicing, mRNA export, protein synthesis, and ribosomal biogenesis [1–3]. Early in vitro studies revealed that, despite their conceptual similarity to the better studied DNA helicases, many DExD/H-box proteins appear not to be processive enzymes in terms of unwinding RNA duplexes [4]. Given that RNA duplexes in vivo are rarely longer than ~10 contiguous base pairs and that they are often stabilized by proteins [5–7], it seems plausible that DExD/H-box proteins may function as ribonucleoprotein particle ATPases (RNPases) to remodel RNA–protein complexes [8, 9]. Several studies [10, 11] appear to support such a hypothesis. However, it remains a formidable challenge to mechanistically

Ren-Jang Lin (ed.), *RNA-Protein Complexes and Interactions: Methods and Protocols*, Methods in Molecular Biology, vol. 1421, DOI 10.1007/978-1-4939-3591-8_15, © Springer Science+Business Media New York 2016

dissect how DExD/H-box proteins accomplish such a feat, especially in light of the diversity and complexity of the cellular RNPs.

We have attempted to resolve this issue by using yeast Prp28p as a model system. Prp28p is an evolutionarily conserved DExD/H-box splicing factor that facilitates the U1/U6 switch at the 5′ splice site (5′ss) during pre-mRNA splicing [6, 10, 12], which is accomplished by the spliceosome, arguably the most complex RNP in the cell [13–15]. Unlike ribosome, spliceosome is built anew for each pre-mRNA through a highly dynamic process, which demands a series of coordinated conformational changes that culminate in the formation of the spliceosome catalytic center [15, 16]. At least eight DExD/H-box proteins [3], including Prp28p, are known to participate in the sequential conformational changes of the splicing machinery, in which dense protein contacts on RNAs are virtually assured. Understanding how Prp28p interacts and exerts its function on spliceosome is therefore of central importance to the splicing field and to the RNA field as a whole.

We have previously shown that Prp28p can be made dispensable in the presence of mutations that alter specific components of the U1 snRNP [10, 12]. These data thus suggest a model that Prp28p counteracts the stabilizing effect by those proteins on U1 snRNP/5′ss interaction. To uncover how Prp28p works within the complex spliceosomal milieu, we adapted an experimental strategy that was previously developed for studying the transcription preinitiation complex [17, 18], which was in turn based on the invention by Schultz and colleagues [19]. Briefly, we strategically replaced selected amino-acid residues along the body of Prp28p by p-benzoyl-phenylalanine (BPA) using nonsense suppressor tRNA and its cognate aminoacyl tRNA synthetase in vivo. We then prepared splicing extracts from these yeast strains for spliceosomal assembly and for UV-activated crosslinking reactions. This approach allowed us to uncover a number of spliceosomal proteins that interacts with Prp28p in a splicing-dependent manner.

Here we describe such an example in which Prp8p [20], a crucial protein located at the heart of the spliceosome, was identified as a physical and functional partner of Prp28p. We suggest this approach may very well be applicable for studying other DExD/H-box proteins functioning in their physiologically relevant environments, thereby shedding new lights on the longstanding question as to how DExD/H-box proteins function in vivo.

2   Materials (*see* Note 1)

2.1   Plasmid and Strain Construction

1. Plasmid pCA8093 (2-μm/*LEU2*) (Chang lab).

2. Plasmid pLH157 [17, 18].

3. Yeast strains with *prp28Δ::kanMAX4* and carrying pCA8009 (*PRP28/CEN/URA3*) (Chang lab).

4. *p*-Benzoyl-L-phenylalanine (BPA) (Bachem).

5. BPA (0.25 mM)-containing medium devoid of tryptophan and leucine (*see* **Note 2**).

6. BPA medium containing 5-fluoroorotic acid (5-FOA) (*see* **Note 2**).

2.2 Extract, RNA, and Protein Preparations

1. YPD (1 % yeast extract, 2 % peptone, 2 % dextrose) containing 0.25 mM BPA (*see* **Note 2**).

2. Glass dounce homogenizer (Wheaton).

3. Plasmid pAdML-M3 [21].

4. Plasmids pM3-Act-wt, pM3-Act-5′ss, pM3-Act-BPS, pM3-Act-3′ss, and pM3-Act-ΔI (Chang lab).

5. Ribo MAX™ Large Scale RNA Production System-SP6 (Promega).

6. RNA elution buffer: 0.3 M sodium acetate, pH 5.2, 2 mM EDTA, 0.1 % SDS.

7. Phenol/chloroform solution (50 % phenol, 50 % chloroform, 0.05 % 8-hydroxyquinoline, 0.1 M NaCl, 1 mM EDTA, 0.01 M sodium acetate, pH 6.0).

8. MS2-MBP fusion protein expressing *E. coli* strain (Chang lab).

9. Buffer A: 20 mM Tris–HCl, pH 7.5, 200 mM NaCl.

10. 5 mM sodium phosphate, pH 7.0.

11. Constant Cell Disruption System (Constant Systems, Ltd).

12. JA-25.50 rotor (Beckman-Coulter).

13. Amylose Resin (New England Biolab).

14. Wash Buffer: 20 mM HEPES-KOH, pH 7.9, 150 mM NaCl, 0.05 % NP-40.

15. Amylose Elution Buffer: 5 mM sodium phosphate, pH 7.0, 15 mM maltose.

16. Heparin Elution Buffer: 20 mM HEPES-KOH, pH 7.9, 100 mM KCl, 0.5 mM dithiothreitol, 0.2 mM phenylmethylsulfonyl fluoride, 15 % glycerol.

2.3 Photocross-linking Reactions and Affinity Isolation of Cross-Linked Products

1. MS2-binding site containing pre-mRNA.

2. MS2-MBP fusion protein.

3. 5× splicing buffer (300 mM potassium phosphate [pH 7.4], 15 % PEG8000, 10 mM spermidine).

4. Splicing extracts.

5. Parafilm.

6. Metal block.

7. CL-1000 Ultraviolet Crosslinker (UVP) with 365-nm UV light tubes.

8. Amylose Resin (New England Biolabs).

9. G-150 buffer (20 mM HEPES-KOH [pH 7.9], 1.5 mM MgCl$_2$, 150 mM KCl).

10. G-150 wash buffer (20 mM HEPES-KOH [pH 7.9], 1.5 mM MgCl$_2$, 150 mM KCl, 5 % glycerol, 0.01 % NP40).

11. 2× SDS-PAGE loading buffer (0.125 M Tris–HCl [pH 6.8], 4 % SDS, 10 % 2-mercaptoethanol, 0.004 % bromophenol blue, 20 % glycerol).

2.4 Detection and Identification of the Crosslinked Products

1. Anti-HA11 monoclonal antibody (Covance).

2. Anti-Prp28p polyclonal antibody (Chang lab).

3. Anti-Prp8p polyclonal antibody (gift from Dr. S.-C. Cheng).

4. HRP-conjugated anti-rabbit IgG (Invitrogen).

5. HRP-conjugated anti-mouse IgG (Invitrogen).

6. Immobilon Western Chemiluminescent HRP Substrate (Millipore).

7. SuperSignal West Femto Maximum Sensitivity Substrate (Thermo).

8. Fujifilm Super RX-N Blue X-Ray Film (Fuji).

9. FOCUS™-FAST Silver Kit (G-Biosciences).

10. Restore™ Western Blot Stripping Buffer (Thermo).

11. RNase A (20 mg/mL) (Sigma).

12. G150 elution buffer (20 mM HEPES-KOH [pH 7.9], 1.5 mM MgCl$_2$, 150 mM KCl, 5 % glycerol, 0.01 % NP40, 12 mM maltose).

13. Trichloroacetic acid to 10 % (v/v).

3 Methods

3.1 Plasmid and Strain Construction

3.1.1 Site Selection for BPA Replacement

For immuno-detection of Prp28p and its crosslinked products, yeast strains capable of expressing Prp28p with specific BPA replacements were first constructed. There are several considerations for choosing which amino-acid residues for BPA replacement. We first selected hydrophilic amino-acid residues that are not strictly conserved among DExD/H-box proteins, arguing that they are more likely to situate on Prp28p's surface and that their BPA replacements are less likely to significantly impact on Prp28p's function. We then used the structural information of Vasa [22], another DExD/H-box protein, to computationally model Prp28p structure and to guide our final selections.

3.1.2 Plasmids for Expressing BPA-Containing Prp28p In Vivo

Plasmid pCA8093 (2-µm/*LEU2*) was used to express a tagged version of Prp28p, which contains an HA epitope at its C-terminus. Expression of *PRP28-HA* on pCA8093 is driven by a strong promoter from the *TDH3* gene, instead by the native *PRP28* promoter. We employed standard PCR-based site-specific mutagenesis on pCA8093 to alter the selected codons within *PRP28* open-reading-frame (ORF) into TAG (amber) stop codon (*see* **Note 3**).

To enable BPA replacement at the designated position as defined by the introduced TAG stop codon within *PRP28* ORF, pLH157 (*TRP1*) [17, 18] was used. This plasmid harbors genes encoding the *E. coli* nonsense suppressor tRNA and its cognate aminoacyl tRNA synthetase, respectively.

3.1.3 Construction of Yeast Strains

Yeast strains derived from BY4705 (*MATα ade2Δ::hisG his3Δ200 leu2Δ0 lys2Δ0 met15Δ0 trp1Δ63 ura3Δ0*) or BJ2168 (*MAT a leu2 trp1 ura3-52 prb1-1122 pep4-3 prc1-407 gal2*) can both support photo-crosslinking experiments. BY4705 is a designer deletion strain derived from the S288C strain background [23] and BJ2168 is a protease-deficient designer strain [24]. All the derived strains contain the following three key features: (1) chromosomal deletion of *PRP28* (i.e., *prp28Δ::kanMAX4*); (2) a pCA8093-derived plasmid as the sole provider for Prp28p, in which a designated amino-acid residue is to be replaced by BPA; (3) pLH157 (*TRP1*). Construction of these strains was done as follows. A haploid *prp28Δ* yeast strain containing pCA8009 (*PRP28/CEN/URA3*) was co-transformed by pLH157 and a version of the pCA8093-derived plasmid. Transformants were initially selected on BPA (0.25 mM)-containing solid medium devoid of tryptophan and leucine. Preparation of BPA-containing medium was described [17, 18] (*see* **Note 2**). Transformants were then picked and streaked for single colonies on BPA medium containing 5-fluoroorotic acid (5-FOA) [25], but again without tryptophan and leucine. In this procedure, 5-FOA served to counterselect pCA8009, thus affording isolation of yeast cells that produced a version of BPA-modified Prp28p in the absence of the original wild-type Prp28p (*see* **Note 4**). In what follows, we will describe experiments using a yeast strain expressing a version of Prp28p (i.e., Prp28p-K136[BPA]) in which the lysine residue at position 136 was replaced by BPA.

3.2 Extract, RNA, and Protein Preparations

3.2.1 Preparation of Splicing Extracts and In Vitro Splicing Assay

Yeast cells were typically grown in 2 L of YPD containing 0.25 mM BPA at 30 °C to 3–4 OD_{600}. Yeast splicing extracts from BJ2168-derived strains were prepared according to the standard dounce protocol [26, 27] or by liquid-nitrogen-grinding method [27, 28]. Splicing extracts from BY4705-derived strains were prepared by the liquid-nitrogen-grinding method. All extracts were assayed for their ATP-dependent splicing activity using a standard splicing assay [26, 27]. Extracts exhibiting robust splicing activity at 2 mM

ATP, but no splicing activity at 0.05 mM ATP, were chosen for subsequent experiments [29].

3.2.2 Production of Aptamer-Tagged Pre-mRNAs and MS2-MBP Protein

The transcription template for producing an actin transcript tagged with three MS2 RNA aptamers, which was derived from pAdML-M3 [21], has been documented [14]. The actin M3-Act-wt pre-mRNA thus produced contains three binding sites for MS2 coat protein at its 5′ end (120 nucleotides), which is followed by actin exon 1 (88 nucleotides), intron (309 nucleotides), and truncated exon 2 (169 nucleotides). For producing actin transcript variants, we further constructed four transcription templates: (1) M3-Act-5′ss is mutated at the 5′ splice site (GUAUGU to AUUUGU); (2) M3-Act-BPS harbors a branch-site mutation (UACUAAC to UACUACC); (3) M3-Act-3′ss has changes at the 3′ splice site (AGAG to ACAC); and (4) M3-Act-Δi bears a precise intron deletion (*see* **Note 5**).

Non-radioactively labeled RNA transcripts can be synthesized using Ribo MAX™ Large Scale RNA Production System-SP6. Briefly, 5 μg of purified *Sac*I-linearized plasmid DNA is first treated with DNA polymerase I Large (Klenow) Fragment for converting the 3′ overhangs into blunt ends and used in a large-scale SP6-transcription reaction. DNA template is then removed by DNaseI digestion. The produced RNA transcripts are isolated by phenol/chloroform extraction, ethanol precipitation, and dissolved in 20 μL of nuclease-free water. The sample is then electrophoresed on an 8-M-urea-5 % polyacrylamide gel (29:1). After electrophoresis, a gel slice containing the full-length transcript is excised and placed into 400 μL of RNA elution buffer in a microfuge tube. The position of the full-length transcript in the gel can be determined according to the location of migration of a [^{32}P]-labeled transcript, which served as a marker on the same gel. Elution of the RNA transcript is achieved by shaking the microfuge tube at 37 °C overnight. The next day, the mixture is extracted with 350 μL of phenol/chloroform solution. After centrifugation, the aqueous layer is recovered and mixed with 700 μL of 100 % ethanol and incubated at –80 °C for 1 h. The precipitated RNA is then collected by centrifugation, washed with 1 mL of 70 % ethanol, vacuum dried, and dissolved in 50 μL of nuclease-free water. The purified RNA is adjusted to 10 ng/μL and stored at –80 °C for later use.

MS2-MBP fusion protein is overexpressed in *E. coli* and purified [30]. Standard molecular-biology protocol is used to transform the MS2-MBP-expressing plasmid (1 μl; 50 ng/μl) into the chemically competent BL21 Rosetta cells (50 μl). After heat shock for 1 min at 42 °C, cells are placed on ice, and 800 μL of LB medium was added into the microfuge tube, which is then incubated at 37 °C with gentle shaking for 1 h. A 400-μL aliquot is removed and inoculated into 25 mL of LB containing ampicillin (75 μg/mL) and chloramphenicol (30 μg/mL) and grown overnight at 37 °C. The next day, the 25-mL pre-culture is added into 1 L of LB containing both ampicillin and chloramphenicol for growth to 0.3–0.6 OD_{600} (about 2–3 h). To induce for MS2-MBP expression, 400 μL of 0.25 M IPTG is added and the culture is grown for 2.5 h. All procedures thereafter are done at 4 °C. Cells are harvested, washed with ice-cold water, transferred into a 50-mL Falcon tube, and recollected. The cell pellet is resuspended in 25 mL of Buffer A and cells are disrupted using a Constant Cell Disruption System. Cell debris and particulates are removed by low-speed centrifugation at 4000 rpm ($1935 \times g$) for 15 min. The resulting supernatant is recovered and aliquoted into pre-chilled microfuge tubes, which are centrifuged at 13,000 rpm ($16,000 \times g$) in a microcentrifuge for 5 min. The cell lysate is then carefully pooled for subsequent purification of the MS2-MBP fusion protein. Three columns (25 mL), each containing 2 mL (bed volume) of Amylose Resin, are packed and pre-equilibrated with 20 mL of Buffer A. The pooled lysate is split into three aliquots and added separately into three columns for mixing with the amylose beads overnight by rotation in a head-over-tail fashion. The next day, the columns are drained by gravity, each washed first with 20 mL of Wash Buffer, and then with 20 mL of 5 mM sodium phosphate (pH 7.0). Elution of the MS2-MBP protein is done by adding 4.0 mL of Amylose Elution Buffer into each column. The eluant is then applied into 1 mL (bed volume) of Heparin Sepharose, which is packed in a 10-mL column and pre-equilibrated in 10 mL of 5 mM sodium phosphate (pH 7.0). The flow-through fraction is then re-applied in a sequential manner into two subsequent Heparin Sepharose columns. The three columns are each washed with 10 mL of 5 mM sodium phosphate (pH 7.0). To elute the bound MS2-MBP protein, 500 μL of Heparin Elution Buffer is added to each column. A total of ten fractions, each consisting of 2–3 drops of the eluant are collected from each column. MS2-MBP proteins should be eluted mostly in the first five fractions.

3.3 Photocross-linking Reactions and Affinity Isolation of Cross-Linked Products

3.3.1 Spliceosomal Assembly

1. MS2-MBP was pre-bound to M3-Act-wt transcript in a 40-μL reaction consisting of 20 μL (490 fmole) of M3-Act-wt transcript (in nuclease-free water) and 20 μL (15.6 pmole) of MS2-MBP (in Heparin Elution Buffer). The binding reaction was incubated on ice for 30 min.

2. Spliceosome was then assembled onto the MS2-MBP-bound M3-Act-wt transcript in a 200-μL reaction by sequential addition of 20 μL of nuclease-free H_2O (Promega), 40 μL of 5× splicing buffer containing various concentrations of ATP (*see* below), and 100 μL of splicing extract. All reagents were kept on ice before use. After gentle mixing by pipeting, the reaction was incubated at 25 °C for 30 min.

3. Final concentrations of the key components in the spliceosomal assembly reaction are as follows: 2.45 nM M3-Act-wt transcript,

78 nM MS2-MBP, 50 % yeast splicing extract, 60 mM potassium phosphate (pH 7.4), 3 % PEG8000, and 2 mM spermidine. Note that the final concentration of ATP may vary from 0, 0.02, 0.05, 0.2, to 2 mM for the purpose of enriching a particular stage of spliceosomal assembly [14, 29] (*see* **Note 6**).

3.3.2 UV Crosslinking

1. After spliceosomal assembly, the reaction content was transferred onto the surface of a Parafilm lying on top of a metal block, which was in turn settled in a bucket of ice.

2. The whole setup was then moved into a CL-1000 Ultraviolet Crosslinker. The reaction content (i.e., the 200-μL droplet) was placed directly under the 365-nm UV light tubes, which were about 5 cm above the Parafilm surface.

3. The sample was UV irradiated for 20 min.

3.3.3 Affinity Purification of the Assembled Spliceosome

1. A large aliquot (50 μL for four reactions) of amylose resin was pre-washed once in excessive volume of G-150 buffer. After centrifugation and removal of supernatant, the resin was resuspended in 1 mL of G-150 buffer and aliquoted (230 μL each) into microfuge tubes. After brief centrifugation, the supernatant was removed to yield a bed volume of 10-μL resin (by calculation) in each tube.

2. The UV-irradiated sample (200 μL) was transferred into a resin-containing tube, which was then placed on a mixing platform rotating in a 360° manner for 1–2 h at 4 °C.

3. After binding, the resin was washed four times with ice-cold G-150 wash buffer which contains glycerol and NP40. Each wash consisted of the following steps: addition of 1 mL of wash buffer, tube inversion for four times, brief centrifugation at 4 °C, and removal of supernatant. Tubes were kept on ice between each step of four washes.

4. The washed resin was resuspended in 20 μL of 2× SDS-PAGE loading buffer and boiled for 5 min. After centrifugation at top speed for 2 min, 20 μL of supernatant was removed and resolved by SDS-7.5 % PAGE.

3.4   Detection and Identification of the Crosslinked Products

3.4.1 Immuno-Detection of Prp28p-Crosslinked Products

1. Standard Western blotting analysis was used to detect Prp28p and its covalently crosslinked products. In a typical Western procedure, anti-Prp28p polyclonal antibody (1:2000 dilution) and HRP-conjugated anti-rabbit IgG (1:10,000 or 1:40,000 dilution, depending on the HRP substrate used subsequently) were employed. Alternatively, mouse monoclonal anti-HA11 antibody (1:2000 dilution) and HRP-conjugated anti-mouse IgG (1:10,000 or 1:40,000 dilution) were used.

2. Immobilon Western Chemiluminescent HRP Substrate was chosen for developing the Western blot when the secondary

antibody was used in 1:10,000 dilution. For more sensitive detection, SuperSignal West Femto Maximum Sensitivity Substrate was employed along with 1:40,000 dilution of the secondary antibody. Fujifilm Super RX-N Blue X-Ray Film was used for detecting chemiluminescence. The exposure time was typically from 5 to 10 min.

3. Figure 1a shows a representative result from the Prp28p-K136[BPA] experiment.

3.4.2 Demonstration of Splicing-Dependent Crosslinking Events

1. To establish the splicing relevance of the crosslinking event, two criteria must be met. The first is to show that the appearance of the crosslinked product is ATP-dependent. This was done by repeating the experiment using different concentrations of ATP (Fig. 1b). In this case, very little crosslinked

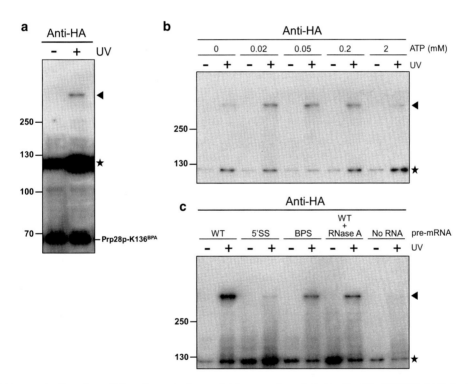

Fig. 1 Demonstration of a splicing-dependent crosslinking event mediated by Prp28p-K136[BPA]. (**a**) Detection of a UV-dependent crosslinked product. Prp28p, which was tagged by HA, and its crosslinked product (arrowhead; top band), were both immuno-detected by anti-HA11 antibody. A cross-reacted protein by anti-HA11, which appeared in reaction with (+) or without (−) UV irradiation, is marked (*). Pre-stained molecular weight standards are indicated to the left. (**b**) Detection of ATP-dependent crosslinked product. Spliceosomal assembly reactions with ATP concentrations ranging from 0 to 2 mM were performed for detecting the crosslinked product. (**c**) Detection of splicing-substrate-dependent crosslinked product. UV-crosslinking experiments using M3-Act pre-mRNA variants are shown. M3-Act-wt transcript (WT); M3-Act-5'ss transcript (5'SS); M3-Act-BPS transcript (BPS); M3-Act-wt transcript with RNase A addition (WT + RNase A); and no addition of M3-Act transcript (No RNA)

product was detected in the absence of ATP and in the presence of 2 mM ATP. The former is consistent with the fact that, without ATP, spliceosome assembly does not go beyond binding of U1 snRNP to pre-mRNA, which is an ATP-independent process. The latter is in line with our observation that, at 2 mM ATP, Prp28p is no longer associated with the spliceosome (our unpublished data).

2. The second criterion for establishing the splicing relevance is to show that the signal of the crosslinked product is significantly reduced in reactions with no added pre-mRNA or with pre-mRNAs containing intron splice-site mutations. We developed four different transcript variants (described above) to achieve this goal. The results from such an experiment employing M3-Act-5'ss and M3-Act-BPS transcripts are shown (Fig. 1c). Furthermore, addition of 2 μL of RNase A (20 mg/mL) did not abolish the crosslinked signal, arguing that the observed crosslinking event was unlikely due to an indirect interaction bridged by RNA.

3.4.3 Identification of the Crosslinked Product

1. To identify the crosslinked product, we scaled up the spliceosomal assembly reaction from 200 μL to 20 mL in a 50-mL Falcon tube. The content was then transferred into a plastic petri dish (100 mm dia.) for UV irradiation. Subsequent scale-up affinity purification of spliceosome was also done in a 50-mL Falcon tube. The eventual washed resin was resuspended in 1 mL of wash buffer and transferred into a 2-mL Eppendorf tube. After brief centrifugation and removal of the wash buffer, 10 μL of RNase A (20 mg/mL) was added to the resin. After brief vortexing to thoroughly mixing the content, the tube was incubated at 37 °C for 30 min.

2. To enhance recovery of the crosslinked product, the resin-bound proteins were sequentially eluted three times with G150 elution buffer containing maltose. Each elution consisted of the following steps: addition of 1 mL of G150 elution buffer, vigorous vortexing for 5 min at room temperature, brief centrifugation, and removal of the eluant. Proteins in the pooled eluant were precipitated by adding trichloroacetic acid to 10 % (v/v) and then separated by SDS-7.5 % PAGE.

3. To control for experimental background, two samples, one from the UV-irradiated reaction and another from the mock (i.e., no UV irradiation) reaction, were loaded in two separate lanes on the same gel for SDS-PAGE. After silver-staining the gel (*see* **Note 7**), regions corresponding to the location of the crosslinked product in both lanes were cut out separately and subjected to standard mass-spectrometry analysis. In this case, comparison of the two sets of data revealed that the protein crosslinked to Prp28p-K136[BPA] is most likely Prp8p, a splicing factor situated at the heart of the spliceosome.

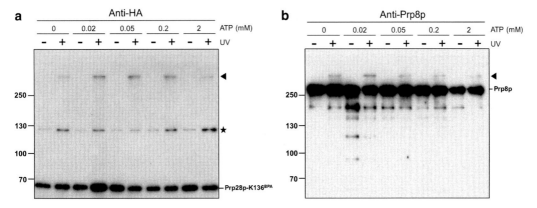

Fig. 2 Identification of Prp8p as an interaction partner with Prp28p-K136^{BPA}. (**a**) Shows an expanded image from the experiment described in Fig. 1b, with signals corresponding to Prp28p-K136^{BPA} alone close to the bottom included. (**b**) Validation of Prp8p within the crosslinked product. To authenticate the prediction by mass spectrometry analysis that Prp8p is present in the Prp28p-K136^{BPA}-crosslinked product, the blot used in (**a**) was stripped off signals and antibodies and re-probed with anti-Prp8p antibody. The presence of a band co-migrating exactly with the Prp28p-K136^{BPA}-crosslinked product and above the Prp8p signals confirms that Prp28p-K136^{BPA} and Prp8p are covalently linked in a single species

4. To confirm the mass-spectrometry prediction, we performed Western blotting analysis using a series of crosslinking reactions done in various ATP concentrations (Fig. 2). The blot was initially reacted with anti-HA11 antibody to visualize Prp28p-HA and its crosslinked product (Fig. 2a). After stripping off the original signals and antibodies (*see* **Note 8**), the same blot was reacted with anti-Prp8p antibody (1:2000) to probe for the presence of Prp8p. The majority of Prp8p, which has a calculated molecular mass of 280 kD, was found to migrate at ~260 kD as reported previously [31] (Fig. 2b). Importantly, a minor fraction of Prp8p was found to precisely co-migrate with the crosslinked product (Fig. 2b), establishing that Prp28p indeed physically interacts with Prp8p in a splicing-dependent manner.

4 Notes

1. Information regarding vendors and suppliers provided herein is for convenience rather than a strict requirement. Many commonly used reagents are available from various reliable sources.

2. The composition and preparation of standard YPD and synthetic complete (SC) media, either with or without agar, are well documented [32]. To prepare BPA-containing medium, dissolve 67.4 mg of BPA in 10 mL of 1 N HCl, then transfer the solution into 1 L of autoclaved and pre-cooled medium.

The introduced acidity is then neutralized by addition of 10 mL of 1 N NaOH into the medium. For plasmid-selection purpose, specific amino acids in the SC medium are dropped out according to the selectional markers of the plasmids present in the yeast strain used in the study. To prepare agar plates that simultaneously contain both BPA and 5-FOA, the following procedures are used. In Flask 1, seven grams of yeast nitrogen base (without amino acids, but with ammonium sulfate) is dissolved in 360 mL of water. In Flask 2 (2 L by volume), 20 g of Bacto-agar is placed in 500 mL of water. Flasks 1 and 2 are autoclaved separately. Flask 3 contains the following ingredients: 2 g of 5-FOA, 10 mg of adenine sulfate, 100 mg of uracil, and 100 mL of 10× amino-acid mix (*see* below). A stir bar is dropped into Flask 3, which is placed on a stirring hot plate for continuous mixing at a temperature between 75 and 80 °C until all compounds are dissolved into solution. The Flask 3 solution is filter-sterilized before use. Solutions from Flasks 1 and 3 are poured into Flask 2, in which 40 mL of sterile 50 % glucose is also added. The content in Flask 2 is then allowed to cool down and then mixed by gentle hand-swirling of the flask without causing air bubbles. Finally, 10 mL of BPA (in 1 N HCl) was added and the acidity is neutralized as described above. The filter-sterilized 10× amino-acid mix (100 mL) is prepared according to the standard recipe for making SC medium, except that the concentrations for all the needed amino acids are increased by tenfolds.

3. The introduced suppressor tRNA can only decode TAG stop codon. As a result, the native stop codon for *PRP28*, which is TAA, cannot be recognized by the BPA-charged suppressor tRNA. For genes which unfortunately end with TAG stop codon, changing it into either TAA or TGA is needed prior to altering any selected internal codon into TAG. For our study, we generated a total of 42 TAG-containing versions of *PRP28* for strain constructions.

4. Strains that fail to grow or have significant growth defect upon 5-FOA counter-selection were discarded. We chose to work only with strains that exhibit a growth rate comparable to that of the isogenic parental strain. Depending on the fitness of strain, the later immuno-detection of Prp28p-crosslinked product may be improved by using SuperSignal West Femto Maximum Sensitivity Substrate (Thermo), a more sensitive reagent.

5. The biochemical consequences for using various M3-Act transcripts are as follows: (1) M3-Act-5′ss completely blocks spliceosomal formation; (2) M3-Act-BPS partially inhibits spliceosomal assembly (*see* below); (3) M3-Act-3′ss permits the formation of an active spliceosome that can only execute

the first chemical step; and (4) M3-Act-Δi, because it is intronless, cannot form spliceosome [33]. The intron in our M3-Act transcripts (5′ss, BPS, and 3′ss) contains a native branch-site-like sequence UACUA<u>A</u><u>G</u> just upstream of the authentic branch site UACUA<u>A</u><u>C</u>. The UACUA<u>A</u>G sequence can substitute, albeit inefficiently, for the authentic UACUA<u>A</u>C box in the splicing process [34]. As a result, our version of M3-Act-BPS, in which the authentic branch site was mutationally inactivated, still permitted some spliceosomal assembly, although to a lesser extent, and thus yielding a reduced crosslinking signal.

6. Varying ATP concentration in splicing reaction is a common practice for selectively enriching specific forms of the splicing machinery along the spliceosomal assembly pathway [14, 29]. At 0.02 mM ATP, spliceosomal assembly is advanced mostly only to the U1/U2-bound stage. Whereas at 0.05 mM ATP, the predominant form of spliceosome contains all five snRNPs, but remains pre-catalytic. At 2 mM ATP, spliceosome progresses to a stage by which U1 and U4 are no longer present and the spliceosomal catalytic center is established for executing both steps of the splicing chemistry. By varying ATP concentration, we have shown that Prp28p only transiently interacts with spliceosome during splicing. At 2 mM ATP, Prp28p is no longer associated with the spliceosome, suggesting that its primary action occurs prior to the catalytic steps.

7. We have used the FOCUS™-FAST Silver Kit from the G-Biosciences for gel staining and de-staining. Typically, gels are placed in a sterile petri dish and soaked in 20 mL of fixing solution (30 % ethanol and 10 % acetic acid) for 30 min, washed twice with 10 % ethanol, each for 5 min, and washed three times with double-distilled water for 5 min each. Gels are then soaked in diluted FOCUS™ Silver Stain for 25 min, rinsed with double-distilled water for 15 s. The rinsed gel is then soaked in FOCUS™ Developer and placed on a rotation table for gentle rocking until the desired bands are visualized. To stop the developing reaction, the gel is transferred into a new petri dish and soaked in 20 mL of 2 % acetic acid for 10 min. A gel slice corresponding to the desired band is excised from the gel using a clean razor blade, transferred into a microfuge tube, washed three times with 1 mL of double-distilled water for 5 min each. To the washed gel slice, 60 µL of SilverOUT™ reagent is added into the tube, vortexed for 10 s, and then incubated for 10 min at the room temperature. The de-stained gel slice is further washed by 1 mL of double-distilled water, vortexed, and incubated for 10 min. After removing the water from the tube, the completely de-stained gel slice is sent for mass spectrometry analysis.

8. We use Restore™ Western Blot Stripping Buffer (Thermo) for blot stripping. Blot is placed in 10 mL of Restore™ Western Blot Stripping Buffer and incubated at room temperature with gentle shaking for 10 min. After that, the blot is washed twice in 20 mL of TBST buffer (50 mM Tris–HCl [pH 7.6], 150 mM NaCl, 0.05 % Tween 20) for 5 min each. The stripped blot is then tested for proper stripping by re-probing with the secondary antibody. If no signal is detected, the blot can then be re-employed for another primary antibody.

Acknowledgments

We thank C.-M. Lin and L.-C. Chang for assistances during the earlier phase of the project; R. Reed for providing pAdML-M3; J. Vilardell for pMS2-MBP; H.-T. Chen for pLH157 and BY4705 strain; S.-C. Cheng for anti-Prp8p and BJ2168 strain; and H.-T. Chen and S.-C. Cheng for insightful discussions. This project has been supported by grants from Ministry of Science and Technology (101-2311-B-001-005 and 102-2311-B-001-029), Thematic Projects (Academia Sinica; AS-99-TP-B20 and AS-103-TP-B12), and Academia Sinica.

References

1. Tanner NK, Linder P (2001) DExD/H box RNA helicases: from generic motors to specific dissociation functions. Mol Cell 8(2):251–262

2. Bleichert F, Baserga SJ (2007) The long unwinding road of RNA helicases. Mol Cell 27(3):339–352

3. Chang TH, Tung L, Yeh FL, Chen JH, Chang SL (2013) Functions of the DExD/H-box proteins in nuclear pre-mRNA splicing. Biochim Biophys Acta 1829(8):764–774. doi:10.1016/j.bbagrm.2013.02.006

4. Jankowsky E (2011) RNA helicases at work: binding and rearranging. Trends Biochem Sci 36(1):19–29. doi:10.1016/j.tibs.2010.07.008, S0968-0004(10)00141-6 [pii]

5. Gutell RR, Larsen N, Woese CR (1994) Lessons from an evolving rRNA: 16S and 23S rRNA structures from a comparative perspective. Microbiol Rev 58(1):10–26

6. Staley JP, Guthrie C (1999) An RNA switch at the 5′ splice site requires ATP and the DEAD box protein Prp28p. Mol Cell 3(1):55–64

7. Zhang D, Rosbash M (1999) Identification of eight proteins that cross-link to pre-mRNA in the yeast commitment complex. Genes Dev 13(5):581–592

8. Schwer B (2001) A new twist on RNA helicases: DExH/D box proteins as RNPases. Nat Struct Biol 8(2):113–116

9. Will CL, Luhrmann R (2001) RNP remodeling with DExH/D boxes. Science 291(5510):1916–1917

10. Chen JY, Stands L, Staley JP, Jackups RR, Latus LJ, Chang T-H (2001) Specific alterations of U1-C protein or U1 small nuclear RNA can eliminate the requirement of Prp28p, an essential DEAD box splicing factor. Mol Cell 7(1):227–232

11. Jankowsky E, Gross CH, Shuman S, Pyle AM (2001) Active disruption of an RNA–protein interaction by a DExH/D RNA helicase. Science 291(5501):121–125

12. Hage R, Tung L, Du H, Stands L, Rosbash M, Chang TH (2009) A targeted bypass screen identifies Ynl187p, Prp42p, Snu71p, and Cbp80p for stable U1 snRNP/Pre-mRNA interaction. Mol Cell Biol 29(14):3941–3952. doi:10.1128/MCB.00384-09, MCB.00384-09 [pii]

13. Jurica MS, Moore MJ (2003) Pre-mRNA splicing: awash in a sea of proteins. Mol Cell 12(1):5–14

14. Fabrizio P, Dannenberg J, Dube P, Kastner B, Stark H, Urlaub H, Luhrmann R (2009) The evolutionarily conserved core design of the catalytic activation step of the yeast spliceosome. Mol Cell 36(4):593–608. doi:10.1016/j.molcel.2009.09.040

15. Wahl MC, Will CL, Luhrmann R (2009) The spliceosome: design principles of a dynamic RNP machine. Cell 136(4):701–718. doi:10.1016/j.cell.2009.02.009, S0092-8674(09)00146-9 [pii]

16. Staley JP, Guthrie C (1998) Mechanical devices of the spliceosome: motors, clocks, springs, and things. Cell 92(3):315–326

17. Chen HT, Warfield L, Hahn S (2007) The positions of TFIIF and TFIIE in the RNA polymerase II transcription preinitiation complex. Nat Struct Mol Biol 14(8):696–703. doi:10.1038/nsmb1272

18. Mohibullah N, Hahn S (2008) Site-specific cross-linking of TBP in vivo and in vitro reveals a direct functional interaction with the SAGA subunit Spt3. Genes Dev 22(21):2994–3006. doi:10.1101/gad.1724408

19. Chin JW, Cropp TA, Anderson JC, Mukherji M, Zhang Z, Schultz PG (2003) An expanded eukaryotic genetic code. Science 301(5635):964–967

20. Grainger RJ, Beggs JD (2005) Prp8 protein: at the heart of the spliceosome. RNA 11(5):533–557

21. Zhou Z, Licklider LJ, Gygi SP, Reed R (2002) Comprehensive proteomic analysis of the human spliceosome. Nature 419(6903):182–185

22. Sengoku T, Nureki O, Nakamura A, Kobayashi S, Yokoyama S (2006) Structural basis for RNA unwinding by the DEAD-box protein drosophila vasa. Cell 125(2):287–300. doi:10.1016/j.cell.2006.01.054

23. Brachmann CB, Davies A, Cost GJ, Caputo E, Li J, Hieter P, Boeke JD (1998) Designer deletion strains derived from Saccharomyces cerevisiae S288C: a useful set of strains and plasmids for PCR-mediated gene disruption and other applications. Yeast 14(2):115–132

24. Jones EW (1991) Tackling the protease problem in Saccharomyces cerevisiae. Methods Enzymol 194:428–453

25. Sikorski RS, Boeke JD (1991) In vitro mutagenesis and plasmid shuffling: from cloned gene to mutant yeast. Methods Enzymol 194:302–318

26. Lin R-J, Newman AJ, Cheng S-C, Abelson J (1985) Yeast mRNA splicing in vitro. J Biol Chem 260(27):14780–14792

27. Stevens SW, Abelson J (2002) Yeast pre-mRNA splicing: methods, mechanisms, and machinery. Methods Enzymol 351:200–220

28. Umen JG, Guthrie C (1995) A novel role for a U5 snRNP protein in 3′ splice site selection. Genes Dev 9:855–868

29. Tarn W-Y, Lee K-R, Cheng S-C (1993) Yeast precursor mRNA processing protein PRP19 associates with the spliceosome concomitant with or just after dissociation of U4 small nuclear RNA. Proc Natl Acad Sci U S A 90:10821–10825

30. Jurica MS, Moore MJ (2002) Capturing splicing complexes to study structure and mechanism. Methods 28(3):336–345

31. Stevens SW, Abelson J (1999) Purification of the yeast U4/U6.U5 small nuclear ribonucleoprotein particle and identification of its proteins. Proc Natl Acad Sci U S A 96(13):7226–7231

32. Sherman F (2002) Getting started with yeast. Methods Enzymol 350:3–41

33. Vijayraghavan U, Parker R, Tamm J, Iimura Y, Rossi J, Abelson J, Guthrie C (1986) Mutations in conserved intron sequences affect multiple steps in the yeast splicing pathway, particularly assembly of the spliceosome. EMBO J 5:1683–1695

34. Cellini A, Parker R, McMahon J, Guthrie C, Rossi J (1986) Activation of a cryptic TACTAAC box in the *Saccharomyces cerevisiae* actin intron. Mol Cell Biol 6:1571–1578

Chapter 16

Evolution of Cell-Type-Specific RNA Aptamers Via Live Cell-Based SELEX

Jiehua Zhou and John J. Rossi

Abstract

Live cell-based SELEX (Systematic Evolution of Ligand EXponential enrichment) is a promising approach for identifying aptamers that can selectively bind to a cell-surface antigen or a particular target cell population. In particular, it offers a facile selection strategy for some special cell-surface proteins that are original glycosylated or heavily post-translationally modified, and are unavailable in their native/active conformation after in vitro expression and purification. In this chapter, we describe evolution of cell-type-specific RNA aptamers targeting the human CCR5 by combining the live cell-based SELEX strategy with high-throughput sequencing (HTS) and bioinformatics analysis.

Key words Live cell-based SELEX, Cell-type-specific RNA aptamer, High-throughput sequencing (HTS), Human CCR5 protein, Aptamer-mediated targeted siRNA delivery

---

1 Introduction

Nucleic acid aptamers are single-stranded DNA or RNA molecules, which can be selected from a combinatorial DNA or RNA library through SELEX technology [1, 2]. Because of their superior characteristics, such as small size, high stability (dehydrated form), lack of immunogenicity, facile chemical synthesis, and adaptable modification, aptamer and aptamer-functionalized agents have been used extensively for targeted therapeutics, molecular diagnostics, in vivo imaging and tracking systems, biosensor systems, and bio-marker discovery [3–5]. To select and identify cell-type-specific, nucleic acid aptamers, two typically selection procedures have been applied [2]: (1) traditional purified membrane protein-based SELEX and (2) live cell-based SELEX. Although most cell-type-specific aptamers have been successfully generated using the purified protein-based selection procedure, it is largely limited to target some single pass receptors. It is challenging to develop aptamers for many multiple transmembrane spanning receptors, or some special cell-surface proteins that are original glycosylated or heavily post-translationally

Ren-Jang Lin (ed.), *RNA-Protein Complexes and Interactions: Methods and Protocols*, Methods in Molecular Biology, vol. 1421, DOI 10.1007/978-1-4939-3591-8_16, © Springer Science+Business Media New York 2016

modified, and are unavailable in their native/active conformation after in vitro expression and purification. Live cell-based SELEX techniques have provided a promising alternative for generating aptamers that can recognize particular target membrane proteins under their native conditions [6, 7].

In principle, live cell-based SELEX methodology relies on the differences between the target cells population (positive cells that express the target of interest) and the control cell population (negative cells that do not express the target protein). Therefore, the selection procedure generally includes positive selection against the target cells and counter selection against related non-target cells to remove nonspecific binding [8]. For example, by using this method, Giangrande group successfully enriched for 2′-fluoropyrimidine RNA aptamers that could selectively bind and be internalized by breast cancer cells that expressed human EGFR2 (HER2) [9].

One of the major challenges of cell-based SELEX is the nonspecific binding and/or uptake of nucleic acids to dead cells, resulting in the delay of target-specific sequence enrichment or even failure of aptamer selection [2, 5]. Selection procedures, especially those that adversely affect cell health and damage cells, such as cell culture, washing condition, and detachment of cells, should be carefully controlled in order to avoid causing damage or death to fragile cells. On the other hand, the complexity of target cell-surface proteins or low expression of the desired targets generally increases the number of selection rounds needed (~14 cycles and up to 25 cycles) to achieve a high-affinity aptamers. The additional enzymatic amplification steps can also introduce biases and artifacts. Previous studies have demonstrated that once maximal molecular enrichment is achieved, further selection rounds not only increase the cost and time but also attenuate the efficiency of the selection as well as the efficacy of the aptamers [2]

In this chapter, we describe detailed practical procedures for live cell-based SELEX. Considering the limited resource and the risk of changed conformation of purified CCR5 protein after purification, we utilized live cell-based SELEX methodology for generating cell-type-specific 2′-fluoropyrimidine RNA aptamers directed to human CCR5. Particularly, the tricky steps and problems often encountered in the selection are highlighted and discussed in detail to improve the success and efficiency of the SELEX.

2 Materials

2.1 Cell Culture and Cell-Surface Target Protein Detection by Flow Cytometry

1. U373-Magi cells, U373-Magi-CCR5E cells, CEM-NKr cells, CEM-NKr-CCR5 cells (AIDS Research and Reference Reagent Program).

2. Complete medium for U373-Magi cells: 90 % DMEM (Dulbecco's Modified Eagle Medium) supplemented with

10 % FBS (Fetal Bovine Serum), 0.2 mg/mL G418 and 0.1 mg/mL hygromycin B. Store at 4 °C.

3. Complete medium for U373-Magi-CCR5E cells: 90 % DMEM supplemented with 10 % FBS, 0.2 mg/mL G418, 0.1 mg/mL hygromycin B and 1.0 μg/mL puromycin.

4. Cell stripper solution (Cellgro, Mediatech Inc.). Store at 4 °C.

5. Washing buffer: DPBS (Dulbecco's Phosphate Buffered Saline) (Cellgro, Mediatech Inc.). Store in aliquots at 4 °C.

6. Binding buffer: DPBS (pH 7.0–7.4) with Ca^{2+} and Mg^{2+} (pH 7.0–7.4, 1 mM $CaCl_2$; 2.7 mM KCl; 1.47 mM KH_2PO_4; 1 mM $MgCl_2$; 136.9 mM NaCl; 2.13 mM Na_2HPO_4) (Cellgro, Mediatech Inc.). Store in aliquots at 4 °C.

7. APC-conjugated mouse Anti-human CD195 antibody (BD Pharmingen). Store at 4 °C.

8. Flow cytometry instrument.

2.2 Preparation of RNA Library and Aptamer Selection

1. The starting DNA library contained 20 nucleotides of random sequences (synthesized and obtained commercially). Make stock solution with water and store in aliquots at –20 °C (*see* **Note 1**).

2. Taq PCR polymerase and buffer. Store in aliquots at –20 °C.

3. dNTP. Make 10 mM dNTP mixture and store in aliquots at –20 °C.

4. QIAquick Gel purification Kit (QIAGEN).

5. DuraScribe T7 transcription Kit (EPICENTRE Biotechnologies). Store at –20 °C (*see* **Note 2**).

6. DNase I. Store at –20 °C.

7. RNase free Bio-Spin 30 Columns (Bio-Rad). Store at 4 °C.

8. Acid phenol:chloroform (5:1) solution (pH 4.5, Ambion). Phenol and chloroform are human health hazards. Take appropriate measures to prevent exposure.

9. Chloroform:Isopropanol (24:1) solution.

10. Glycogen (20 mg/mL). Store in aliquots at –20 °C.

11. RNA refolding and binding buffer: DPBS (pH 7.0–7.4) with Ca^{2+} and Mg^{2+} (pH 7.0–7.4, 1 mM $CaCl_2$; 2.7 mM KCl; 1.47 mM KH_2PO_4; 1 mM $MgCl_2$; 136.9 mM NaCl; 2.13 mM Na_2HPO_4) (Cellgro, Mediatech Inc.). Store in aliquots at 4 °C.

12. Yeast tRNA. Make 10 mg/mL in water and store in aliquots at –20 °C.

13. TRIzol reagent (Invitrogen). This reagent containing phenol and guanidine is a human health hazard. Take appropriate measures to prevent exposure. Store in aliquots at 4 °C.

14. ThermoScript RT-PCR system (Invitrogen) (*see* **Note 3**).

15. 2× iQ SyberGreen MasterMix (BIO-RAD) for PCR. Store at −20 °C.

2.3 Illumina High-Throughput Sequencing (HTS) and Data Analysis

1. The relevant PCR primer. Prepare 5 mM stock solution in water and store in aliquots at −20 °C (*see* **Note 4**).

2. Forty percent acrylamide/bis-acrylamide solution (AccuGel 19:1, National Diagnostics). This is a neurotoxin when unpolymerized. Take appropriate measures to prevent exposure.

3. *N,N,N,N'*-Tetramethyl-ethylenediamine (TEMED). Store at 4 °C.

4. Ammonium persulfate: prepare 10 % solution in water and immediately freeze in single use (150 μL) aliquots at −20 °C.

5. 10×TBE.

6. Running buffer (1×TBE): dilute 10×TBE with water. Store at room temperature.

7. Illumina HiSeq2000 instrument.

2.4 Flow Cytometry and Live-Cell Confocal Microscopy

1. Silencer siRNA Labeling Kit (Ambion). Store at −20 °C.

2. Trypsin/EDTA 1×solution. Store in aliquots at 4 °C.

3. Peripheral blood mononuclear cells (PBMCs) from healthy volunteers. Isolate PBMCs from whole blood by centrifugation through Ficoll-Hypaque solution (Histopaque-1077, Sigma) and deplete for CD8 T cells (cytotoxic/suppressor cells) using Dynabeads CD8 (Invitrogen, CA). Culture PBMCs in T-cell active medium (BioE) in a humidified 5 % CO_2 incubator at 37 °C.

4. 35 mm plates for live-cell confocal microscopy (Glass Bottom Dish, MatTek).

5. Poly-lysine.

6. Hoechst 33342 (Molecular Probes): prepare 0.15 mg/mL solution in water and store in aliquots at 4 °C.

7. Zeiss LSM 510 Meta Inverted 2 photon confocal microscope.

2.5 CCR5 Knockdown Experiment

1. The CCR5 siRNA and relevant PCR primer. Prepare 5 mM stock solution in water and store in aliquots at −20 °C (*see* **Note 5**).

2. Lipofectamine 2000 (Invitrogen).

3. STAT-60 reagent (TEL-TEST). This reagent containing phenol and guanidine is a human health hazard. Take appropriate measures to prevent exposure. Store in aliquots at 4 °C.

4. DNA-free DNA removal kit (Ambion). Store at −20 °C.

5. RNase inhibitor (Promega). Store at −20 °C.

6. Moloney murine leukemia virus reverse transcriptase (MML-RT) and random primers (Invitrogen). Store at –20 °C.

7. 2× iQ SyberGreen MasterMix (BIO-RAD) for PCR. Store at –20 °C.

8. CFX96 Touch™ Real-Time PCR Detection System (Bio-Rad).

3   Methods

By combining the live cell-based SELEX strategy with high-throughput sequencing (HTS) and bioinformatics analysis, 2′-fluoropyridine-modified RNA aptamers directed to human CCR5 can be successfully isolated. The high-throughput sequencing (HTS) technology and bioinformatics analysis facilitate the rapid identification of individual RNA aptamers and monitor the library evolution. Multiple SELEX libraries from different rounds of selection can simultaneously be analyzed by HTS. Millions of sequence reads obtained from each round can provide comprehensive and in-depth information, such as the basic sequences, total reads, the complexity of the library, the enrichment factor, frequency of each unique sequences, and distribution of each nucleotide at random region, which allow us to better understand the selection progression and the molecular evolution. One of the best candidates (G-3) efficiently binds CCR5 and is internalized into human CCR5 expressing U373-Magi-CCR5E cells, CEM-NKr-CCR5 cells, and primary PBMCs. The selection procedures, progression, and aptamer characterization are described in this section.

3.1 Detection of the Cell-Surface Target Protein (CCR5)

1. Wash the adherent cell lines (U373-Magi cells and U373-Magi-CCR5E cells) with pre-warmed DPBS, and detach cells with Cell stripper from the plates (*see* **Note 6**).

2. After washing cells twice with 500 μL binding buffer containing DPBS (pH 7.0–7.4) and Ca^{2+} and Mg^{2+}, resuspend cell pellets in binding buffer and incubate at 37 °C for 30 min. And then pellet cells and resuspend in 100 μL of pre-warmed binding buffer (*see* **Note 7**).

3. Add APC mouse anti-human CD195 antibody for cell-surface staining (*see* **Note 8**). After incubation at room temperature for 30 min in the dark, wash cells three times with 500 μL of pre-warmed binding buffer, and resuspend cells in 350 μL of DPBS pre-warmed to 37 °C and immediately analyze by flow cytometry. An example result is shown in Fig. 1.

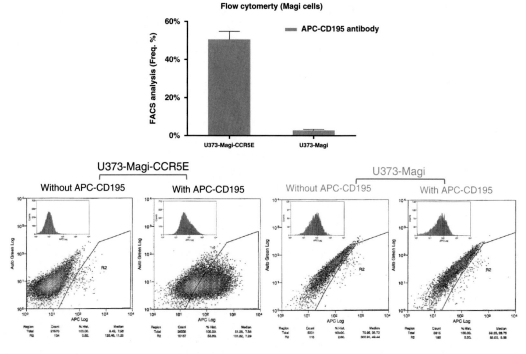

Fig. 1 Cell-surface expression of CCR5 on U373-Magi-CCR5E cells and U373-Magi cells was measured by flow cytometry using non-permeabilized cells stained with CCR5 antibody conjugated to APC (APC anti-human CD195). (Reprinted from [17], Figure S1)

3.2 Generation of RNA Library by In Vitro Transcription

1. Amplify the initial single-stranded DNA library pool (0.4 μM) by PCR using 3 μM each of 5′- and 3′-primers, along with 2 mM $MgCl_2$ and 200 μM of each dNTP. Conduct 10 PCR cycles as follows: 93 °C, 3 min (the initial step); 93 °C, 1 min, 63 °C, 1 min, 72 °C, 1 min; 72 °C, 7 min (the last step); 4 °C, keep! (*see* **Note 9**).

2. Recover the resulting PCR products using a QIAquick Gel purification Kit.

3. Transcribe RNA library from its PCR-generated DNA templates using the DuraScription Kit. In the transcription reaction mixture, replace CTP and UTP with 2′-F-CTP and 2′-F-UTP (*please refer to* Subheading 2) to produce RNA that is resistant to RNase A degradation. Set up a 20 μL transcription reaction, by mixing:

   (a) 1 μg of DNA template.

   (b) 2 μL 10×buffer.

   (c) 2 μL ATP (50 mM).

   (d) 2 μL GTP (50 mM).

   (e) 2 μL 2′-F-dCTP (50 mM).

   (f) 2 μL 2′-F-dUTP (50 mM).

   (g) 2 μL DTT (100 mM).

   (h) 2 μL T7 RNA polymerase.

Incubate at 37 °C for 6 h (*see* **Note 10**), and purify with Bio-Spin 30 Columns following phenol extraction and ethanol precipitation. And resuspend RNA pellet into water.

4. Remove the template DNA by DNase I digestion. Set up a 200 μL reaction, by mixing:

    (a) 10 μg of the transcribed RNA pool.

    (b) 20 μL 10× reaction buffer.

    (c) 5 μL DNase I.

    Incubate at 37 °C for 1 h (*see* **Note 11**). After phenol extraction and ethanol precipitation, resuspend RNA pellet into RNase-free water.

5. Refold RNA pool in binding buffer containing DPBS (pH 7.0–7.4) and Ca²⁺ and Mg²⁺, by heating mixture at 65 °C for 5 min and then slowly cooling to 37 °C. Continue incubation at 37 °C for 10 min.

3.3 Live Cell-Based SELEX

1. In principle, perform a SELEX procedure as described by Tuerk and Gold [10]. In this case, conduct a modified cell-based SELEX described by Thiel et al. [9, 11–13]. The detailed selection procedure and conditions are summarized in Fig. 2 and Table 1 (*see* **Note 12**).

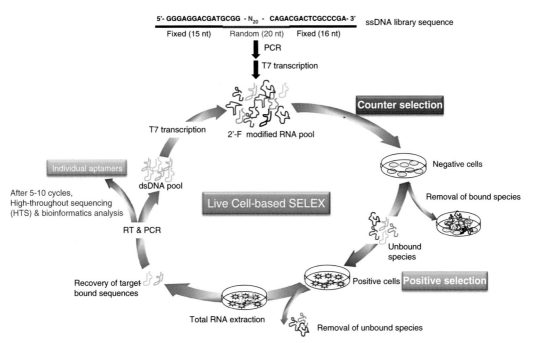

Fig. 2 Schematic of live cell-based SELEX procedure for evolution of RNA aptamers. It consists of four main steps: (1) counter selection by incubating library with negative cells that do not express the target protein; (2) positive selection by incubating recovered unbound sequences with positive cells expressing the target protein; (3) recovery of target-bound sequences; and finally (4) re-amplification of recovered species and make new RNA pool for next selection round. After 5–10 selection cycles, individual aptamer sequences are identified through barcode-based high-throughput, Illumina Deep Sequencing (HTS) and bioinformatics analysis. (Reprinted from [17], Fig. 1a)

Table 1
The selection condition. The numbers of cells, plate size, medium volume, the amount of RNA pool and tRNA, washing condition, and incubation time are indicated here. (Reprinted from [17], Table S1)

SELEX rounds	Positive cells (plate size and medium volume)	Negative cells (plate size and medium volume)	RNA pool (incubation time)	RNA work Con.	Competitor tRNA	Washing
1	3×10^6 cells (15 cm, 15 mL)	3×10^6 cells (15 cm, 15 mL)	4 nmol (30 min)	333 nM	0	2×12 mL
2	3×10^6 cells (15 cm, 15 mL)	3×10^6 cells (15 cm, 15 mL)	4 nmol (25 min)	333 nM	2.5 nmol	3×12 mL
3	1.5×10^6 cells (10 cm, 12 mL)	3×10^6 cells (10 cm, 12 mL)	2.5 nmol (25 min)	208 nM	5 nmol	4×12 mL
4	1.5×10^6 cells (10 cm, 12 mL)	3×10^6 cells (10 cm, 12 mL)	2.5 nmol (20 min)	208 nM	15 nmol	5×12 mL
5	7.5×10^5 cells (6 cm, 8 mL)	2.25×10^6 cells (10 cm, 8 mL)	1.5 nmol (20 min)	188 nM	15 nmol	6×8 mL
6	7.5×10^5 cells (6 cm, 8 mL)	2.25×10^6 cells (10 cm, 8 mL)	1.5 nmol (15 min)	188 nM	20 nmol	6×8 mL
7	3×10^5 cells (3.5 cm, 5 mL)	1.5×10^6 cells (6 cm, 5 mL)	0.8 nmol (15 min)	160 nM	20 nmol	7×5 mL
8	3×10^5 cells (3.5 cm, 5 mL)	1.5×10^6 cells (6 cm, 5 mL)	0.8 nmol (10 min)	160 nM	40 nmol	8×5 mL
9	1.5×10^5 cells (3.5 cm, 5 mL)	1.2×10^6 cells (6 cm, 5 mL)	0.5 nmol (10 min)	100 nM	40 nmol	9×5 mL

Note
To avoid nonspecific interaction between nucleic acids and the cell surface, the yeast tRNA (100 μg/mL) as a competitor was first incubated with non-targeted cells or targeted cells at 37 °C for 25 min and then ready for selection step
Counter-selection: RNA pool was incubated with U373-Magi negative cells at 37 °C for 30 min

2. For the first cycle of selection, 24 h before selection, seed U373-Magi cells (CCR5-negative) and U373-Magi-CCR5E cells (CCR5-positive) at equal density (5×10^6 cells per plate, respectively) on 150 mm tissue culture dish with 25 mL complete culture medium.

3. On the day of selection, wash U373-Magi cells and U373-Magi-CCR5E cells with 15 mL pre-warmed binding buffer to remove dead cells, and add 15 mL pre-warmed binding buffer supplemented with 100 μg/mL yeast tRNAs (*see* **Note 13**).

4. After 25 min incubation at 37 °C, remove the buffer from U373-Magi cells and add the refolded, initial RNA library (4 nmol of 0-RNA library in 15 mL refolding buffer). Incubate at 37 °C for 30 min and collect the supernatants from U373-Magi cells. Discard the U373-Magi cells (*see* **Note 14**).

5. Following **step 4**, after 25 min incubation at 37 °C, remove the buffer from U373-Magi-CCR5E cells and add the supernatants

from U373-Magi cells (15 mL from **step 5**). Incubate at 37 °C for 30 min. Wash the U373-Magi-CCR5E cells twice with 12 mL pre-warm binding buffer to remove unbound sequences and cell-surface RNAs with weak binding (*see* **Note 15**).

6. Recover the cell-surface bound RNA with strong binding affinity and internalized RNA sequences by TRIZol extraction. After ethanol precipitation, resuspend RNA pellet into RNase-free water (*see* **Note 16**).

7. Reversely transcribe the recovered RNA pool by using the ThermoScript RT-PCR system. Add 1 μL of 10 μM gene-specific primer (reverse primer) and 2 μL of 10 mM dNTP mixture into 9 μL of the recovered RNA, and heat at 65 °C for 5 min. Immediately, chill the reaction on the ice. Subsequently, add the following reagents:

   (a) 4 μL 5 × First strand buffer.

   (b) 1 μL 0.1 M DTT.

   (c) 1 μL RNaseOut (40 U/μL).

   (d) 1 μL ThermoScript RT (15 U/μL).

   (e) 1 μL RNase-free water.

   Incubate the reaction (total volume of 20 μL) at 50 °C for 60 min. Heat the mixture at 85 °C for 5 min to inactivate reverse transcriptase and chill on ice. The cDNA is ready for PCR (*see* **Note 17**).

8. Amplify the cDNA by PCR using 2 μM each of 5′- and 3′-primers, along with 2 mM MgCl₂ and 200 μM of each dNTP. Conduct 15 PCR cycles as follows: 94 °C, 5 min; 94 °C, 1 min, 63 °C, 1 min, 72 °C, 1 min; 72 °C, 7 min; 4 °C, keep! Run 2.5 % agarose gel to verify PCR product (*see* **Note 18**)

9. Recover the resulting PCR products using a QIAquick Gel purification Kit and transcribe it to new RNA pool as described in Subheading 3.2.

10. After purification, refold the new RNA pool for the next selection round (counter selection and positive selection) as described in Subheading 3.3 (*see* **Note 15**)

3.4 Monitor the Progress of SELEX or Detect Binding Affinity by qRT-PCR

1. Apply quantitative real-time PCR (qRT-PCR) methods to monitor SELEX progress.

2. 24 h before selection, seed U373-Magi cells (CCR5-negative) and U373-Magi-CCR5E cells (CCR5-positive) at equal density (3×10^4 cells per well, respectively) on a 48-well plate with 250 μL complete culture medium.

3. On the day of selection, wash U373-Magi cells and U373-Magi-CCR5E cells with 250 μL pre-warmed binding buffer to remove dead cells, and add 250 μL pre-warmed binding buffer supplemented with 100 μg/mL yeast tRNAs.

4. After 15 min incubation at 37 °C, remove the buffer from both cell lines and add the refolded RNA library or RNA aptamer candidate (0.1 nmol of RNA library in 250 μL refolding buffer) supplemented with 1 nmol yeast tRNA.

5. Incubate at 37 °C for 15 min. Wash the cells six times with 250 μL pre-warmed binding buffer to remove unbound RNA and cell-surface RNAs with weak binding.

6. Recover the cell-surface bound RNA with strong binding affinity and internalized RNA sequences by TRIZol extraction. And reversely transcribe those recovered RNA pools as described in Subheading 3.3.

7. Analyze the resulting cDNA by qRT-PCR using $2 \times iQ$ SyberGreen Mastermix and specific primer sets at a final concentration of 400 nM (triplex assay). An example result is shown in Fig. 3 (*see* **Note 19**).

Add the following reagents in one 25 μL qPCR reaction:

(a) 1 μL cDNA reaction mixture.

(b) 12.5 μL $2 \times iQ$ SyberGreen Mastermix.

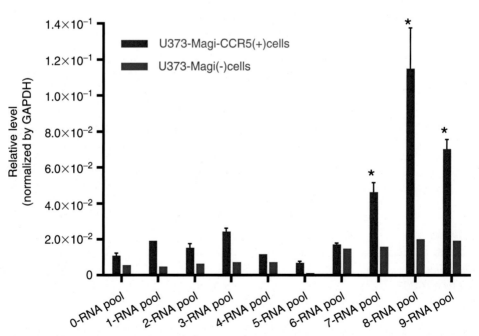

Fig. 3 The progress of SELEX. Progression of the selection was monitored using quantitative RT-PCR (qRT-PCR) and normalizing to GAPDH gene. Nine rounds of live cell-based SELEX were performed to enrich for RNA aptamers that bind and internalize into U373-MAGI-CCR5E (CCR5⁺) cells. Nonspecific aptamers were removed by pre-clearing against U373-MAGI (CCR5⁻) cells. From the seventh selection round, selective binding/internalization of RNA pools was observed in U373-MAGI-CCR5E (CCR5⁺) cells. (Reprinted from [17], Fig. 1b)

(c) 2 µL primer mixture (5 mM).

(d) 9.5 µL RNase-free water.

Standard PCR program: 1 cycle of 94 °C for 5 min; 35–40 cycles of 94 °C for 30 s; 55–62 °C (mostly 60) for 30 s; 72 °C for 30 s; 1 cycle of 72 °C for 5 min.

3.5 Illumina High-Throughput Sequencing (HTS) and Data Analysis

1. After nine rounds of SELEX, send the RNA libraries for selection rounds 5, 6, 7, 8, and 9 for Illumina high-throughput sequencing analysis.

2. Reversely transcribe 1 µg of RNA library using RT primer and amplify the resulting cDNA by 8 PCR cycles. Gel purify the product and dilute with 1 mL of hybridization buffer. Use a final DNA concentration of 10 pM for single read flow cell cluster generation (*see* **Note 4**).

3. Bioinformatics analysis (*see* **Note 20**). Count unique reads in each sample and identify the most frequent 1000 unique sequences in each sample. For alignment and grouping analysis, classify the top 40 unique sequences into six major groups (groups 1–6) based on the similarity in the sequences and secondary structures. Predict their secondary structures by MFold RNA (http://mfold.rna.albany.edu/?q=mfold/RNA-Folding-Form). An example result is shown in Table 2 and Fig. 4.

Table 2
Bioinformatics analysis of high-throughput sequence data from selection rounds. (a) The total reads and useful reads were defined as follows. The 3′-fixed oligo sequence and 3′-Solexa adapter were identified and trimmed from each reads. The reads with 20-base (random domain) after processing were considered as usable reads and retained for further analysis. The most frequent 1000 unique sequences were identified and listed here for the clarity. The molecular enrichment at each round was calculated by the formula (total reads of top 1000 unique sequences at round X/round 5). (b) Bioinformatics analysis of RNA aptamers to identify related sequence and structure groups. After alignment of top 40 sequences, six groups of RNA aptamers were identified. The representative RNA aptamers and the reads of each group are listed here. Only the random sequences of the aptamer core regions (5′–3′) are indicated. Groups 2, 4, and 5 (G-2, G-4, and G-5 aptamers) shared a conserved sequence, which is comprised of 10 nucleotides UUCGUCUG(U/G)G. (Reprinted from [17], Tables 1 and 2)

(a)	Round 5	Round 6	Round 7	Round 8	Round 9
Total reads	48,175,338	37,973,185	36,503,548	27,876,693	30,723,210
Useful reads	35,003,576	24,043,982	25,650,155	17,994,328	16,273,964
Total reads of top 1000 unique sequences	1,553,154	8,877,699	18,721,730	15,831,299	15,473,124
Molecular enrichment (fold)	1	5.72	12.05	10.19	9.96

(continued)

Table 2
(continued)

(b)

Groups	RNA	20-nt random sequence	Reads of each group				
			Round 5	Round 6	Round 7	Round 8	Round 9
1	G-1	AUCGUCUAUUA GUCGCUGGC	144,560	1,405,945	4,385,441	2,572,509	8,228,896
2	G-2	UCCUUGGCUU UUCGUCUGUG	447,271	2,700,796	5,191,611	4,572,814	3,151,265
3	G-3	GCCUUCGUUU GUUUCGUCCA	409,975	2,033,703	3,150,589	3,452,540	1,427,892
4	G-4	UCCCGGCUCG UUCGUCUGUG	129,752	695,110	1,564,701	2,568,638	1,148,903
5	G-5	UUCGUCAU UU UUCGUCUGGG	208,652	1,282,740	3,133,297	1,591,731	701,763
6	G-6	CCUUUCGUCU GUUUCUGCGC	69,760	158,855	233,840	159,454	94,378
Others		Orphan sequences	20,151	42,315	0	0	0
		Total Reads of all groups	1,430,121	8,319,464	17,659,479	14,917,686	14,753,097

Fig. 4 Bioinformatics analysis of high-throughput sequence data from selection rounds. (**a**) Distribution of frequencies of top 1000 unique sequences at each round. The most frequent 1000 unique sequences were identified at each selection round. From Round 7, enrichment saturation was observed. (**b**) The molecular enrichment at each round. The molecular enrichment at each round was calculated by the formula (total reads of top 1000 sequences in round X/round 5). From Round 5 to Round 9, the molecular diversity was significantly converged, suggesting some specific sequences have been enriched. (**c**) Sequence logo for each select round. Bioinformatics analysis of RNA aptamers to identify related sequence and structure families. Through the alignment of primary sequences, the distributions within each round were identified at the 20-nt random domain. From Round 7, highly represented sequences were observed. (**d**) The frequency of each group at each selection round. After alignment of top 40 sequences, six groups of RNA aptamers were identified. The percent frequency of each group at each selection round was calculated by the formula (the reads of each group/the useful reads at each round). (Reprinted from [17], Figure S2a, b and Fig. 1c, d)

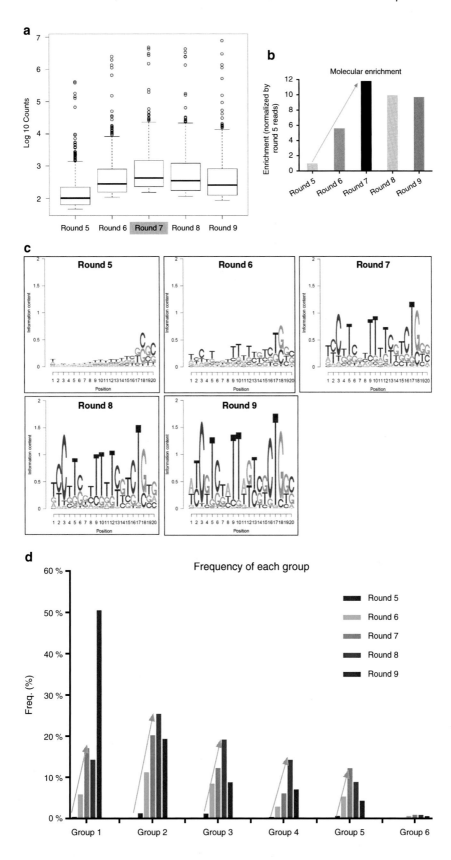

3.6 Cell-Surface Binding and Internalization Studies by Flow Cytometry

1. Generate 2'-fluoropurimidine modified RNA aptamers (G-1, G-2, G-3, G-4, G-5, and G-6) as described in Subheading 3.2 (*see* **Note 21**)

2. Generate fluorescent aptamers using the Silencer siRNA Labeling Kit (*see* **Note 22**). Add the following reagents in order:

   (a) 22.5 μL nuclease-free water.

   (b) 5 μL 10× Labeling Buffer.

   (c) 15 μL RNA (5 μg).

   (d) 7.5 μL Labeling Dye.

   Incubate at 37 °C for 1 h.

3. After incubation, add 5.0 μL (0.1 vol) 5 M NaCl and 125 μL (2.5 vol) cold 100 % EtOH, and mix thoroughly. Incubate at –20 °C for 60 min. And then centrifuge at top speed in a microcentrifuge (12,000×g) at 4 °C for 20 min. Remove supernatant and wash pellet with 175 μL of 70 % EtOH. Air dry pellet in the dark. And then suspend labeled RNA in 15 μL of nuclease-free water.

4. Measure the absorbance of the labeled RNA at 260 nm and at the absorbance maximum for the fluorescent dye (Cy3 at 550 nm).

5. Calculate the base:dye ratio and RNA concentration according to the calculator provided by http://www.genelink.com/tools/gl-BDratiores.asp (*see* **Note 23**).

6. Refold the mixture in refolding buffer as described above (*see* Subheading 3.2).

7. Cell-surface binding assay with adherent cell lines: Wash U373-Magi cells or U373-Magi-CCR5E cells with pre-warmed DPBS, and detach cells with Cell stripper from the plates (*see* **Note 6**).

8. Cell-surface binding assay with suspension cells: directly pellet the CEM-NKr cells, CEM-NKr-CCR5 cells, or PBMCs from the flask.

9. After washing cells twice with 500 μL binding buffer containing DPBS (pH 7.0–7.4) and Ca^{2+} and Mg^{2+}, resuspend cell pellets in binding buffer and incubate at 37 °C for 30 min. And then pellet cells and resuspend in 100 μL of pre-warmed binding buffer (*see* **Note 7**).

10. Add the refolded Cy3-labeled experimental RNAs and incubate at RT for 30 min in the dark (*see* **Note 24**).

11. Wash cells three times with 500 μL of pre-warmed binding buffer, and resuspend cells in 350 μL of DPBS pre-warmed to 37 °C and immediately analyze by flow cytometry. An example result is shown in Fig. 5.

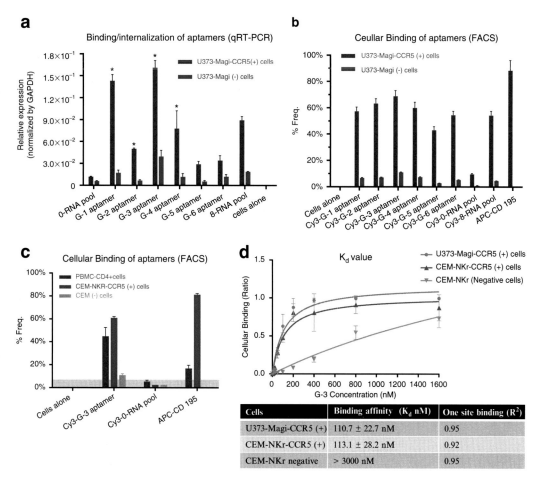

Fig. 5 Cell-type-specific binding and internalization studies of individual RNA aptamers. (**a**) Cell-type-specific binding/internalization was evaluated by qRT-PCR. Six representative RNA aptamers from each groups were incubated with U373-Magi-CCR5E (CCR5 positive) cells or U373-Magi negative cells. The total RNA was isolated for cDNA synthesis followed by qPCR amplification. The RNA aptamers showed selective binding/internalization to CCR5 expressing cells. The 0-RNA pool was used as negative controls. Data represent the average of three replicates. Cell-surface binding of Cy3-labeled RNAs was assessed by flow cytometry. The 0-RNA pool was used as negative control. Data represent the average of three replicates. (**b**) Cy3-labeled RNAs were tested for binding to U373-Magi-CCR5 positive cells and U373-Magi negative cells. The selected aptamers showed cell-type-specific binding affinity. APC-CD195 antibody was used to stain cellular surface CCR5. (**c**) One of the best RNA aptamer, G-3, was chosen for further binding affinity test with PBMC-CD4+ cells, CEM-NKr-CCR5 positive cells, CEM negative cells. (**d**) Cell-surface binding constant (K_d) of G-3 aptamer was evaluated. The U373-Magi-CCR5 positive cells, CEM-NKr-CCR5 positive cells, and CEM negative cells were incubated with the increasing amounts of Cy3-labeled G-3 aptamer. The binding affinity was analyzed by flow cytometry assay. The calculated K_d determinations are indicated. (Reprinted from [17], Fig. 2a, b, c and d)

3.7 Cell-Surface Binding and Internalization Studies by Live-Cell Confocal Microscopy

1. Prepare and refold Cy3-labeled aptamers as described above (*see* Subheading 3.6).

2. For the adherent cell lines: Before one day of assay, grow the U373-Magi cells or U373-Magi-CCR5E cells in 35 mm plate pre-treated with poly-lysine seeding at 0.3×10^6 in 2 mL of complete medium to allow about 70–80 % confluence in 24 h (*see* **Note 25**). On the day of the experiments, wash cells with 1 mL of pre-warmed binding buffer and then incubate with 1.5 mL of pre-warmed complete growth medium for 30–60 min in a humidified 5 % CO_2 incubator at 37 °C.

3. For the suspension cells: on the day of experiments, pellet the cells CEM-NKr cells, CEM-NKr-CCR5 cells, or PBMCs from the flask. Wash the cells twice with pre-warmed binding buffer and grow the cells in 35 mm plate pre-treated with poly-lysine seeding at 1×10^6 in 1.5 mL of complete medium. Incubate the cells for 30–60 min in a humidified 5 % CO_2 incubator at 37 °C for attaching on the dish surface.

4. Add the Cy3-labeled aptamer at a 67 nM final concentration into the media and incubate for live-cell confocal microscopy in a 5 % CO_2 microscopy incubator at 37 °C. Collect the images every 15 min using a Zeiss LSM 510 Meta Inverted 2 photon confocal microscope system under water immersion at $40 \times$ magnification.

5. After 4–5 h of incubation and imaging, stain the cells with 0.15 mg/mL Hoechst 33342 (nuclear dye for live cells) for 15 min at 37 °C. Collect the images are as described above. An example result is shown in Fig. **6.**

3.8 Validate the Target Protein as Human CCR5

1. Transfect the CCR5 siRNA and control non-silencing siRNA NC-1 to U373-Magi-CCR5E cells using Lipofectamine 2000 (*see* **Notes 5** and **26**).

2. After 48 h of transfection, isolate total RNAs with STAT-60, and then treat the total RNAs with DNase I to remove genomic DNA.
   Mix the following reagents:
   (a) 8 μL nuclease-free water.
   (b) 1.5 μL $10 \times$ DNase Buffer.
   (c) 4 μL RNA (2 μg).
   (d) 0.5 μL RNain inhibitor.
   (e) 1.0 μL RNase-free DNase I.

   Incubate at 37 °C for 1 h, heat at 80 °C for 10 min to inactivate DNase I, and immediately chill the reaction on the ice.

3. Reversely transcribe the cDNA. Add 2 μL of 50 ng/μL random primer and 1 μL of 10 mM dNTP into the reaction mixture,

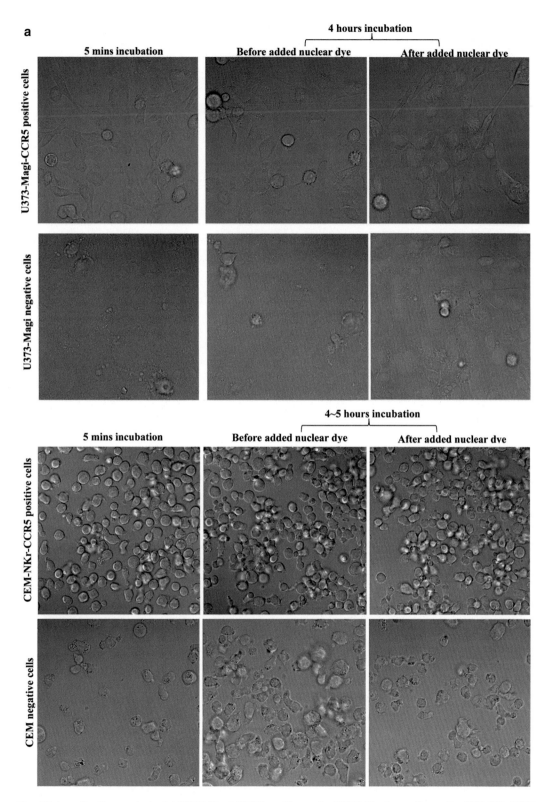

Fig. 6 Internalization analysis. (**a**) U373-Magi-CCR5 positive cells, U373-Magi negative cells, CEM-NKr-CCR5 positive cells, CEM negative cells, or (**b**) PBMC-CD4+ cells were grown in 35 mm plates treated with polylysine and incubated with a 67 nM concentration of Cy3-labeled G-3 aptamer in complete culture media for real-time live-cell confocal microscopy analysis. The images were collected using 40 × magnification. (**c**) Localization analysis. (Reprinted from [17], Fig. 2f, g and h)

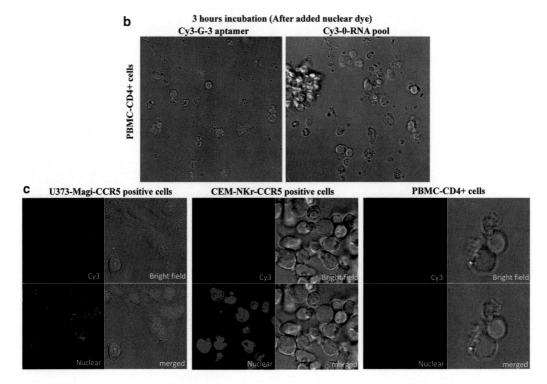

Fig. 6 (continued)

and heat at 65 °C for 5 min. Immediately, chill the reaction on the ice. Subsequently, add the following reagents:

(a) 5 μL 5×First strand buffer.

(b) 2.5 μL 0.1 M DTT.

(c) 0.5 μL RNasin inhibitor.

(d) 1.0 μL MMLV-RT.

Incubate the reaction (total volume of 27 μL) at 25 °C for 10 min and at 37 °C for 1 h. Heat the mixture at 70 °C for 15 min to inactivate reverse transcriptase and chill on ice. The cDNA is ready for qRT-PCR analysis.

4. Analyze expression of the target genes by quantitative RT-PCR using 2× iQ SyberGreen Mastermix and specific primer sets at a final concentration of 400 nM (triplex assay) as described in Subheading 3.4 (*see* **Note 19**). An example result is shown in Fig. 7.

5. Analyze expression of the CCR5 protein by flow cytometry. After 48 h of transfection, wash the cells and detach with Cell stripper as described in Subheading 3.6. An example result is shown in Fig. 7.

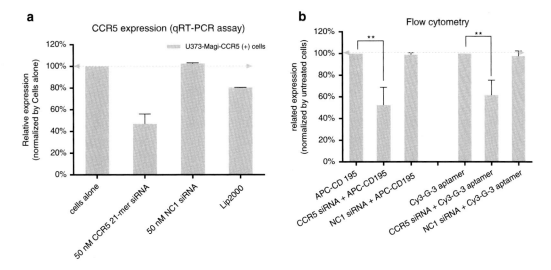

Fig. 7 Knockdown of CCR5 reduced binding affinity of aptamers. CCR5 siRNA was transfected into U373-Magi-CCR5 positive cells. After 48 h post-transfection, (**a**) CCR5 gene expression was detected by qRT-PCR and (**b**) cell-surface binding of Cy3-labeled G-3 aptamer was assessed by flow cytometry. A scrambled siRNA (NC1) was used as negative control. Data represent the average of three replicates. (Reprinted from [17], Figure S3e and Fig. 2e)

4   Notes

1. Common chemicals can be purchased from various reliable sources such as Sigma-Aldrich. Suppliers for some reagents are provided for convenience, and in most cases they can also be purchased from other reputable suppliers. Unless stated otherwise, all solutions should be prepared in autoclaved water that has a resistivity of 18.2 MΩ-cm, which is referred to as "water" in the text.

2. In order to obtain the best performance, avoid repeated freeze–thaw steps and finish the T7 transcription Kit within 6 months which contains 2′-F-dCTP and 2′-F-dUTP. 2′-F-dCTPand2′-F-dUTPare2′-fluorine-2′-deoxyribonucleoside-5′-triphosphates.

3. This special reverse transcriptase is good for RNAs with highly secondary structures. The cDNA synthesis was conducted at a higher temperature (55–60 °C).

4. The sample preparation and deep sequencing can be performed by a DNA sequencing core facility. The following is a protocol we often use: 1.0 μg of RNA pool is reverse-transcribed using RT primer (5′-CAG ATT GAT GGT GCC TAC AGT CGG GCG UGT CGT CTG-3′), and then is subjected to PCR

amplification for 8 cycles using primers JH5 (5′-AAT GAT ACG GCG ACC ACC GAC AGG TTC AGA GTT CGA TCG GGA GGA CGA TGC GG-3′) and RT/index primer (5′-CAG ATT GAT GGT GCC TAC AGT CGG GCG UGT CGT CTG-3′). The PCR product is subjected to 6 % TBE PAGE gel purification with size selection (for targeted smRNAs of 51 nt). The purified library is followed by the second round of PCR amplification for 4 cycles with primers PE-mi-index primer (5′-CAA GCA GAA GAC GGC ATA CGA GAT <u>NNNNNN</u> CAG ATT GAT GGT GCC TAC AG-3′) and R2 (5′-AAT GAT ACG GCG ACC ACC GA-3′) then followed by 6 % TBE PAGE gel purification with size selection (for targeted smRNAs of 51 nt). The purified library is quantified using qPCR with a forward primer (5′-CAA GCA GAA GAC GGC ATA CG-3′) and a reverse primer (5′-AAT GAT ACG GCG ACC ACC GA-3′). The quantified, denatured PCR product from smRNA library is loaded in 1 mL of hybridization buffer to a final DNA concentration of 10 pM then used for single read flow cell cluster generation and 40 cycle (40 nt) sequencing on an Illumina HiSeq2000.

5. The CCR5 siRNA (sense: 5′ P-CUC UGC UUC GGU GUC GAA A dTdT-3′; Antisense: 5′ P-UUU CGA CAC CGA AGC AGA G dTdT-3′) has been demonstrated previously to knockdown CCR5 expression. Primers are as follows: CCR5 forward primer: 5′-AAC ATG CTG GTC ATC CTC AT-3′; CCR5 reverse primer: 5′-AAT AGA GCC CTG TCA AGA GT-3′. GAPDH expression is used for normalization of the qPCR data.

6. To avoid the damage of the cells, the cells are gently and briefly incubated with a small amount of Cell Stripper solution for detachment.

7. Washing step: spin cells at 1000 rpm in a microcentrifuge for 5 min at 25 °C.

8. 1 μL antibody should be enough for 2×10^5 cells in 100 μL reaction system.

9. To preserve the abundance of the original DNA library, PCR was limited to 10 cycles. PCR Forward primer: 5′-<u>TAA TAC GAC TCA CTA TAG GG</u>A GGA CGA TGC GG-3′ (32 mer); PCR Reverse primer: 5′-TCG GGC GAG TCG TCT G-3′ (16 mer). Initial DNA pool template: 5′-GGG AGG ACG ATG CGG - N_{20}- CAG ACG ACT CGC CCG A-3′ (51 nt).

10. Mix all the reagents at room temperature and finally add T7 RNA polymerase. Increase the amount of DNA template (1 μg to 4 μg) and reaction time (overnight incubation) will improve yield of RNA.

11. The work concentration of RNA in the reaction is 50 ng/ μL. The template DNA pool should be completely removed to avoid its contamination during SELEX procedure.

12. In principle, live cell-based SELEX methodology relies on the differences between the target cells population (positive cells that express the target of interest) and the control cell population (negative cells that do not express the target protein) [8]. Therefore, the selection procedure generally includes positive selection against the target cells and counter selection against related non-target cells to remove nonspecific binding.

13. Dead cells nonspecifically take up nucleic acids and cause nonspecific binding and/or internalization of nucleic acids during cell-based SELEX, thereby resulting in the delay of target-specific sequence enrichment or even failure of aptamer selection. In this study we therefore carefully controlled the selection procedures, especially those that would probably adversely affect cell health and damage cells, such as cell culture, washing condition, detachment of cells, etc. For example, the cells used for selection are healthy, with >95 % cells alive. A non-enzymatic cell disassociation solution is used here to detach a monolayer of cells. tRNA is used as a competitor to reduce nonspecific binding.

14. This is a counter-selection step. Counter-selection step is performed per cycle to minimize nonspecific binding with the non-targeted cells (negative cells). The supernatants with unbound sequences from negative cells plate are the pre-cleared 0-RNA pool which is transferred to the targeted cells for positive selection.

15. This is a positive selection step. In order to increase the stringency of selection, the density and number of target cells and incubation time were reduced and competitor tRNA and washing times were progressively increased during the selection procedure. The selection conditions for each round are summarized in Table 1. In our study, we totally conducted nine selection rounds.

16. In order to improve recovery yield, 1 μL of glycogen is added into ethanol precipitation step.

17. A no-RT, negative control needs to be set up. The cDNA can be stored at −20 °C or used for PCR immediately.

18. The amount of cDNA for PCR should be optimized to reduce nonspecific PCR product and increase PCR efficiency. Generally, we use 2, 1, 0.5, 0.02, 0.002 μL of original cDNA reaction mixture for a 50-μL PCR reaction. Under our experimental condition, 0.5 μL of original cDNA reaction mixture is optimal.

19. Allow all reagents to reach room temperature before use. Label test tubes to be used for the preparation of standards and specimens. If the entire 96-well plate will not be used, remove surplus strips from the plate frame. Place surplus strips and desiccant into the Re-sealable plastic bag (provided by the manufacturer), seal and store at 4 °C. Primers were as follows: RNA pool Forward primer: 5′-<u>TAA TAC GAC TCA CTA TAG GG</u>A GGA CGA TGC GG-3′ (32 mer); RNA pool Reverse primer: 5′-TCG GGC GAG TCG TCT G-3′ (16 mer). GAPDH expression was used for normalization of the qPCR data. GAPDH forward primer: 5′-CAT TGA CCT CAA CTA CAT G-3′; GAPDH reverse primer: 5′-TCT CCA TGG TGG TGA AGA C-3′. Remember to include a non-template control (NTC), a standard positive control for quantification (a 3–10-fold dilution curve in duplicates) and a "no RT" negative control for each cDNA sample.

20. Reads processing and data analysis were conducted. In brief, the processing principles were as follows. Bases after Ns in each read were considered low quality and were removed. The 3′-fixed oligo and 3′-Solexa adapter were identified and trimmed from each reads. The reads with 20-base after processing were considered as usable reads and retained for further analysis. Unique reads in each sample were counted. The most frequent 1000 unique sequences were identified in each sample. The most frequent 1000 unique sequence in round 9 were obtained and matched to the other four samples (top 1000 unique reads) and their frequencies were recorded. The consensus sequence of round 9 was used to compare to the reads in each round. For alignment and grouping analysis, the top 40 sequences were divided into 6 groups according to their predicted secondary structures by MFold RNA and QuickFold RNA.

21. The RNA aptamers sequences are listed in Table 2b.

22. Fluorescent dye reagent and Cy3-labeled RNA are sensitive to light and should limit the exposure to light for the entire procedure.

23. Detected by spectrophotometry, the base:dye ratio can be calculated, which should be lower than 200. The lower the base:dye ratio, the more dye molecules are present on the RNA, indicating an effective labeling. Dye molecules (Cy3, Cy5, or fluorescein) covalently attach to the guanine base of the RNA. The dye molecule can also attach to the adenosine base of RNA, however, this reaction is very rare.

24. The different work concentrations (0–1600 nM) of Cy3-labeled experimental RNA aptamers are used for cell-surface binding experiment by flow cytometry. The dissociation

constants were calculated using nonlinear curve regression with a Graph Pad Prism 6.0.

25. The surface of the plate for confocal microscopy is very smooth and is treated with poly-lysine to increase the attachment of the cells on the dish.

26. The siRNA transfection is performed in a 6-well plate according to the manufacturer's instruction. The CCR5 knockdown is confirmed by flow cytometry (APC-CD195 antibody staining) and qRT-PCR assay. In parallel, the cells are stained with the Cy3-labeled G-3 aptamer.

Acknowledgments

We thank James O. McNamara II and Paloma H. Giangrande for helpful discussions. We thank Mayumi Takahashi for providing primary PBMCs. The authors would like to thank City of Hope DNA sequencing core and Bioinformatics Core facility (Harry Gao, Xiwei Wu and Jinhui Wang) for Solexa Deep sequencing and data processing. The following reagents were obtained through the NIH AIDS Research and Reference Reagent Program, Division of AIDS, NIAID, NIH: the U373-Magi and U373-Magi-CCR5E cell lines were obtained from Dr. Michael Emerman [14]; CEM-NKr and CEM-NKr-CCR5 cell lines were obtained from Dr. Peter Cresswell [15, 16].

This work was supported by National Institutes of Health grants (grant numbers R01AI29329, R01AI42552, and R01HL07470) to J.J.R. Funding for open access charge: National Institutes of Health.

Conflict of Interest declaration: J.J.R. and J.Z. have an issued patent entitled "Cell-type-specific aptamer-siRNA delivery system for HIV-1 therapy". USPTO, No. US 8, 222, 226 B2, issued date: July 17, 2012. J. J. R. and J. Z. have a patent pending on "Cell-specific internalizing RNA aptamers against human CCR5 and used therefore", The United States Patent, and application number: 62/025,368, filed on July 16, 2014.

References

1. Mayer G (2009) The chemical biology of aptamers. Angew Chem Int Ed Engl 48:2672–2689

2. Ozer A, Pagano JM, Lis JT (2014) New technologies provide quantum changes in the scale, speed, and success of SELEX methods and aptamer characterization. Mol Ther Nucleic Acids 3:e183

3. Sundaram P, Kurniawan H, Byrne ME et al (2013) Therapeutic RNA aptamers in clinical trials. Eur J Pharm Sci 48:259–271

4. Ni X, Castanares M, Mukherjee A et al (2011) Nucleic acid aptamers: clinical applications and promising new horizons. Curr Med Chem 18:4206–4214

5. Zhou J, Rossi JJ (2014) Cell-type-specific, aptamer-functionalized agents for targeted disease therapy. Mol Ther Nucleic Acids 3:e169

6. Fang X, Tan W (2010) Aptamers generated from cell-SELEX for molecular medicine: a chemical biology approach. Acc Chem Res 43:48–57

7. Guo KT, Paul A, Schichor C et al (2008) CELL-SELEX: novel perspectives of aptamer-based therapeutics. Int J Mol Sci 9:668–678

8. Cerchia L, Giangrande PH, McNamara JO et al (2009) Cell-specific aptamers for targeted therapies. Methods Mol Biol 535:59–78

9. Thiel KW, Hernandez LI, Dassie JP et al (2012) Delivery of chemo-sensitizing siRNAs to HER2+–breast cancer cells using RNA aptamers. Nucleic Acids Res 40:6319–6337

10. Tuerk C, Gold L (1990) Systematic evolution of ligands by exponential enrichment: RNA ligands to bacteriophage T4 DNA polymerase. Science 249:505–510

11. Cerchia L, Duconge F, Pestourie C et al (2005) Neutralizing aptamers from whole-cell SELEX inhibit the RET receptor tyrosine kinase. PLoS Biol 3:e123

12. Cerchia L, Esposito CL, Jacobs AH et al (2009) Differential SELEX in human glioma cell lines. PLoS One 4:e7971

13. Thiel WH, Bair T, Peek AS et al (2012) Rapid identification of cell-specific, internalizing RNA aptamers with bioinformatics analyses of a cell-based aptamer selection. PLoS One 7:e43836

14. Vodicka MA, Goh WC, Wu LI et al (1997) Indicator cell lines for detection of primary strains of human and simian immunodeficiency viruses. Virology 233:193–198

15. Howell DN, Andreotti PE, Dawson JR et al (1985) Natural killing target antigens as inducers of interferon: studies with an immunoselected, natural killing-resistant human T lymphoblastoid cell line. J Immunol 134:971–976

16. Lyerly HK, Reed DL, Matthews TJ et al (1987) Anti-GP 120 antibodies from HIV seropositive individuals mediate broadly reactive anti-HIV ADCC. AIDS Res Hum Retroviruses 3:409–422

17. Zhou J, Satheesan S, Li H et al (2015) Cell-specific RNA aptamer against human CCR5 specifically targets HIV-1 susceptible cells and inhibits HIV-1 infectivity. Chem Biol 22: 379–390

Chapter 17

mCarts: Genome-Wide Prediction of Clustered Sequence Motifs as Binding Sites for RNA-Binding Proteins

Sebastien M. Weyn-Vanhentenryck and Chaolin Zhang

Abstract

RNA-binding proteins (RBPs) are critical components of post-transcriptional gene expression regulation. However, their binding sites have until recently been difficult to determine due to the apparent low specificity of RBPs for their target transcripts and the lack of high-throughput assays for analyzing binding sites genome wide. Here we present a bioinformatics method for predicting RBP binding motif sites on a genome-wide scale that leverages motif conservation, RNA secondary structure, and the tendency of RBP binding sites to cluster together. A probabilistic model is learned from bona fide binding sites determined by CLIP and applied genome wide to generate high specificity binding site predictions.

Key words RNA-binding protein (RBP), CLIP tag clusters, motif, Binding site prediction, mCarts

1 Introduction

RNA-binding proteins (RBPs) bind to short, degenerate sequences in RNA to regulate a number of post-transcriptional processes including alternative splicing, alternative polyadenylation, and RNA stability [1–3]. These short sequences have traditionally made it difficult to computationally predict RBP binding sites on a genome-wide scale with high specificity. As a result, previous approaches using consensus sequence [4] or position weight matrices (PWMs) [5] to search for functional binding sites have low discriminative power.

Recently, the development of HITS-CLIP (crosslinking and immunoprecipitation combined with high-throughput sequencing) and its variants has made it possible to map direct, in vivo RBP binding sites genome wide in a given condition [6–8]. In brief, UV light is used to crosslink protein-RNA complexes in direct contact in cells or tissue. The protein of interest is isolated along with its bound RNA fragments, which are then purified in very stringent conditions and sequenced in depth.

Ren-Jang Lin (ed.), *RNA-Protein Complexes and Interactions: Methods and Protocols*, Methods in Molecular Biology, vol. 1421, DOI 10.1007/978-1-4939-3591-8_17, © Springer Science+Business Media New York 2016

HITS-CLIP has resulted in a wealth of information about RBP function and provided new insights into RNA regulation. However, a HITS-CLIP experiment only provides a snapshot reflecting the experimental conditions rather than the complete landscape of protein-RNA interactions. Some transcripts may not be expressed under the conditions in which the HITS-CLIP experiment was performed, and thus will not be detected as bound by the RBP [9, 10]. In addition, some binding sites might escape the detection by HITS-CLIP due to technical issues that limit the complexity and depth of CLIP libraries [9]. Finally, while CLIP data provide evidence of protein-RNA interactions, they do not directly provide a mechanism for the specific recognition, which is coded in the RNA sequences and structures.

Several algorithms have been developed to take advantage of the HITS-CLIP data to generate genome-wide RBP binding profiles [11–13]. We previously developed mCarts which predicts clusters of RBP motif sites by integrating several intrinsic and extrinsic features of functional protein-RNA interactions, including the number and clustering of individual motif sites, their accessibility as determined by RNA secondary structures, and cross-species conservation [14]. A hidden Markov model (HMM) framework learns quantitative and subtle rules of these features from in vivo RBP binding sites derived from HITS-CLIP data and generates genome-wide predictions of new sites with high specificity and sensitivity. These predictions extend the RBP interaction map observed from HITS-CLIP data, providing a broader picture of RBP binding and regulation. This protocol describes the process of installing the mCarts software, training the mCarts model, and predicting RBP binding sites in the mouse genome for the RBP Nova, a neuron-specific RBP that binds to YCAY (Y = C/U) motifs to regulate alternative splicing [6].

Although downstream analysis after obtaining predicted binding sites is beyond the scope of this protocol and varies depending on specific RBPs, typical steps include cross-validation using CLIP data or other known list of binding sites and correlation of RBP binding with altered RNA splicing or gene expression upon RBP perturbation [14]. For example, we previously showed that a high validation rate was achieved when a subset of top candidate alternative exons with strong YCAY clusters as predicted by mCarts were tested for Nova-dependent splicing by RT-PCR. In addition, on a genome-wide scale, alternative exons with evidence of Nova binding from both CLIP data and mCarts predicted motif sites are more likely to have Nova-dependent splicing than those exons that are supported by CLIP or bioinformatics predictions alone. These analyses suggest that CLIP and mCarts data are complementary to each other and the combination of the two can help identify direct Nova targets more accurately.

2 Materials

2.1 Computer

1. This analysis requires a personal computer or cluster running a UNIX-based operating system (Linux or Mac OS X) with sufficient memory (8 GB RAM; 16 GB recommended) to run the software and enough storage for the data files (30 GB for mouse and 40 GB for human) and the CLIP data (up to a few GB, but varies).

2. The software package provides a set of command line tools implemented in perl and C++ and relies on several standard Unix tools such as awk and sort. Updates to the software, documentation, and protocol can be found at http://zhanglab.c2b2.columbia.edu/index.php/MCarts.

3. Commands that should be entered into the terminal (*see* **Note 1**) will be identifiable by a different typeface. The beginning of a command is indicated by a "$" (which should not be entered on the command line). Commands often span multiple lines in the text, but they should be entered as a single line. For example:

```
$ perl ~/src/script.pl -option filename
```

2.2 The mCarts Software

1. Download the mCarts software and the required czplib perl libraries from the links provided at http://zhanglab.c2b2.columbia.edu/index.php/MCarts_Documentation#Download.

2. Install mCarts (v1.2.x or later) by running the following commands:

```
$ tar xzvf mCarts.v1.x.x.tgz
$ cd mCarts
$ make
```

The make command requires the GCC compiler as well as the Boost (http://www.boost.org) and popt libraries (http://rpm5.org/files/popt/), which are installable from your distribution's package manager.

3. Add mCarts to your path (optionally, add this command to your .bash_profile; *see* **Note 2**):

```
$ export PATH=~/src/mCarts:$PATH
```

4. Add czplib to the perl libraries path by running the following command (this needs to be added to your .bash_profile if SGE/OGE is installed on your system; otherwise, optionally add this command to your .bash_profile; *see* **Note 2**):

```
$ tar xzvf czplib.v1.x.x.tgz
$ export PERL5LIB=~/src/plib
```

5. Install the perl modules Math::CDF and Bio::SeqIO (e.g., using CPAN).

2.3 The CIMS Software (for CLIP Data Processing)

1. Download the CIMS package from the link provided at http://zhanglab.c2b2.columbia.edu/index.php/CIMS_Documentation#Download (suggested location is ~/src).

2. Decompress the CIMS package by running the following command:

```
$ tar xzvf CIMS.v1.x.x.tgz
```

2.4 The Reference Library Files

1. Download the reference library files for the organism corresponding to your CLIP data from the link provided at http://zhanglab.c2b2.columbia.edu/index.php/MCarts_Documentation#Download (suggested location is ~/data/):
mm10 library files: mCarts_lib_data_mm10.tgz.
hg19 library_files: mCarts_lib_data_hg19.tgz.
For running the protocol with the sample data, download the mm10 database.

2. Decompress the library files:

```
$ tar xzvf mCarts_lib_data_mm10.tgz.
```

2.5 The RepeatMasker Database

A copy of the database as it stood at publication time is available at http://zhanglab.c2b2.columbia.edu/index.php/MCarts_Documentation#Download. Using this file as you follow the protocol will ensure that your results match ours, but we recommend using the latest version when performing your own analysis.

1. Go to the UCSC Genome Browser (http://genome.ucsc.edu).

2. Click "Tools > Table browser" at the top of the page.

3. Set "genome" to your organism of interest (for this protocol, select "mouse").

4. Set "assembly" to the assembly matching your dataset ("mm10").

5. Set "group" to "Variation and Repeats" (mouse) or "Repeats" (human).

6. Set "track" to "RepeatMasker."

7. Set "output format" to "BED - browser extensible data."

8. Set the name to something memorable, e.g., "mm10.rmsk.bed."

9. Click "Get output."

10. Click "get BED" to download the file (suggested location is ~/data/).

2.6 The CLIP Data

1. For following along with the protocol, download and decompress the sample Nova CLIP data, Nova_CLIP_uniq_mm10.bed, from the link provided at http://zhanglab.c2b2.colum-

bia.edu/index.php/MCarts_Documentation#Download (*see* **Note 3** for details about this dataset):

```
$ gunzip Nova_CLIP_uniq_mm10.bed.gz
```

2. Alternatively, provide a BED file for an RBP of your choosing. This file should be in BED format and should contain only unique CLIP tags that represent independent captures of protein-RNA interactions. If this is the case, the CLIP data must have been mapped and filtered properly with removal of PCR duplicate tags (*see* **Note 4**).

3 Methods

This protocol assumes that the mCarts software is located in the directory ~/src/, that the mCarts library files are located in ~/data/, and that the CLIP data file is in the current working directory. Adjust the paths accordingly. As you are progressing through the protocol, you can compare the number of lines in each file with those provided in Table 1.

Table 1
The number of lines expected in each file (obtained using *wc -l filename*)

Filename	Lines
Nova_CLIP_uniq_mm10.bed	4,401,394
Nova_CLIP_uniq_mm10.cluster.0.bed	1,766,040
Nova_CLIP_uniq_mm10.cluster.bed	249,393
Nova_CLIP_uniq_mm10.tag.exact.bedGraph	5,112,140
Nova_CLIP_uniq_mm10.cluster.PH.detail.txt	249,393
Nova_CLIP_uniq_mm10.cluster.PH.center.bed	249,393
Nova_CLIP_uniq_mm10.cluster.PH.center.ext50.bed	249,393
Nova_CLIP_uniq_mm10.cluster.PH.center.ext50.normsk.bed	142,786
mm10.exon.uniq.ext1k.bed	272,046
Nova_CLIP_uniq_mm10.cluster.PH.center.ext50.normsk.ext1k.bed	61,028
Nova_CLIP_uniq_mm10.cluster.PH15.center.ext50.normsk.ext1k.bed	7,700
↘CLIP.pos.bed	7,700
mm10.exon.uniq.ext1k.noCLIP.bed	112,798
↘CLIP.neg.bed	112,798

3.1 Generate the Positive Training File by Identifying Regions with Strong CLIP Tag Clusters

Since Nova is known to be an important splicing factor, we will limit the CLIP tag clusters to exons and flanking intronic sequences for training.

1. Identify CLIP tag clusters by grouping overlapping CLIP tags (this step is slightly different from our previous method to generate Nova clusters. *See* **Note 3** for discussion comparing this to previous Nova results):

```
$ perl ~/src/CIMS/tag2cluster.pl -v -s -maxgap "-1"
Nova_CLIP_uniq_mm10.bed Nova_CLIP_uniq_mm10.cluster.0.bed
```

2. Select the clusters containing > 2 tags:

```
$ awk '$5>2' Nova_CLIP_uniq_mm10.cluster.0.bed >
Nova_CLIP_uniq_mm10.cluster.bed
```

3. Create a bedGraph file, which is used to determine the CLIP tag coverage at each position in the genome:

```
$ perl ~/src/CIMS/tag2profile.pl -ss -exact -of bedgraph -n Nova -v
Nova_CLIP_uniq_mm10.bed Nova_CLIP_uniq_mm10.tag.exact.bedGraph
```

4. Determine the peak heights of the clusters:

```
$ perl ~/src/CIMS/extractPeak.pl -s --no-match-score 0 -of detail -v
Nova_CLIP_uniq_mm10.cluster.bed Nova_CLIP_uniq_mm10.tag.exact.bedGraph
Nova_CLIP_uniq_mm10.cluster.PH.detail.txt
```

5. Determine the center position of the clusters:

```
$ awk '{print
$1"\t"int(($8+$9)/2)"\t"int(($8+$9)/2)+1"\t"$4"\t"$7"\t"$6}'
Nova_CLIP_uniq_mm10.cluster.PH.detail.txt >
Nova_CLIP_uniq_mm10.cluster.PH.center.bed
```

6. Extend the cluster centers 50 nt in each direction:

```
$ awk '{print $1"\t"$2-50"\t"$3+49"\t"$4"\t"$5"\t"$6}'
Nova_CLIP_uniq_mm10.cluster.PH.center.bed
> Nova_CLIP_uniq_mm10.cluster.PH.center.ext50.bed
```

7. Remove clusters that overlap with repetitive regions:

```
$ perl ~/src/CIMS/tagoverlap.pl -big -region mm10.rmsk.bed -r -v
Nova_CLIP_uniq_mm10.cluster.PH.center.ext50.bed
Nova_CLIP_uniq_mm10.cluster.PH.center.ext50.normsk.bed
```

8. Extend the known exons by 1000 nt in each direction:

```
$ perl ~/src/CIMS/bedExt.pl -l -1000 -r 1000 -chrLen
~/data/mCarts_lib_data_mm10/chrLen.txt -v
~/data/mCarts_lib_data_mm10/mm10.exon.uniq.bed mm10.exon.uniq.ext1k.bed
```

9. Determine which exonic regions (exons ± 1000 nt) contain CLIP clusters:

```
$ perl ~/src/CIMS/tagoverlap.pl -region mm10.exon.uniq.ext1k.bed -ss --
keep-score --keep-tag-name --complete-overlap --non-redundant -v
Nova_CLIP_uniq_mm10.cluster.PH.center.ext50.normsk.bed
Nova_CLIP_uniq_mm10.cluster.PH.center.ext50.normsk.ext1k.bed
```

10. Select the top clusters based on peak height (PH):

```
$ awk '$5>=15'
Nova_CLIP_uniq_mm10.cluster.PH.center.ext50.normsk.ext1k.bed >
Nova_CLIP_uniq_mm10.cluster.PH15.center.ext50.normsk.ext1k.bed
```

This results in 7700 regions spanning 770,000 nucleotides. *See* **Note 5** for information about picking cluster threshold.

11. Create a symbolic link to the positive region file, which makes future commands clearer and easily reusable with a different training file:

```
$ ln -s Nova_CLIP_uniq_mm10.cluster.PH15.center.ext50.normsk.ext1k.bed
CLIP.pos.bed
```

3.2 Generate the Negative Training File by Filtering Out Any Regions with CLIP Tags

1. Select exonic regions (exons ± 1000 nt) that contain no CLIP tags (n.b. tags, not clusters):

```
$ perl ~/src/CIMS/tagoverlap.pl -big -region Nova_
CLIP_uniq_mm10.bed -
ss --keep-score -r -v mm10.exon.uniq.ext1k.bed
mm10.exon.uniq.ext1k.noCLIP.bed
```

This results in 112,798 regions spanning 252,523,292 nucleotides.

2. Create a symbolic link to the negative region file, which makes future commands clearer and easily reusable with a different training file

```
$ ln -s mm10.exon.uniq.ext1k.noCLIP.bed CLIP.neg.bed
```

3.3 Train the mCarts Model

1. Run the mCarts training:

```
$ mCarts -ref mm10 -f CLIP.pos.bed -b CLIP.neg.bed
-lib
~/data/mCarts_lib_data_mm10  -w  YCAY  --min-site  3
--max-dist 30 --train-
only -v Nova_HMM_D30_m3
```

The whole genome is divided into a number of smaller splits for parallelization. Individual jobs are submitted to the queuing system when it is detected (Oracle Grid Engine (OGE), formerly known as Sun Grid Engine or SGE, is currently supported); jobs are run locally otherwise, in which case 24–36 h of runtime should be expected. Additional details on mCarts are worth noting (*see* **Note 6**).

If the program finished without errors, the following files should be created in the *Nova_HMM_D30_m3* directory:

– BLS (directory)

– formatted (directory)

– model.txt

- params.txt
- train_neg.txt
- train_pos.txt

2. To visualize the model, open the *models.txt* file (located in the *Nova_HMM_D30_m3* output directory) in Microsoft Excel. For each of the following categories, create a line graph comparing the positive to the negative regions for distance (distance between neighboring motif sites). There is a long tail for the distance parameters, so visualizing the score for all 1000 nt is not necessary (try ~100 nt). For conservation_0 (intron), conservation_1 (CDS), conservation_2 (5′ UTR), conservation_3 (3′ UTR), and accessibility, create a scatterplot comparing the positive and negative regions, using the "#" row for the *x*-axis. The "#" row indicates Branch Length Score (BLS) for conservation and degree of single strandedness for accessibility. The results for Nova are shown in Fig. 1.

3.4 Run the Model on the Whole Genome

1. Run the Nova model on the mm10 genome:

```
$ mCarts -v --exist-model ./Nova_HMM_D30_m3
```

If the program finished without errors, the following additional files should be created in the *Nova_HMM_D30_m3* directory:

- cluster.bed
- out (directory)
- qsub (directory; only if SGE is available)
- scripts (directory; only if SGE is available)
- scripts.list (only if SGE is available)

2. Convert the motif cluster BED file into a bedGraph file:

```
$ perl ~/src/CIMS/tag2profile.pl -ss -exact -weight -of bedgraph -n
"Nova_motif" -v ./Nova_HMM_D30_m3/cluster.bed
./Nova_HMM_D30_m3/cluster.bedGraph
```

3.5 Visualizing and Interpreting the Results

1. From the plots generated by the model training, we observe the following:
   - YCAY motifs are clustered more closely in positive training regions.
   - Positive regions are more accessible (more single stranded).
   - Positive regions have higher conservation in the 5′ UTR, CDS, intron, and 3′ UTR.

2. The cluster.bedGraph file generated by mCarts can be loaded into a genome browser such as the UCSC Genome Browser.

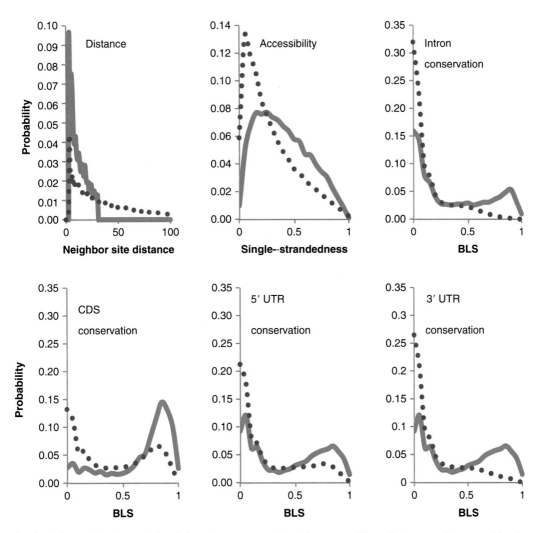

Fig. 1 Features of positive (*solid line*) and negative (*dashed line*) Nova YCAY clusters as determined by the mCarts model

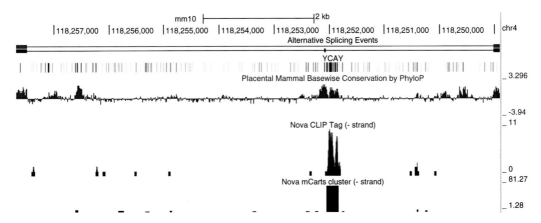

Fig. 2 Exon 6 of Ptprf contains a cluster of highly conserved YCAYs. The motif cluster predicted by mCarts matches these and the binding profile determined by HITS-CLIP

This allows for the visualization of RBP binding clusters and their associated scores (Fig. 2). Figure 2 shows exon 6 of *Ptprf*, which contains 22 highly conserved YCAY elements and whose inclusion has been previously shown to be activated by Nova [15].

4   Notes

1. This protocol assumes familiarity with the UNIX command line. There are many great introductory resources available (e.g., ref. [16, 17]), but instruction in its use is beyond the scope of this protocol.

2. Unix-based operating systems contain a special file, ~/.bash_profile, which is automatically executed upon starting the bash shell. To avoid having to add software to your path manually each time you open a new terminal window, you can add the commands directly to ~/.bash_profile. Simply edit the file and add the commands of interest, then reload the profile manually using:

```
$ . ~/.bash_profile
```
Note the "." at the beginning.

3. The sample CLIP data we provide for this protocol is from ref. [18]. It consists of 4,401,528 unique tags originally mapped to mm9. We used the LiftOver utility (*see* **Note 4**) to translate the coordinates to mm10, resulting in 4,401,394 unique tags. Another important detail to note is that the results presented in this protocol will differ slightly from those presented in previous work [14, 18] due to the use of a different clustering algorithm. The method described here is more straightforward and has been successfully used in subsequent work [19].

4. Regarding data pre-processing, stringent mapping and filtering of CLIP data are critical for defining robust RBP binding sites. Detailed discussion of CLIP data processing is beyond the scope of this protocol, but readers are referred to the CIMS software package we developed [8]. It is often the case that the raw CLIP data for your RBP of interest was aligned to an earlier version of the reference genome. For example, the Nova data in this protocol was previously mapped to mm9. To convert the mm9 coordinates to mm10 coordinates, we use the LiftOver utility developed by the UCSC Genome Browser group (https://gcnome-store.ucsc.edu) [20]. The required chain files can be downloaded from UCSC as well (http://hgdownload.cse.ucsc.edu/downloads.html). For converting mm9 to mm10, download and unzip http://hgdownload.cse.ucsc.edu/goldenPath/mm9/liftOver/mm9ToMm10.over.chain.gz, then execute the following command:

```
$ liftOver Nova_CLIP_unique_tag_mm9.bed mm9ToMm10.over.chain
Nova_CLIP_uniq_mm10.bed Nova_CLIP_unique_tag_mm9mm10.unmapped
```

In some cases, such as this one, the BED file contains track lines that LiftOver can't handle (you will get an error). To get rid of these lines:

```
$ grep -v "track" file.bed > file.noheader.bed
```

5. To focus the model training on the most robust clusters, we pick the set of clusters with the greatest peak height (PH). The cutoff value depends on specific datasets, but in our experience based on cross-validation analyses, the exact value does not greatly affect the outcome. We generally pick a threshold where at least 5000–6000 confident clusters are obtained to reduce the variation in parameter estimation. The following command provides a summary of the peak heights, listing (1) the PH, (2) the number of clusters with that PH, and (3) the cumulative number of clusters at that peak height:

```
$ cut -f5 Nova_CLIP_uniq_mm10.cluster.PH.center.ext50.normsk.ext1k.bed
| sort -nr | uniq -c | awk 'BEGIN{cumul=0} {print $2"\t"$1"\t"$1+cumul;
cumul=$1+cumul}'
```

For this dataset, we choose to set the cutoff at 15, which corresponds to 7700 clusters.

6. In this mCarts protocol, we run the analysis with the following parameters:

- ref mm10: the reference genome being used.
- f CLIP.pos.bed: the foreground (positive) training set.
- b CLIP.neg.bed: the background (negative) training set.
- lib ~/data/mCarts_lib_data_mm10: the location of the mCarts library files for mouse.
- w YCAY: the motif we are searching for (IUPAC code is allowed).
  mCarts currently does not accept "U" in the motif so be sure to provide a "T" instead (e.g., "TGCATG" instead of "UGCAUG").

  min-site 3: the minimum number of sites in a cluster.

  max-dist 30: the maximum distance between neighboring sites in a cluster.

  train-only: only train for now; we will test in the next step.

- v: verbose; print out what the software is doing.

The full mCarts documentation is available at http://zhanglab.c2b2.columbia.edu/index.php/MCarts_Documentation and a full description of the methodology in ref. [14].

As of this writing, the direct software links are as follows:

mCarts: http://sourceforge.net/p/mcarts/.

czplib: http://sourceforge.net/p/czplib/.

CIMS: http://sourceforge.net/p/ngs-cims/.

Acknowledgements

The authors would like to thank Lauren E. Fairchild and Huijuan Feng for their assistance in testing the protocol and for providing feedback on the manuscript. This work was supported by grants from the National Institutes of Health (NIH) (R00GM95713) and the Simons Foundation Autism Research Initiative (297990 and 307711) to C.Z.

References

1. Licatalosi DD, Darnell RB (2010) RNA processing and its regulation: global insights into biological networks. Nat Rev Genet 11:75–87

2. Ray D, Kazan H, Cook KB et al (2013) A compendium of RNA-binding motifs for decoding gene regulation. Nature 499:172–177

3. Cook KB, Kazan H, Zuberi K et al (2011) RBPDB: a database of RNA-binding specificities. Nucleic Acids Res 39:D301–D308

4. Chasin LA (2007) Searching for splicing motifs. Adv Exp Med Biol 623:85–106

5. Galarneau A, Richard S (2005) Target RNA motif and target mRNAs of the Quaking STAR protein. Nat Struct Mol Biol 12:691–698

6. Licatalosi DD, Mele A, Fak JJ et al (2008) HITS-CLIP yields genome-wide insights into brain alternative RNA processing. Nature 456:464–469

7. König J, Zarnack K, Rot G et al (2010) iCLIP reveals the function of hnRNP particles in splicing at individual nucleotide resolution. Nat Struct Mol Biol 17:909–915

8. Moore MJ, Zhang C, Gantman EC et al (2014) Mapping Argonaute and conventional RNA-binding protein interactions with RNA at single-nucleotide resolution using HITS-CLIP and CIMS analysis. Nat Protoc 9:263–293

9. Darnell RB (2010) HITS-CLIP: panoramic views of protein–RNA regulation in living cells. Wiley Interdiscipl Rev RNA 1:266–286

10. Blencowe BJ, Ahmad S, Lee LJ (2009) Current-generation high-throughput sequencing: deepening insights into mammalian transcriptomes. Genes Dev 23:1379–1386

11. Maticzka D, Lange SJ, Costa F et al (2014) GraphProt: modeling binding preferences of RNA-binding proteins. Genome Biol 15:R17

12. Cereda M, Pozzoli U, Rot G et al (2014) RNAmotifs: prediction of multivalent RNA motifs that control alternative splicing. Genome Biol 15:R20

13. Han A, Stoilov P, Linares AJ et al (2014) De novo prediction of PTBP1 binding and splicing targets reveals unexpected features of its RNA recognition and function. PLoS Comput Biol 10:e1003442

14. Zhang C, Lee K-Y, Swanson MS et al (2013) Prediction of clustered RNA-binding protein motif sites in the mammalian genome. Nucleic Acids Res 41:6793–6807

15. Jelen N, Ule J, Živin M et al (2007) Evolution of nova-dependent splicing regulation in the brain. PLoS Genet 3:e173–e1847

16. Stein LD (2002) Unix survival guide. John Wiley, Hoboken, NJ

17. Buffalo V (2015) Bioinformatics data skills. O'Reilly Media, Sebastopol, CA

18. Zhang C, Frias MA, Mele A et al (2010) Integrative modeling defines the Nova splicing-regulatory network and its combinatorial controls. Science 329:439–443

19. Weyn-Vanhentenryck SM, Mele A, Yan Q et al (2014) HITS-CLIP and integrative modeling define the rbfox splicing-regulatory network linked to brain development and autism. Cell Rep 6:1139–1152

20. Hinrichs AS, Karolchik D, Baertsch R et al (2006) The UCSC genome browser database: update 2006. Nucleic Acids Res 34:D590–D598

Chapter 18

Design of RNA-Binding Proteins: Manipulate Alternative Splicing in Human Cells with Artificial Splicing Factors

Yang Wang and Zefeng Wang

Abstract

The majority of human genes undergo alternative splicing to produce multiple isoforms with distinct functions. The dysregulations of alternative splicing have been found to be closely associated with various human diseases; thus new approaches to modulate disease-associated splicing events will provide great therapeutic potentials. Here we report protocols for constructing novel artificial splicing factors that can be designed to specifically modulate alternative splicing of target genes. By following the method outlined in this protocol, it is possible to design and generate artificial splicing factors with diverse activities in regulating different types of alternative splicing. The artificial splicing factors can be used to change splicing of either minigenes or endogenous genes in cultured human cells, providing a new strategy to study the regulation of alternative splicing and function of alternatively spliced products.

Key words Alternative splicing, Artificial splicing factors, SR proteins, hnRNP proteins, RRM, RS domain, Glycine-rich domain

1 Introduction

The majority of genes in higher eukaryotes are transcribed as pre-mRNAs containing multiple introns that are removed to generate mature mRNA through RNA splicing. As an essential step of gene expression, splicing involves a series of precisely controlled reactions catalyzed by a dynamic ribonucleoprotein complex known as the spliceosome. Most human genes undergo alternative splicing (AS) to produce multiple spliced isoforms [1], providing a major mechanism to regulate gene functions and increase the coding complexity of human genome. The choice of different spliced isoforms is tightly regulated in different tissues and developmental stages, and the disruption of splicing regulations is a common cause of a variety of human diseases [2]. Therefore, modulation of alternative splicing may have broad applications in both basic molecular research and translational research.

Ren-Jang Lin (ed.), *RNA-Protein Complexes and Interactions: Methods and Protocols*, Methods in Molecular Biology, vol. 1421, DOI 10.1007/978-1-4939-3591-8_18, © Springer Science+Business Media New York 2016

Generally, the splicing process is regulated by multiple *cis*-elements that serve as splicing enhancers or silencers to control core splicing events. Based on the activities and relative locations of these *cis*-elements, they are classified as exonic splicing enhancers (ESEs) or silencers (ESSs), and intronic splicing enhancers (ISEs) or silencers (ISSs). The splicing regulatory *cis*-elements are recognized by various *trans*-acting protein factors (i.e., splicing factors) to either promote or suppress the use of nearby splice sites [3, 4]. Many splicing factors have modular organization, with separate sequence-specific RNA-binding modules to recognize their targets and splicing effector module to change splicing. The best known examples are the members of the Serine/Arginine-rich (SR) protein family, which contain N-terminal RNA recognition motifs (RRMs) that bind to *exonic splicing enhancers* (ESEs) in pre-mRNAs and C-terminal RS domains that promote exon inclusion [5]. As another example, the hnRNP protein hnRNP A1 binds to *exonic splicing silencers* (ESSs) through its RRM domains and inhibits exon inclusion through a C-terminal Glycine-rich domain [6].

Because the majority (~90 %) of all human genes undergo alternative splicing, engineering novel splicing factors to specifically modulate splicing offers tremendous promise for both basic and applied research. This can be achieved in principle by a modularly designed protein factor that contains a target recognition module and a splicing effector module. However, natural splicing factors usually contain RNA recognition motif (RRM) or K homology (KH) domains that recognize short RNA elements with moderate affinities. It is impractical to engineer an RNA recognition module using these domains due to their weak RNA-binding affinity and the absence of a predictive RNA recognition "code." By using the unique RNA recognition module of PUF domains (named for *Drosophila* Pumilio and *C. elegans fem-3* binding factor), we have, for the first time, engineered novel artificial splicing factors that can specifically recognize almost any given pre-mRNA target [7]. These *artificial* factors are engineered by combining different splicing regulatory domains to a modified PUF domain that specifically recognize different RNA sequences. The canonical PUF domain contains eight repeats of three α-helices, each recognizing a single base in an 8-nt RNA target. The amino acids at the certain positions of the second α-helix form specific hydrogen bonds with the edge of RNA base and thus determine the RNA-binding specificity (Fig. 1a). The code for base recognition of PUF repeat is surprisingly simple (Fig. 1a), which can direct the mutagenesis of PUF domain to generate modified PUFs that recognize any possible 8-base (reviewed in Ref. 8). By choosing different functional domains, such *engineered splicing factors* (ESFs) can function as either splicing activator or splicing inhibitor and can

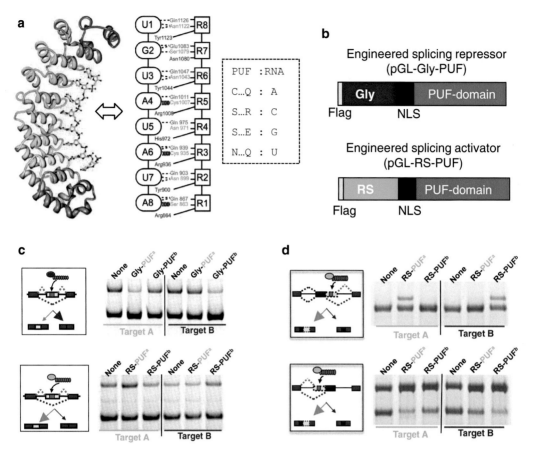

Fig. 1 Design of ESFs and their activity in modulating exon skipping. (**a**) Specific binding between PUF domain and RNA targets is illustrated with RNA-PUF structure and a schematic diagram. The PUF binding code for each of the four RNA bases is shown at *right*, which is used to design PUF mutations. (**b**) Modular domain organization of ESFs. The C-terminal RNA-binding module (PUF domain) is fused with a functional module (Gly-rich domain or RS domain) to produce novel splicing repressors or activators with designed specificity. In addition, a NLS and the FLAG epitope tag are included to facilitate nuclear localization and detection of ESFs. Expression of the ESFs is driven by CMV promoters (*arrow*). (**c**) Gly-PUF ESFs were co-expressed with exon skipping reporters, and splicing pattern was assayed by RT-PCR. The modified PUFᵃ and PUFᵇ specifically bind to 8-mer targets A and B respectively (*in same colors*). All combinations were used so the PUF-target pairs of different color serve as controls. The effects of RS-PUF on exon skipping (**c**), competing 5′ or 3′ splice site reporter (**d**) were assayed by methods similar to panel (**c**)

regulate various types of alternative splicing events associated with splicing reporters or endogenous genes.

Here we describe in detail how to design and construct ESFs and how to use them to manipulate alternative splicing. We focus on its application in cultured human cells and the common assay of ESF activities of splicing modulation. Using the Bcl-x gene as example, we also describe how to manipulate the splicing of endogenous genes and analyze the downstream effect of splicing changes.

2 Materials

2.1 Reagents

Information regarding vendors and suppliers provided herein is for convenience rather than a strict requirement. Many commonly used reagents are available from various reliable sources.

1. Plasmids encoding ESFs (pGl-RS-PUF, pGl-Gly-PUF).

2. Splicing reporter plasmid (pGZ3-NRE, pEZ-1B-A6G, pEZ-2F-GU).

3. Lentiviral expression vector pWPXLd-Gly-PUF.

4. QuickChange Site-Directed Mutagenesis kit (Stratagene).

5. Cultured cells (HEK293T cells, MDA-MB-231 cells, A549 cells, HeLa cells).

6. Dulbecco's modified Eagle medium (DMEM; Invitrogen).

7. Opti-MEM I (Invitrogen).

8. Lipofectamine 2000 transfection reagent (Invitrogen).

9. TRIzol reagent (Invitrogen) ! CAUTION TRIzol reagent includes phenol, which can cause burns. Wear gloves when handling.

10. 70 % ethanol.

11. DNase I.

12. Superscript III reverse transcriptase (Invitrogen).

13. Cy5-dCTP (GE Healthcare Life Sciences).

14. Primers used for RT and PCR: *see* Table 1

15. Acrylamide (Sigma) ! CAUTION Neurotoxic; Take precautions to avoid contact and ingestion.

Table 1
Primers used in constructing ESF expression vectors

Primer name	Sequence	Notes
Pum-F1	CAC<u>GGATCC</u>TCCCCCCCCAAGAAAAAGAGGA AGGTA<u>TCTAGA</u>GGCCGCAGCCGCCTTTTG	Encodes NLS between BamHI and XbaI sites
Pum-R1	GTG<u>GTCGAC</u>TTACCCTAAGTCAACACC	Encodes a stop codon and SalI site
ASF-RS-F	CAC<u>GCTAGC</u>ATGGACTACAAGGACGACGATGACA AGGGT<u>CTCGAG</u>AGAAGTCCAAGTTATGGAAG	Encodes an N-terminal FLAG tag after NheI site
ASF-RS-R	CAC<u>GGATCC</u>CCGTACGAGAGCGAGATCTG	Contains BamHI site for cloning

Restriction enzyme digestion sites are *underlined*

16. Bis-acrylamide (Fluka) ! CAUTION Neutrotoxic; wear gloves when handling.

17. 10 % 1× TBE PAGE gel (*see* Subheading 2.2).

18. PBS (phosphate balance saline).

19. FITC-conjugated goat anti-mouse IgG.

20. Vector shield's mounting medium with DAPI.

21. SDS-PAGE gel running buffer (*see* Subheading 2.2).

22. 2× SDS-PAGE sample buffer (*see* Subheading 2.2).

23. 12 % SDS-PAGE gel (*see* Subheading 2.2).

24. Antibodies: Caspase-3 (Cell signaling); PARP (Cell Signaling); Bcl-x (BD Bioscience); Actin (Sigma); Alpha-tubulin (Sigma); FLAG M5 (Sigma).

25. Nitrocellulose membrane (Amersham-Pharmacia).

26. ECL Western Blotting Detection Reagents (GE Healthcare).

27. Cell proliferation reagent WST-1 (ROCHE).

28. Cisplatin (Sigma).

29. Paclitaxel (Sigma).

30. TNF-alpha (GenScript).

31. TRAIL (GenWay).

2.2 Reagent Setup

1. *Cultured cells*: The human embryonic kidney cell line 293 T, lung cancer cell line A549, breast cancer cell line MDA-MB-231, and human cervical cancer cell line HeLa were grown in Dulbecco's modified Eagle's medium (DMEM) supplemented with 10 % fetal bovine serum.

2. *30 % (wt/vol) Acrylamide (29:1) solution*: Add 29 g of acrylamide and 1 g of bis-acrylamide into water to a final volume of 100 ml.

3. *2× SDS-PAGE sample buffer*: 0.2 % bromophenol blue (wt/vol), 20 % glycerol (vol/vol), 100 mM Tris–HCl, pH 6.8, 4 % SDS (wt/vol), 715 mM 2-mercaptoethanol.

4. *5× SDS-PAGE running buffer*: Adding 15.14 g of Tris, 93.8 g of glycine, and 5 g of SDS to 1 l H_2O. Adjust the pH to 8.3 and store at room temperature.

5. 12 % SDS-PAGE separating gel.

	Separating gel (10 ml)
H_2O (ml)	3.3
30 % Acrylamide (wt/vol) (29:1) (ml)	4.0
1.5 M Tris–HCl (pH 8.8) (ml)	2.5
10 % SDS (wt/vol) (ml)	0.1

	Separating gel (10 ml)
10 % APS (wt/vol) (ml)	0.1
TEMED (ml)	0.004

	Stacking gel (5 ml)
H$_2$O (ml)	3.4
30 % Acrylamide (wt/vol) (29:1) (ml)	0.83
1 M Tris–HCl (pH 6.8) (ml)	0.63
10 % SDS (wt/vol) (ml)	0.05
10 % APS (wt/vol) (ml)	0.05
TEMED (ml)	0.005

6. 10 % 1× TBE PAGE gel (10 ml).

H$_2$O (ml)	3.4
30 % Acrylamide (wt/vol) (29:1) (ml)	3.3
5× TBE	2.0
10 % APS (wt/vol) (ml)	0.10
TEMED (ml)	0.004

3   Methods

3.1   Construction of ESF Expressing Plasmids (See Note 1)

Generate ESF expression constructs by using the pCI-neo vector. We started with an expression construct that encodes from the N-terminus to the C-terminus: FLAG epitope, Gly-rich domain of hnRNP A1 (residues 195–320 of NP_002127), and the MS2 coat protein (gift of Dr. R. Breathnach from Institut de Biologie-CHR [6]).

3.1.1   Engineered Splicing Repressor

1. The fragment encoding the MS2 coat protein fragment is removed using *Bam*HI/*Sal*I digestion and replaced with a fragment encoding a NLS (PPKKKRKV) and the PUF domain of human Pumilio1, which is generated by PCR amplification using primers Pum-F1 and Pum-R1 (*see* Table 1).

2. The resulting construct expresses a Gly-PUF type ESF (pGL-Gly-PUF) under the control of a CMV promoter (Fig. 1b, top panel).

3.1.2   Engineered Splicing Activator

1. To make an expression construct for an RS-PUF type ESF (pGL-RS-PUF) (Fig. 1b, bottom panel), remove the fragment encoding the FLAG/Gly-rich domain with *Nhe*I/*Bam*HI digestion,

and replace it with a fragment that encodes the RS domain of ASF/SF2 protein with an N-terminal FLAG epitope, which is amplified using primers ASF-RS-F and ASF-RS-R (*see* Table 1).

3.1.3 ESF Variants

1. To generate constructs for different ESFs with mutated PUF domains that recognize different RNA targets (pGL-Gly/RS-PUF), we design the PUF mutations according to the binding code of PUF repeat [8] and introduce the point mutations in consecutive steps using a QuikChange Site-Directed Mutagenesis kit [9].

2. The additional functional domains of ESF (RS domains/fragment or Gly-rich domains/fragment) are amplified by PCR (or synthesized oligonucleotides for short fragments), and cloned between XhoI and BamHI sites.

3. The RS domains are residues 123–238 of 9G8 (NP001026854), residues 180–272 of SRp40 (NP008856), residues 117–221 of SC35 (NP003007), and an (RS)$_6$ sequence.

4. The Gly-rich domains are residues 203-353 of hnRNP A2 (NP112533), residues 211–378 of hnRNP A3 (NP919223), and the short peptide of GYGGGGPGYGNQGGGYGGG.

3.2 Construction of Splicing Reporter (See Note 2)

3.2.1 To Assess the Effect of ESFs on Inclusion/Skipping of a Target Exon

1. A modular reporter (pGZ3) is used for insertion of splicing regulatory sequences (each recognized by different ESFs, shown in different colors in Fig. 1c, d) into the test exon (Exon 12 of the human IGF-II mRNA-binding protein 1, IGF2BP1, Ensembl ID ENSG00000159217).

2. The test exon, together with its flanking introns, is inserted between two GFP exons in pEGFP-C1 vectors as described previously [10].

3. To insert target sequences of PUF domains into this reporter vector, synthesize and anneal oligonucleotides containing the candidate sequences (designated by N$_8$) flanked by *Xho*I and *Apa*I sites.

4. The resulting DNA fragments are digested and ligated into the *Xho*I/*Apa*I digested vector (inside the test exon).

5. The inclusion of the test exon is assayed by body-labeled RT-PCR, as described previously, with primers corresponding to the first and third exon (two GFP exons) of the reporter minigene [11].

3.2.2 To Assess the Effect of ESFs on the Alternative Use of 5′ and 3′ ss

1. Use reporters with competing 5′ (pEZ-1B) and 3′ ss (pEZ-2F) [11].

2. The corresponding target sequences of PUF domains (NRE, A6G, GU) are inserted into these reporters using either *Xho*I/*Apa*I sites (for the competing 3′ ss reporter) or *Xho*I/*Eco*RI sites (for the competing 5′ ss reporter) as described above.

3.3 Construction of Lentiviral Expression Vectors

1. The full-length ESFs are PCR amplified from original expression vectors and integrate into the lentiviral expression vector pWPXLd between MluI and SpeI sites.

2. Lentiviruses are generated by co-transfecting HEK293 cells with packaging vectors pPAX2 and pMD2.G with either pWPXLd-Gly-PUF (531), pWPXLd-Gly-PUF (wt) (Control), or pWPXLd-GFP (mock) using the standard calcium phosphate precipitation method.

3. The titer of lentivirus is determined by infecting HEK293 cells with serial dilutions of the virus preparation.

3.4 Analysis of Exon Inclusion or Alternative Use of Splice Sites of Minigene Reporters

3.4.1 Co-transfection of ESF Plasmids and Splicing Reporter Plasmids

1. Seed 2×10^5 HEK293T cells into each well of a 24-well plate and grow the cells overnight in a humidified incubator at 37°C and 5 % CO_2.

2. Mix Lipofectamine 2000 by gently inverting bottles before use, then dilute 2 µl lipofectamine 2000 in 50 µl Opti-MEM I medium. Mix gently and incubate for 5 min at room temperature.

3. Dilute 0.04 µg of pGL-Gly-PUF expression vectors and 0.2 µg of pGZ3 reporter plasmids or 0.4 µg of pGL-RS-PUF expression vectors and 0.2 µg of pGZ3 reporter plasmids or 0.4 µg of pGL-RS-PUF expression vectors and 0.2 µg of pEZ-1B or pEZ-2F reporter plasmids in 50 µl Opti-MEM I medium in a sterile tube.

4. After 5 min incubation, mix the diluted Lipofectamine 2000 with the diluted plasmids gently and incubate for 20 min at room temperature to allow complex formation to occur.

5. After incubation, add the transfection mixture to each well and incubate for at least 12 h in the incubator.

3.4.2 Isolation of RNA

1. After 12 h incubation, discard medium from each well and wash it with 500 µl PBS per well.

2. Discard PBS and add 200 µl trypsin to each well, incubate in the incubator for 5 min. Then add 1 ml medium to stop the digestion and transfer the cells to a sterile 1.5 ml tube.

3. Centrifuge tubes for 3 min at $5000 \times g$ and discard the medium. Then add 500 µl TRIzol reagent per tube to lyse cells by repetitive pipetting. Incubate the homogenized samples for 5 min.

4. For each TRIzol treated sample, add 0.1 ml of chloroform per 500 µl of TRIzol reagent. Invert tubes for 15 s and incubate for 3 min at room temperature.

5. Centrifuge tubes for 15 min at $12,000 \times g$ at 4°C and transfer the aqueous phase to a fresh tube and use 0.25 ml of isopropanol per 0.5 ml of TRIzol reagent used for the initial homogenization.

6. Mix rigorously and incubate at room temperature for 10 min and then centrifuge at $12,000 \times g$ for 10 min at 4°C. The RNA precipitate is usually visible after centrifugation on the bottom of the tube.

7. Discard the supernatant and wash the RNA pellet with 0.5 ml of 75 % ethanol per 0.5 ml of TRIzol reagent used before.

8. Mix rigorously and centrifuge at $7500 \times g$ for 5 min at 4 °C. Remove the supernatant and dry the RNA pellet. Then dissolve RNA in 50 μl RNase-free water.

9. Add 2 μl DNase I, 7 μl buffer, and 11 μl H_2O to each 50 μl RNA solution and incubate at 37 °C for 1 h and then heat inactivate DNase at 70 °C for 15 min.

3.4.3 Detection of Splice Variants by RT-PCR

1. For each sample, add the following components to a 0.2 ml nuclease-free tube:

Primer: Gexon 3r(10 μM) or Oligo dT	1 μl
RNA (2 μg)	5 μl
dNTP (10 mM)	1 μl
H_2O	3 μl

2. Heat mixture to 65 °C for 5 min and incubate on ice for at least 2 min to prevent reformation of the secondary structure, then add the following components to the same tube.

5× Fist strand buffer	4 μl
DTT (0.1 M)	1 μl
Superscript III RT	1 μl
H_2O	4 μl

3. Mix by pipetting gently up and down and incubate at 50 °C for 60 min and then inactivate the reaction by heating at 70 °C for 15 min.

4. For each sample, add the following to a PCR tube to do the body-labeled PCR.

10× PCR buffer	2.5 μl
10 mM dNTP mix	0.5 μl
Gexon 1f (10 μm) or Bcl-x-fwd (10 μm)	1 μl
Gexon 3r (10 μm) or Bcl-x-rev (10 μm)	1 μl
Taq DNA polymerase (5U/μl)	0.25 μl
Cy5-dCTP	0.5 μl
cDNA	2 μl

5. Heat reaction to 94 °C for 2 min to denature and perform 25 cycles of PCR with 94 °C 30 s, 60 °C 30 s, and 72 °C 30 s, then followed by 72 °C 7 min.

6. PCR products were resolved by electrophoresis through a 10 % PAGE gels with 1× TBE buffer and scanned with a Typhoon 9400 scanner. The amount of each splicing isoform is measured with ImageQuant 5.2 software.

3.5 Analysis of Endogenous Bcl-x Splicing (See Note 3) and Measure Its Effects on Apoptosis (See Note 4)

3.5.1 Transfection of ESF Plasmids

1. Plate 2×10^5 HeLa cells to each well of a 24-well plate and then grow cells overnight in an incubator at 37 °C and 5 % CO_2.

2. After 12 h, transfect cells with 2 μg of pGL-Gly-PUF(wt), 0.2, 1, 2 μg of pGL-Gly-PUF(531) respectively.

3. 24 h later, harvest the cells, 1/3 of the cells for the RNA isolation and PCR analysis (*see* Subheadings 3.4.2 and 3.4.3) and 2/3 for the protein isolation.

3.5.2 Western Blot Analysis of Apoptotic Protein Markers

1. For the western blot analysis, the total cell pellets were boiled in 2× SDS-PAGE loading buffer for 10 min and then resolved by 12 % SDS-PAGE gel and transferred the proteins onto a nitrocellulose membrane.

2. Block the membrane with 5 % milk for 1 h at room temperature and incubate the membrane with Caspase-3 (1:1000), PARP (1:1000), beta-actin (1:5000) primary antibodies respectively at 4 °C overnight.

3. Wash the membrane with PBS containing 0.1 % Tween 20 (PBS-T) for 5 min at room temperature three times on a rocking shaker, then incubate the membrane with HRP-linked secondary antibodies for 1 h at room temperature.

4. Wash the membrane with PBS-T three times at room temperature and then develop the membrane using ECL Western Blotting detection reagents.

3.5.3 Detection of the ESFs (See Note 5) and DNA Fragmentation (See Note 6)

1. For the immunofluorescence assay measuring the apoptosis, 5×10^5 HeLa cells were seeded onto poly-lysine coated glass coverslips in a 6-well plate and transfected with pGL-Gly-PUF(wt) or pGL-Gly-PUF(531) plasmids using Lipofectamine 2000 (*see* **steps 5–9**).

2. 24 h after transfection, fix the cells on coverslips with 1 ml 4 % formaldehyde in 1× PBS for 20 min at room temperature.

3. Gently wash cells on coverslips by adding 2 ml 1× PBS and incubate for 5 min, then remove the PBS with pipettes. Repeat the wash for three times.

4. Permeabilize cells with 0.2 % Triton X-100 in 1× PBS for 10 min and then wash with 1× PBS three times.

5. Block cells with 3 % BSA in 1× PBS for 10 min and wash with 1× PBS three times.

6. Dilute the FLAG antibody to 1:1000 in 3 % BSA/PBS, and pipet 30 μl of the diluted FLAG antibody into a parafilm sheet.

7. Take out the coverslip with cells, dry the excess buffer carefully with Kimwipes, and put it upside down (the cell side down) on the 30 μl primary solution, incubate for 1 h at room temperature.

8. Add about 500 µl of 1× PBS from the side of coverslips incubated with the FLAG antibody to allow it float on top of the solution, then take it back to the 6-well plate with 1× PBS in the well.

9. Wash it again with 1× PBS three times and 5 min for each time.

10. Dilute the anti-mouse secondary antibody (1:500) in 3 % BSA/PBS; again, use 30 µl for each coverslip; put the coverslip upside down on the secondary antibody solution, incubate for 15 min at room temperature.

11. Remove the coverslips from the parafilm as described in **step 8** and put it back to 6-well plate, wash again with 1× PBS three times and 5 min for each time.

12. Mount the coverslips with mounting medium (with DAPI), absorb the excess medium carefully with Kimwipes, and seal the edge with nail polish.

13. Visualize cells using an Olympus fluorescence microscope and photograph using a digital camera.

3.6 Viral Transduction of ESF and the Measurement of Apoptosis by Flow Cytometry

1. To detect the apoptosis with propidium iodide staining, split 2×10^6 HeLa cells, MDA-MB-231 cells, and A549 cells into 60 mm dishes respectively.

2. 24 h later, refresh the medium and prepare the lentivirus.

3. Dilute ten million lentivirus pWPXld-Gly-PUF(wt), pWPXld-Gly-PUF(531), or pWPXld-GFP stocks into 4 ml of fresh medium to make the ratio of virus vs. cell number equal to 5.

4. Change the medium in the plates with 4 ml of virus containing medium prepared in **step 3**. 12 h after the infection, change the medium.

5. After 24 h of infection, collect cells and stain for 5 min in a PBS solution containing a final concentration of 2 µg/ml propidium iodide (PI).

6. Analyze the PI-stained cells with a FACScalibur fluorescence-activated cell sorter (FACS) using CELLQuest software (*see* **Note 7**).

3.7 Determine the Effect of ESFs on Sensitizing Cancer Cells to Anti-tumor Drugs

1. For the cell viability assay, HeLa cells, MDA-MB-231 cells, and A549 cells were infected with lentivirus expressing GFP (pWPXld-GFP), control ESF (pWPXld-Gly-PUF (wt)), or designer ESF (pWPXld-Gly-PUF (531)) (*see* **steps 1–4** in Subheading 3.6).

2. After 72 h of infection, seed 5×10^4 cells per well in 96-well plates and grow cells overnight.

3. The next morning, add cisplatin (5 µM), paclitaxel (10 nM), TNF-alpha (10 nM), TRAIL (2 µg/ml) to each cell line in the 96-well plates, and incubate at 37 °C in the presence of 5 % CO_2 for 24 h.

4. After 24 h incubation, add 10 μl of proliferation reagent WST-1 to each well. The cells are further incubated with WST-1 reagent for 30 min and the absorbance at 450 nm is measured using a Benchmark microplate reader. At least three independent experiments need to be performed for each sample (*see* **Note 8**).

4   Notes

1. We have demonstrated methods for engineering splicing factors construction, splicing reporters construction, and how to use ESFs to modulate the splicing of endogenous gene and provide therapeutic potentials [7].

2. To provide a principle of how the ESFs work, we have included examples derived from our own work. The ESFs not only can be used to promote or inhibit the inclusion of the cassette exon containing a cognate target sequence (Fig. 1c) but also can affect the usage of alternative splice sites in the reporter system (Fig. 1d).

3. The ESFs can also be applied to specifically modulate RNA splicing of endogenous genes. We demonstrate this potential by targeting Bcl-x that produces two antagonistic isoforms using alternative 5′ splice sites. We design a Gly-PUF (531) ESF to recognize an 8-mer between the two alternative 5′ splice sites and "reprogram" splicing code to increase Bcl-xS isoforms (Fig. 2a). When transfected into HeLa cells where Bcl-xL is the predominant form, the Gly-PUF (531) ESF increases splicing of the Bcl-xS isoform in a dose-dependent manner, whereas the control ESF, Gly-PUF(wt), does not affect the Bcl-xS level (Fig. 2b). The increase of Bcl-xS protein level is also confirmed by Western blotting using a Bcl-x antibody (Fig. 2c).

Fig. 2 (continued) PUF(531) was fused with the Gly-rich domain of hnRNP A1 to inhibit use of the downstream 5′ ss (indicated with the *heavy red arrow*). (**b**) Modulation of Bcl-x 5′ ss usage. HeLa cells were transfected with different amounts of the Gly-PUF(531) expression construct. Gly-PUF(wt) was used as a control. Two isoforms of Bcl-x are detected with RT-PCR using primers corresponding to exons 1 and 3 of the Bcl-x gene. The percentage of Bcl-xS isoform is quantified and shown at the bottom. (**c**) Expression levels of Bcl-xL and Bcl-xS in the presence of ESFs. Samples are in the same order as panel (**b**), and all proteins are detected by Western blots. The expression of ESFs is detected with anti-Flag antibody, and the tubulin level is used as control. The blot was exposed for a longer time for Bcl-xS because the available Bcl-x antibody detects Bcl-xL with much higher sensitivity. (**d**) Cleavage of PARP and caspase 3 in HeLa cells transfected with different amounts of ESF expression constructs. Samples are detected by western blot at 24 h after transfection. (**e**) Localization of ESFs in transfected HeLa cells detected using immunofluorescence microscopy with anti-FLAG antibody. The cells were co-stained with DAPI to show nuclei. Some nuclei, especially in cells transfected with Gly-PUF(531), are fragmented due to apoptosis. (**f**) Percentage of cells undergoing apoptosis (i.e., with fragmented nuclear DNA) from pictures of randomly chosen fields. Data are from Ref. 7

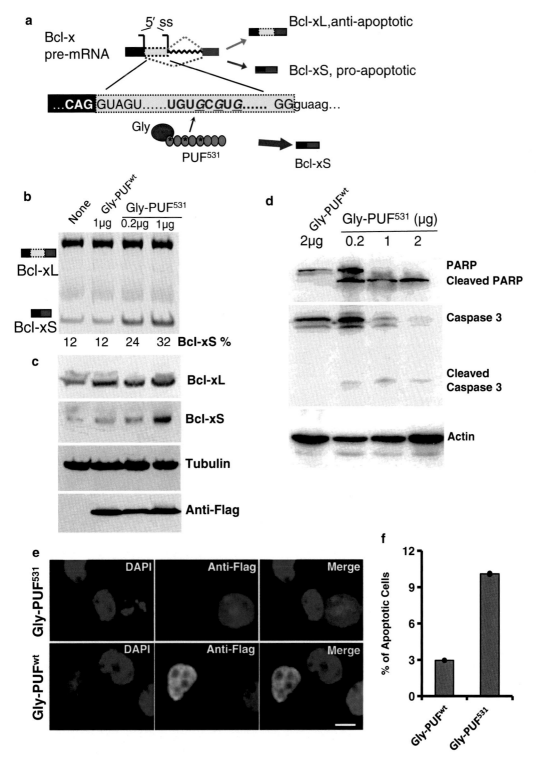

Fig. 2 Use an ESF to modulate splicing of endogenous Bcl-x pre-mRNA. (**a**) Alternative splicing of Bcl-x pre-mRNA. Two alternative 5′ ss in exon 2 of Bcl-x are used to generate Bcl-xL and Bcl-xS. The sequence UGUGCGUG between the two splice sites was chosen as the ESF target, and mutations (*asterisks*) were introduced in wild-type PUF repeats 1, 3, and 5 (Q867E/Q939E/C935S/Q1011E/C1007S) to recognize this sequence. The resulting

4. The increased splicing of the pro-apoptotic Bcl-x isoform can promote cells expressing the designer ESF (Gly-PUF (531)) to undergo apoptosis. The induction of the pro-apoptotic Bcl-xS isoform leads to cleavage of caspase 3 and poly (ADP-ribose) polymerase (PARP), two known molecular markers in the apoptosis pathway (Fig. 2d).

5. Using immunofluorescence microscopy with anti-FLAG antibodies, the ESFs are predominantly localized in the nuclei of the transfected cells (Fig. 2e).

6. Many cells expressing Gly-PUF[531] have fragmented nuclear DNA, indicating that they are undergoing apoptosis (Fig. 2e). Examination of >200 cells from randomly chosen fields indicates that cells transfected with Gly-PUF(531) have more fragmented nuclear DNA (Fig. 2f).

7. The increase of apoptosis can also be observed with flow cytometry of propidium iodide stained cells (not shown).

8. Different cancer cells stably expressing Gly-PUF(531) also have increased chemosensitivity to the treatment of common anti-tumor drugs (Fig. 3).

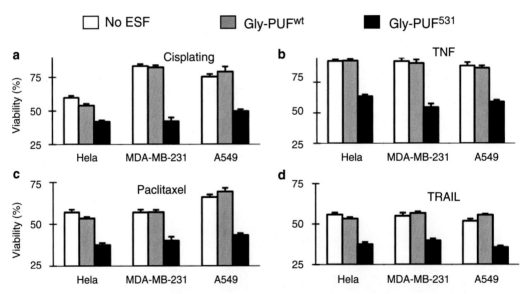

Fig. 3 Effect of ESFs in the chemosensitivities of multiple cultured cancer cell lines. (**a**) Effect of ESFs on cisplatin sensitivity of different cancer cell lines. Cancer cells were infected with lentivirus to express Gly-PUF(531), control ESF (Gly-PUF(wt)), or GFP (as mock infection), and cisplatin was added to a final concentration of 5 μM at 72 h after infection. Cell viability was measured with the WST-1 assay 24 h after drug treatment. All treatments were repeated at least twice, and the means with error bars representing the standard deviation of triplicate samples in one representative experiment are plotted. *White bars* represent cells of mock infection, *grey bars* represent control Gly-PUF(wt) infection, and black bars represent Gly-PUF(531) infection. (**b, c**, and **d**) Effect of ESFs on the sensitivities to TNF-alpha, paclitaxel, and TRAIL in different cancer cell lines. Experimental conditions are the same as described for panel (**b**) except final concentrations of 20 ng/ml TNF-alpha (**b**), 10 nM paclitaxel (**c**), or 100 ng/ml TRAIL (**d**) were used. The significant differences (*P* < 0.05, judged by paired *T*-test) of cell viabilities were observed for all drug treatments between the Gly-PUF(531) and Gly-PUF(wt) infected cells. Data are from Ref. 7

Acknowledgements

This work is supported by NIH grant R01-CA158283 and the Jefferson Pilot award to Z.W. Y.W. is funded by the Young Thousand Talents Program and the National Natural Science Foundation of China (grants 31471235 and 81422038).

References

1. Wang ET, Sandberg R, Luo S, Khrebtukova I, Zhang L, Mayr C, Kingsmore SF, Schroth GP, Burge CB (2008) Alternative isoform regulation in human tissue transcriptomes. Nature 456(7221):470–476. doi:10.1038/nature07509

2. Wang GS, Cooper TA (2007) Splicing in disease: disruption of the splicing code and the decoding machinery. Nat Rev Genet 8(10):749–761. doi:10.1038/nrg2164

3. Black DL (2003) Mechanisms of alternative pre-messenger RNA splicing. Annu Rev Biochem 72:291–336. doi:10.1146/annurev.biochem.72.121801.161720

4. Wang Z, Burge CB (2008) Splicing regulation: from a parts list of regulatory elements to an integrated splicing code. RNA 14(5):802–813, doi:rna.876308

5. Graveley BR, Maniatis T (1998) Arginine/serine-rich domains of SR proteins can function as activators of pre-mRNA splicing. Mol Cell 1(5):765–771

6. Del Gatto-Konczak F, Olive M, Gesnel MC, Breathnach R (1999) hnRNP A1 recruited to an exon in vivo can function as an exon splicing silencer. Mol Cell Biol 19(1):251–260

7. Wang Y, Cheong CG, Hall TM, Wang Z (2009) Engineering splicing factors with designed specificities. Nat Methods 6(11):825–830, doi:nmeth.1379

8. Wei H, Wang Z (2015) Engineering RNA-binding proteins with diverse activities. Wiley Interdiscipl Rev RNA. doi:10.1002/wrna.1296

9. Cheong CG, Hall TM (2006) Engineering RNA sequence specificity of Pumilio repeats. Proc Natl Acad Sci U S A 103(37):13635–13639, doi:0606294103

10. Xiao X, Wang Z, Jang M, Burge CB (2007) Coevolutionary networks of splicing cis-regulatory elements. Proc Natl Acad Sci U S A 104(47):18583–18588, doi:0707349104

11. Wang Z, Xiao X, Van Nostrand E, Burge CB (2006) General and specific functions of exonic splicing silencers in splicing control. Mol Cell 23(1):61–70, S1097-2765(06)00333-9

Chapter 19

Informational Suppression to Probe RNA:RNA Interactions in the Context of Ribonucleoproteins: U1 and 5′ Splice-Site Base-Pairing

Jiazi Tan and Xavier Roca

Abstract

Informational suppression is a method to map specific RNA:RNA interactions by taking advantage of the rules of base complementarity. First, a predicted Watson-Crick base pair is broken by single-nucleotide substitution which disrupts the RNA's structure and/or function. Second, the base pair is restored by mutating the opposing nucleotide, thereby rescuing structure and/or function. This method applies to RNP:RNA interactions such as 5′ splice-site (5′ss) base-pairing to the 5′ end of U1 small nuclear RNA as part of a small nuclear RNP. Our protocol aims to determine the 5′ss:U1 base-pairing register for natural 5′ss, because for distinct 5′ss sequences the nucleotides on each strand can be aligned differently. This methodology includes cloning of a wild-type splicing minigene and introduction of 5′ss variants by PCR mutagenesis. A U1-expression plasmid is mutated to construct "suppressor U1" snRNAs with restored base-pairing to mutant 5′ss in different registers. Cells are transfected with combinations of minigenes and suppressor U1s, and the splicing patterns are analyzed by reverse transcription and semiquantitative PCR, followed by gel electrophoresis. The identity of suppressor U1s that rescue splicing for specific mutations indicates the register used in that 5′ss. We also provide tips to adapt this protocol to other minigenes or registers.

Key words RNA:RNA interactions, Pre-mRNA splicing, U1 snRNA/snRNP, 5′ splice sites, Informational suppression, Base-pairing

1 Introduction

Many RNAs in the cell are assembled with proteins during their biogenesis to form stable and functional ribonucleoprotein (RNP) particles, which include the ribosomal subunits and the small nuclear RNPs (snRNPs) [1, 2]. These particles exert their functions via protein:protein, RNA:protein, or RNA:RNA interactions. While the two former types of interactions are discussed elsewhere in this book, in this chapter we will discuss the method of informational suppression to characterize RNA:RNA interactions in the context of RNPs and pre-messenger RNA (pre-mRNA) splicing.

Ren-Jang Lin (ed.), *RNA-Protein Complexes and Interactions: Methods and Protocols*, Methods in Molecular Biology, vol. 1421, DOI 10.1007/978-1-4939-3591-8_19, © Springer Science+Business Media New York 2016

Pre-mRNA splicing is the two-step *trans*-esterification reaction that excises introns and joins exons in eukaryotic pre-mRNAs [3]. Splicing is catalyzed by a large and dynamic macromolecular complex known as the spliceosome, which is composed of five snRNPs and dozens to hundreds of proteins [4]. An early event in splicing is the U1 snRNP-mediated recognition of the partially conserved motif at the 5′ end of introns termed 5′ splice site (5′ss), which includes the last three exonic nucleotides and the first eight intronic nucleotides [5, 6]. The 5′ end of the U1 small nuclear RNA (snRNA), which is the RNA moiety of the U1 snRNP, base pairs to the 5′ss to a maximum of 11 base pairs with the consensus sequence (Fig. 1). However, in most eukaryotes, the actual 5′ss deviate from the consensus at different positions, and informational suppression experiments are shedding light into the mechanisms of recognition of the high diversity of functional 5′ss sequences [5]. Protein factors also affect 5′ss recognition such as the U1-C polypeptide [7]. Recent crystal structures of the U1 snRNP base-paired to pre-mRNA-like sequences have shown that U1-C contacts the phosphoribose backbone of both strands and stabilizes duplexes with certain mismatches [8]. Nevertheless, this and all previous studies indicate that the major force driving the 5′ss:U1 snRNP interaction is base-pairing.

The term "informational suppression" was originally coined to describe alterations in mRNA translation that suppress the phenotypes caused by nonsense or frameshift mutations [9, 10]. The scope of this term was later expanded to include intra- or intermolecular RNA:RNA interactions in other gene expression steps such as splicing [11]. In our case, the mutation to be suppressed occurs at a particular position of a specific 5′ss sequence, and the suppressor in this case is an exogenous U1 snRNA carrying the allele-specific compensatory mutation that restores base-pairing to the mutant 5′ss. Thus, as opposed to classical genetic suppression screens aiming to identify unknown factors that genetically interact with the mutant gene [8], here we know the two molecules involved. In the experiment presented below, we use suppressor U1 (also termed U1 shift) experiments to characterize the 5′ss:U1 interaction at a single-nucleotide resolution, so as to assess the correct base-pairing register as outlined below.

Back in 1986, experiments with suppressor U1 snRNAs carrying compensatory mutations led to the formal demonstration that 5′ss are recognized by base-pairing to U1 snRNA in mammals [12], and two years later analogous evidence was provided for budding yeast [13, 14]. Since then a number of studies have used this approach, showing that some mutations at certain 5′ss positions in some contexts could be suppressed while others could not [15–17]. Some of these suppressor U1s that rescue splicing at mutant 5′ss are being proposed as potential therapeutic agents when delivered to cells in expression plasmids [18]. In addition, as U6 snRNA base pairs to 5′ss after U1 leaves the spliceosome [19–21], suppressor U6 tests have also been used in the past [19, 20, 22]. In the last few years we have learnt that the 5′ss:U1 interaction is far from simple, as this small helix with a maximum of 11 base pairs can accommodate multiple base-pairing registers. We have recently used suppressor U1s to demonstrate that certain 5′ss sequences are recognized by noncanonical registers, in which certain positions of the 5′ end of U1 base pair to positions of the 5′ss that are different from those in the canonical register originally described by Zhuang and Weiner [5, 12]. Such noncanonical registers include the following (Fig. 1): (1) the shifted register, in which all the base pairs

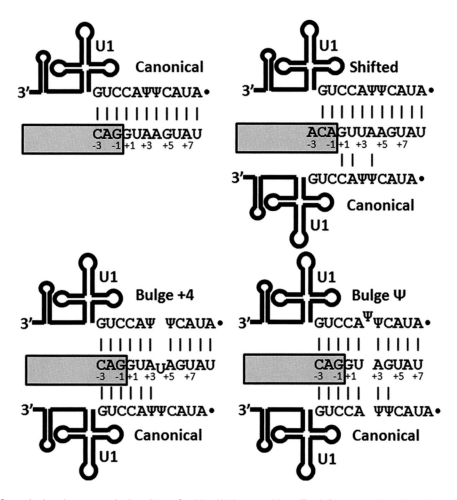

Fig. 1 Canonical and noncanonical registers for 5′ss:U1 base-pairing. *Top left*, canonical register for consensus 5′ss. *Top right*, shifted register. *Bottom left and right*, bulge +4 and bulge pseudouridine register, respectively. Note that the 5′ss comprises the last three exonic nucleotides and the first eight intronic nucleotides, which are respectively numbered −3 to −1 and +1 to +8. Exons are drawn as boxes. The sequence of the 11 nucleotides at the 5′ end of U1 is shown, and the U1 snRNA secondary structure is schematically depicted. Dot indicates the 2,2,7-trimethylguanosine cap at the 5′ end of U1. Ψ represents pseudouridine, a uridine regioisomer

are shifted by one nucleotide downstream in the pre-mRNA, with the first intronic G base-pairing to U1-C9 instead of U1-C8 in the canonical register [23]; and (2) many bulge registers, in which a one-nucleotide bulge at either the 5′ss or U1 alters the positions of several base pairs [24]. These registers substantially modify the stability of the 5′ss:U1 helix, so they are essential for 5′ss prediction and for the molecular diagnosis of 5′ss mutations in genetic disease. In this chapter, we describe the suppressor U1 methods to test such noncanonical registers.

2   Materials

Here we are listing for convenience the reagents and suppliers that we routinely use in our current lab, yet in most cases the reagents cited in the original paper [24] or from other companies can be utilized. The materials can be classified into three categories: reagents for plasmid construction and mutagenesis (for Subheading 3.1–3.3), cell transfection reagents (for Subheading 3.4), and materials for the analysis of splicing by RT-PCR from the plasmid-derived RNAs (for Subheading 3.5).

2.1 Construction of Splicing Minigenes and U1 snRNA Expression Plasmids

1. Backbone plasmids: pcDNA3.1+ (or −) (Life Technologies), and pNS6.U1 [23].

2. Human genomic DNA, 216 µg/mL (Promega).

3. PrimeSTAR® Max DNA Polymerase Premix (Takara Bio). For PCR amplification of genomic fragments.

4. Oligonucleotide primers to construct *RPGR* splicing minigene (Integrated DNA Technologies).
   Rpgr.ex3.Eco.F (5′caagcag *GAATTC*cagGAAATAATAAACT TTACATGTTTGGC3′);
   Rpgr.In3.R (5′tatcttagccccaccccatggtacttcctttgtggtccctgg3′);
   Rpgr.In3.F:
   (5′ccagggaccacaaaggaagtaccatagggtggggctaagata3′);
   Rpgr.In4.R
   (5′caatctagaatgtaactagcaactgatgctgtcccatgatagaatccaag3′);
   Rpgr.In4.F
   (5′cttggattctatcatgggacagcatcagttgctagttacattctagattg3′);
   Rpgr.ex5.Xho.R (5′TACAGAGA *ctcgag*tctcacCAGTTAGGGCAGCTGAAG3′).

   Non-italicized uppercase and lowercase nucleotides indicate exonic and intronic sequences, respectively. The 5′ sequences of the outer primers carrying restriction sites (underlined) and spacers are shown in italics. All stocks are kept at 100 µM (100 pmol/µL) and diluted to 10 µM (10 pmol/µL) for use.

5. Qiaquick® Gel Extraction Kit (Qiagen) and Qiaquick® PCR Purification Kit (Qiagen).

6. 10× NEBuffer 4 (New England Biolabs). The final 1× is composed of 50 mM Potassium Acetate, 20 mM Tris-acetate, 10 mM Magnesium Acetate, 1 mM DTT, pH 7.9 at 25 °C.

7. NEB Bovine Serum Albumin (BSA), 20 mg/mL (100×). Stored in 20 mM Tris–HCl, 100 mM KCl, 0.1 mM EDTA, 50 % Glycerol, pH 8.0 at 25 °C.

8. Restriction enzymes, EcoRI and XhoI, 20 U/μL (New England Biolabs).

9. Agarose (Vivantis).

10. 10× TBE buffer: Combine 108 g Tris–HCl (Promega), 55 g boric acid (Merck) and 9.8 g EDTA (Affymetrix) (pH 8.0). Add water to 1 L volume.

11. NEB Gel loading dye, Blue (6×). 1× buffer consists of 2.5 % Ficoll®-400, 11 mM EDTA, 3.3 mM Tris–HCl, 0.017 % SDS, 0.015 % bromophenol blue, pH 8.0 at 25 °C.

12. Ethidium bromide (Bio-Rad).

13. 100 % ethanol, molecular biology grade (Merck).

14. 3 M sodium acetate, pH 5.0. Dissolve 40.8 g sodium acetate (Merck) in 70 mL of deionized water, adjust pH to 5 by adding glacial acetic acid (Merck), and then add water to 100 mL final volume.

15. Isopropanol (Merck).

16. T4 DNA Ligase, 400 U/μL (New England Biolabs).

17. 10× T4 DNA ligase buffer (New England Biolabs): 500 mM Tris–HCl, 100 mM MgCl$_2$, 10 mM ATP, 100 mM DTT, 250 μg/mL bovine serum albumin (BSA).

18. Chemically competent DH5α *E. coli* cells.

19. LB liquid media or plates supplemented with 70 μg/mL of ampicillin (Merck), henceforth referred to as LB^{Amp}.

20. E.Z.N.A.® Plasmid Mini Kit (Omega Bio-tek).

21. PureLink® HiPure Plasmid Midiprep Kit (Invitrogen).

22. 5× HiFi Buffer (Kapa Biosystems), with 2 mM MgCl$_2$ at 1×.

23. HiFi DNA polymerase 1 U/μL (Kapa Biosystems).

24. dNTP mix with 10 mM for each nucleotide (New England Biolabs).

25. Oligonucleotide primers for PCR mutagenesis of *RPGR* splicing minigene (Integrated DNA Technologies).

Rpgr.ex5.mutag.R   (5′TTGACACCAGGGTGTGGTTC CTTCCAC3′);

Rpgr.+3G.F(5′GGAACCACACCCTGGTGTCAACAGgt**G**ta gtgctcaacctgtatgatttc3′);

Rpgr.+5C.F (5′GGAACCACACCCTGGTGTCAACAGgtat **C**gtgctcaacctgtatgatttc3′);

Rpgr.+6C.F (5′GGAACCACACCCTGGTGTCAACAGgtata**C**tgctcaacctgtatgatttc3′);

Rpgr.+7C.F: (5′GGAACCACACCCTGGTGTCAACAGgtatag**C**gctcaacctgtatgatttc3′);

Rpgr.−2C.F: (5′GGAACCACACCCTGGTGTCAAC**C**Ggtatagtgctcaacctgtatgatttc3′).

Uppercase and lowercase nucleotides indicate exonic and intronic sequences, respectively. Bold nucleotides indicate the mutation in each forward (F) primer. All stocks are kept at 100 μM (100 pmol/μL) and diluted to 10 μM (10 pmol/μL) for use.

26. DpnI, 20 U/μL (New England Biolabs).

27. Oligonucleotide primers for PCR mutagenesis of U1 5′ end (Integrated DNA Technologies).

NS6.mutag.R (5′GAGATCTTCGGGCTCTGCC3′);

NS6.U1.5′-G2 (5′GGCAGAGCCCGAAGATCTC A**G**ACTTACCTG GCAGGGGAGATAC3′);

NS6.U1.5′-G3 (5′GGCAGAGCCCGAAGATCTC AT**G**CTTACCTG GCAGGGGAGATAC3′).

Bold nucleotides indicate the mutation in each forward (F) primer. Sequencing primer is U1_F1 (5′GCTGGAAAGG GCTCGGGAGTGCG3′). All stocks are kept at 100 μM (100 pmol/μL) and diluted to 10 μM (10 pmol/μL) for use.

2.2 Transfection of Cells with U1 and Minigene

1. HEK293T cells (ATCC CRL-11268).

2. Hyclone Dulbecco's Modified Eagle's medium (DMEM) (Thermo Scientific).

3. 10 % (v/v) Fetal Bovine Serum (FBS) (Gibco Life Technologies).

4. Penicillin, 10 U/μL, and streptomycin, 10 μg/μL (Gibco Life Technologies).

5. 12-well plates (Fisher Scientific).

6. Trypsin (Gibco Life Technologies).

7. Trypan Blue, 0.4 % (Gibco Life Technologies).

8. X-tremeGENE 9 DNA Transfection Reagent (Roche).

9. Hyclone Opti-MEM (Gibco Life Technologies).

2.3 Splicing Analysis: RNA Extraction, DNase I Digestion, Reverse Transcription (RT), PCR

1. PureLink® RNA Mini Kit (Ambion Life Technologies).

2. 10× RQ1 RNase-free DNase reaction buffer (Promega), composed of 400 mM Tris–HCl (pH 8.0), 100 mM $MgSO_4$, and 10 mM $CaCl_2$.

3. RQ1 RNase-Free DNaseI (Promega), 1 U/μL.

4. RQ1 DNase stop solution, 20 mM EGTA (pH 8.0).

5. 10× RT Buffer (New England Biolabs). The final 1× buffer is composed of 50 mM Tris–HCl, 75 mM KCl, 3 mM MgCl₂, 10 mM DTT, at pH 8.3.

6. Moloney Murine Leukemia Virus Reverse Transcriptase (New England Biolabs), 200 U/µL.

7. Oligonucleotide primer for reverse transcription (RT) (Integrated DNA Technologies), Oligo-dT with 18 Ts (5′TTTTTTTTTTTTTTTTTT3′). Stock is kept at 100 µM (100 pmol/µL) and diluted to 40 µM (40 pmol/µL) for use.

8. Thermus aquaticus (Taq) DNA polymerase, 5 U/µL (Promega).

9. Oligonucleotide primers for RT-PCR (Integrated DNA Technologies): pcDNA-F (5′GAGACCCAAGCTGGCTA GCGTT3′); pcDNA-R (5′GAGGCTGATCAGCGGGTTTA AAC3′). All stocks are kept at 100 µM (100 pmol/µL) and diluted to 10 µM (10 pmol/µL) for use.

10. T4 polynucleotide kinase (PNK), 10 U/µL (New England Biolabs).

11. 10× T4 polynucleotide kinase (T4 PNK) buffer (New England Biolabs): 700 mM Tris–HCl, pH 7.6, 100 mM MgCl₂, 50 mM DTT.

12. γ-[³²P] adenosine triphosphate (ATP), 6000 Ci/mmol, 10 mCi/mL, 250 µCi (Perkin-Elmer).

13. MicroSpin G-25 columns (GE Healthcare).

14. 5× Colourless GoTaq reaction buffer (with 1.5 mM MgCl₂ at 1×) (Promega).

15. MgCl₂, 25 mM (Promega).

16. GoTaq DNA polymerase, 5 U/µL (Promega).

17. 30 % acrylamide/bis-acrylamide solution 29:1 (Biorad).

18. Vertical electrophoresis tank and plates: adjustable slab gel kit, 16.5 cm wide, height adjusts from 16 to 28 cm (CBS Scientific).

19. Ammonium persulfate (Promega) and tetramethylethylenediamine (TEMED) (Sigma Aldrich).

20. Whatman filter paper (Whatman, GE Healthcare Life Sciences).

21. Plastic wrap.

22. Model 583 gel-dryer (Bio-Rad).

23. Phosphorimaging screen, Typhoon Trio variable mode imager, ImageQuant TL software (all from GE Healthcare Life Sciences).

24. Medical X-ray Film General Purpose Green, Kodak Model 2000 X-Ray Film Processor (both from Kodak).

25. GS-800 Calibrated Densitometer (Bio-Rad).

3 Methods

Here, we describe the construction and analysis of the *RPGR* splicing minigene as published previously [24]. This minigene, encompassing *RPGR* exons 3–5 and shortened introns 3 and 4, was used to validate the bulge +4 register as the mechanism of recognition of the 5'ss in exon 4 (Fig. 2a). Nevertheless, this protocol can be applied to any splicing minigene, and the parameters to adjust these procedures to different minigenes will be discussed in Subheading 4.

3.1 Construction of RPGR Splicing Minigenes

The genomic fragment encompassing *RPGR* exon 4 (63 bp) and its flanking two introns and two exons are cloned into the mammalian expression plasmid pcDNA3.1+ (*see* **Note 1** for standard minigene choice and design, and **Notes 2–4** for alternative strategies). As introns 3 and 4 are long, the final minigene is designed to have these introns with internal deletions only to keep 225 nt on each end of the intron. Thus, the first step is to generate the three-exon two-intron minigene with internal intronic deletions using overlap PCR, with the outer primers containing the EcoRI and XhoI restriction sites for subsequent cloning into the pcDNA3.1+ Multiple Cloning Site (MCS) (Fig. 2b).

1. Three PCR reactions are performed to synthesize the 5', middle and 3' fragments of the minigene, using the following recipe:

   (a) 25 μL of PrimeSTAR® Max DNA Polymerase Premix.

   (b) 1 μL (~200 ng) of human genomic DNA.

   (c) Primer pairs: primers for 5' fragment are Rpgr.ex3.Eco.F and Rpgr.In3.R, in tube #1; for middle fragment are Rpgr. In3.F and Rpgr.In4.R, in tube #2; for 3' fragment are Rpgr. In4.F and Rpgr.ex5.Xho.R in tube #3. A total of 1 μL of each primer (10 μM) is added to obtain a final concentration of 0.2 μM.

   (d) 22 μL of nuclease-free water to a final volume of 50 μL.

2. The PCR program for this reaction is as follows:

   (a) 10 s at 98 °C.

   (b) 5 s at 55 °C (or adjust annealing temperature to the lowest Tm of the two primers minus 5 °C).

   (c) 5 s at 72 °C (5 s per kilo-base-pair).

   (d) Repeat **steps 2–4** a total of 34 times (35 cycles).

   (e) Forever at 4 °C.

3. Upon completion, all the volume of each PCR reaction is mixed with 10 μL of 6× loading dye, and then loaded and

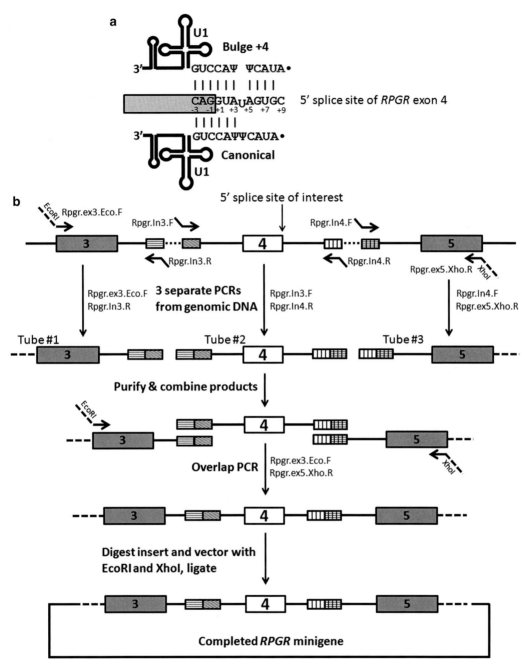

Fig. 2 (**a**) *RPGR* 5′ splice site base-pairing with U1 snRNA in either the bulge +4 or the canonical register. (**b**) *RPGR* minigene cloning strategy by overlap PCR. Exons are represented by *boxes*, introns with *solid lines*, and shortened introns with *dotted lines*. PCR primers are depicted as *horizontal arrows*. Overlapping intronic sequences are marked with the *smaller crosshatched boxes*. These overlapping sequences connect the three initial PCR products in the overlap PCR to generate the exon 3–5 minigene fragment with shortened introns. The restriction sites added in the "outer" primers are indicated as dashed lines. These restriction sites are used to clone the overlap PCR fragment into the pcDNA3.1+ plasmid

separated in a 1.5 % agarose gel in TBE (or TAE) stained with Ethidium Bromide. The size of the bands should be as follows: 356 bp for 5′ fragment, 560 bp for middle, and 431 bp for 3′ fragment.

4. Bands of the correct size are cut out of the gel for DNA extraction using the Qiaquick® Gel Extraction Kit.

   (a) The sliced agarose bands are weighed, and 3 volumes of QG buffer per 1 volume of gel are added to each tube.

   (b) The mixtures are heated at 50 °C until agarose is fully dissolved.

   (c) 1 volume of isopropanol is added to each preparation, and mixtures are pipetted into a spin column placed in a 2 mL collection tube.

   (d) Columns are spun in a microcentrifuge at full speed for 1 min and the flow-through is discarded.

   (e) 750 μL of Wash Buffer (PE) buffer are added to each column, the samples are spun for 1 min, and the flow-through is discarded. Columns are spun for another 2 min at max speed to remove traces of Wash Buffer.

   (f) 50 μL of Elution Buffer (EB) buffer are then added to each column, which is then placed in a clean collection tube and incubated for 1 min. Finally, the tubes are spun for 1 min at maximum speed to elute the DNA.

   (g) The concentration of the eluates of the three purified PCR products is assessed using a spectrophotometer, preferably a Nanodrop.

5. Equal amounts of the three purified PCR products are used for the overlap PCR using the outer primers with the corresponding restriction sites.

   (a) 25 μL of PrimeSTAR® Max DNA Polymerase Premix.

   (b) 10 ng of each PCR product.

   (c) Primers: Rpgr.ex3.Eco.F and Rpgr.ex5.Xho.R, 1 μL of each primer (10 μM) to a final concentration of 0.2 μM of per primer.

   (d) Nuclease-free water to a final volume of 50 μL.

6. The PCR program is same as above but with a slightly longer elongation (10 s in the third step of PCR). To make sure there is enough DNA for cloning, it is advised to make at least 3 reactions of 50 μL each.

7. PCR products of the expected size of 1254 bp are then purified using the Qiaquick® PCR Purification Kit.

   (a) 5 volumes of PB buffer are added to 1 volume of PCR product and mixed.

(b) The mixtures are placed in a column with a collection tube and spun for 1 min in microcentrifuge at full speed.

(c) PE buffer is then added, and from here on this protocol is the same as from the equivalent step in gel extraction.

(d) The concentration is assessed by Nanodrop. A minimum of 1 μg should be obtained.

8. The isolated PCR products and the expression plasmid are then digested with the two restriction enzymes:

(a) 10 μL of 10× NEBuffer 4.

(b) 50 μL of purified PCR product or pcDNA3.1+.

(c) 3 μL of EcoRI.

(d) 3 μL of XhoI.

(e) 1 μL of NEB BSA.

(f) 33 μL of nuclease-free water to a final volume of 100 μL.

(g) Reactions are incubated at 37 °C for 1–3 h.

9. All digested DNA from the PCR product as well as the plasmid are separated in 1 % agarose gel in TBE (or TAE) and gel purified as described above.

(a) Upon elution, the 50 μL of PCR insert and vector DNA are precipitated by adding 0.1 volumes (5 μL) of 3 M sodium acetate, pH 5.0, and then 2.5 volumes (137.5 μL) of 100 % ethanol.

(b) The mixture is kept at –20 °C or –80 °C for at least 1 h, preferably overnight. Then the tubes are spun in a refrigerated microcentrifuge at full speed (>12,000 g) for 30 min at 4 °C. The supernatants are discarded, and 500 μL of 70 % ethanol is added to each tube. Tubes are spun again under the same conditions for 15 min and the supernatants are carefully discarded.

(c) Pellets are air-dried and resuspended in small volume (~7 μL) of nuclease-free water.

(d) After assessing concentration in Nanodrop, the concentration of both insert and vector is adjusted to 100 ng/μL.

10. For ligation, the vector and insert are mixed at a 1:8 molar ratio:

(a) 1 μL of 10× T4 DNA ligase buffer.

(b) ~2 μL of pcDNA3.1+, vector (~200 ng).

(c) ~4 μL of PCR product, insert (~400 ng).

(d) 0.5 μL of T4 DNA ligase.

(e) Top up with nuclease-free water to a final volume of 10 μL.

(f) Reactions are incubated at 16 °C overnight.

11. 4 µL are used to transform 50 µL of chemically competent *Escherichia coli* DH5α cells.

   (a) Bacteria are thawed on ice, plasmids are added and mixtures are kept on ice for 30 min.

   (b) Bacteria are heat-shocked at 42 °C for 45 s, and then ~1 mL of liquid LB with ampicillin (LB^(Amp)) is added after prewarming it at 42 °C. The bacteria are allowed to recover by 1 h incubation at 37 °C.

   (c) The bacteria are precipitated by spinning at $4000g$ and supernatants are removed, leaving ~50 µL which are used to resuspend cells by gentle pipetting.

   (d) Finally, bacteria are plated on standard LB^(Amp) plates.

12. Four or more colonies per plate are grown in liquid LB^(Amp), and minipreps are prepared with the E.Z.N.A.® Plasmid Mini Kit. The presence and size of the insert (1237 bp) is assessed by digesting each miniprep with EcoRI and XhoI (10 µL of final reactions, above protocol is used with 1/10 of the volumes). Digested products are separated on 1.5 % agarose gel. Finally, minipreps with insert of the correct size are sequenced with the pcDNA-F primer to confirm that the clones contain the *RPGR* minigene without spurious mutations.

Once the splicing minigene is constructed, it should be confirmed first that the splicing pattern of the minigene substrate proceeds as expected. To this end, the pcDNA3.1+*RPGR* plasmid is used to transfect HEK293T (or HeLa) cells as outlined below (Subheading 3.4). After transfection, the splicing pattern is analyzed by RT-PCR with pcDNA-F and pcDNA-R primers as per below (Subheading 3.5), and the PCR products are resolved by agarose gel electrophoresis. If the PCR product migrates around the expected size of 444 bp, which corresponds to inclusion of exon 4, then the corresponding band is excised. DNA from the slice is gel-purified and sequenced with pcDNA-F as primer. If the isolated product shows *RPGR* exons 3-4-5 spliced together at the correct boundaries, this new minigene is ready for mutagenesis.

3.2 Mutagenesis of RPGR Splicing Minigene

To test for the mechanism of recognition of the test 5'ss in *RPGR* exon 4, different 5'ss mutations are introduced by PCR mutagenesis. Nucleotides at 5'ss positions −2, +5, +6, and +7 are changed to Cs as this nucleotide does not occur in the entire 5'ss consensus sequence except in position −3 (*see* **Note 5**). In addition, position +3 is also changed to G as this 5'ss mutation has been found associated with X-linked retinitis pigmentosa. The experimental design for PCR mutagenesis is schematically shown in Fig. 3. In brief, the pcDNA3.1+*RPGR* plasmid is used as template for PCR with different forward primers, each one containing one designed mutation, and a common reverse primer that overlaps with the 5' portion of all forward primers.

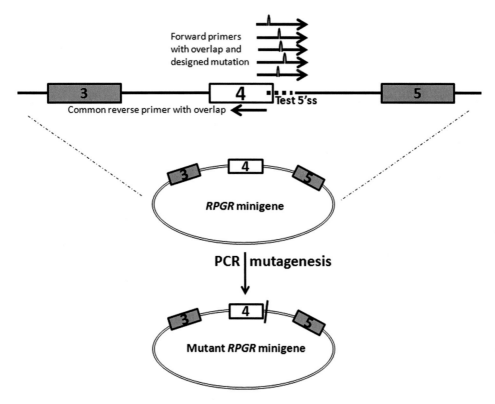

Fig. 3 Schematic of PCR mutagenesis. See text for details

1. The PCR mutagenesis reaction is as follows:
   (a) 5 μL of 5× HiFi Buffer.
   (b) 0.75 μL of dNTPs (to a final concentration of 300 μM each).
   (c) 20 ng of pcDNA3.1+*RPGR* as template.
   (d) 0.75 μL of primer pairs at final concentration of 300 nM each: common reverse primer Rpgr.ex5.mutag.R plus one forward primer from these: Rpgr.+3G.F, Rpgr.–2C.F, Rpgr.+5C.F, Rpgr.+6C.F, Rpgr.+7C.F.
   (e) 0.5 μL (0.5 U) of HiFi DNA polymerase.
   (f) Nuclease-free water to a final volume of 25 μL.
2. The PCR program for this reaction is as follows:
   (a) 2 min at 94 °C.
   (b) 30 s at 94 °C.
   (c) 1 min at 50 °C.
   (d) 5 min at 72 °C.
   (e) Repeat **steps 2–4** a total of 17 times (18 cycles).
   (f) 5 min at 72 °C.
   (g) Forever at 4 °C.

3. Upon completion of the program, the parental unmodified pcDNA3.1+*RPGR* plasmid needs to be removed by digestion before transformation. To this end, 0.75 μL of DpnI is added directly to the PCR tubes, and these reactions are then incubated for 1–2 h at 37 °C, followed by heating at 80 °C for 20 min to inactivate the enzyme.

4. 50 μL of DH5α cells are transformed with 4 μL of the mixture and plated onto LB^Amp plates as described above. 2–4 colonies are cultured in liquid LB^Amp overnight at 37 °C with vigorous shaking.

5. Minipreps are prepared with the E.Z.N.A.® Plasmid Mini Kit, and their concentration is assessed by Nanodrop. All mutations are verified by DNA sequencing using the pcDNA-F primer, and sequencing should also exclude other spurious mutations.

6. Next, LB broth from one confirmed clone for each mutant is sub-cultured in 40 mL of fresh LB^Amp overnight at 37 °C with vigorous shaking. Midipreps are prepared with the PureLink® HiPure Plasmid Midiprep Kit.

7. Each plasmid DNA prepared by midiprep is then analyzed by Nanodrop, which should show a concentration of at least 500 ng/μL, and a 260/280 nm absorbance ratio of at least 1.8 as an indicator of purity.

These mutant *RPGR* splicing minigenes are now ready for splicing analysis upon transfection.

3.3 Mutagenesis of U1 snRNA Expression Plasmids

Suppressor U1 plasmids are eukaryotic gene expression vectors with the U1 snRNA gene, usually under the control of its own promoter and additional regulatory sequences [25]. These plasmids are then mutagenized to introduce compensatory mutations at the 5′ end to rescue base-pairing to the target 5′ss, which is the 5′ss of *RPGR* exon 4 in this case. As the protocol for U1 mutagenesis is analogous to that of *RPGR*, here we only point out the differences between the two mutagenesis schemes.

1. PCR reactions are assembled as in Subheading 3.2, but the *RPGR*-specific primers are replaced by primers NS6.mutag.R paired with either NS6.U1.5′-G2 or NS6.U1.5′-G3. Template is pNS6.U1.

2. PCR program is as in Subheading 3.2, with no changes.

3. DpnI digestion and transformation are performed as in Subheading 3.2. As this is a very low copy plasmid, colonies are cultured in 10 mL of LB^Amp for minipreps.

4. Plasmids are sequenced using pNS6.U1-specific primer U1_F1.

5. Upon confirmation of the correct clone, midipreps are performed using 100 mL of LB^Amp.

3.4 Cell Culture and Transfection

As any cell line can be used to test noncanonical 5′ss:U1 base-airing registers, here we use Human Embryonic Kidney cells transformed with large T antigen (HEK293T), instead of HeLa in the original study [24] (*see* **Note 6** for justification of the use of this transfection-based assay over in vitro methods). HEK293T cells are cultured in Hyclone Dulbecco's Modified Eagle's medium (DMEM) with 10 % (volume/volume or v/v) Fetal Bovine Serum (FBS) and antibiotics (100 U/mL penicillin and 100 mg/mL streptomycin). The protocol for transfection is as follows:

1. Day 1, cells are seeded in 12-well plates:

   (a) Medium on plates is removed by vacuum.

   (b) A total of 4 mL of trypsin (pre-warmed at 37 °C) is added to plates, which are then incubated at 37 °C for 5 min or until cells detach.

   (c) To stop trypsinization, 20 mL of DMEM (pre-warmed at 37 °C) is added, and cells are transferred to a 50 mL tube.

   (d) Tubes are spun at 210g for 5 min at 4 °C. Supernatant is removed and pellets are resuspended in 10 mL of fresh medium.

   (e) Cells are counted in hemacytometer by mixing 10 μL of cells with 10 μL of Trypan Blue.

   (f) More DMEM is added to cells to adjust the volume so that the concentration of cells is 1×10^5 cells/mL.

   (g) A total of 1 mL of medium with cells are added to each well of a 12-well plate, and plates are placed in incubator for 1 day at 37 °C and 5 % CO_2.

2. Day 2, transfection:

   (a) Plated HEK293T cells should be at ~50 % confluence, as qualitatively assessed in an inverted microscope.

   (b) Cells are transfected with 1 μg of total DNA per well, using 3 μL of X-tremeGENE 9 DNA Transfection Reagent diluted in 100 μL of Hyclone Opti-MEM. All the plasmid combinations are summarized in Table 1. Plasmid mixtures consist of:

   A total of 1:12 of pcDNA3.1+RPGR plasmid (~83 ng).
   A total of 10:12 of pNS6.U1 or empty pUC19 as control plasmid (~830 ng).
   A total of 1:12 of empty pUC19 mock plasmid (~83 ng).

   (c) Transfection mixtures are kept for 20 min at room temperature. In the meantime, media from cells are aspirated by vacuum and replaced with fresh DMEM pre-warmed at 37 °C. All transfection mixtures are added to cells dropwise.

3. Day 4, cells are harvested for their RNA around 48 h after transfection, *see* Subheading 3.5.

Table 1

List of DNA samples for transfection

	Splicing minigene	Suppressor U1	Mock plasmid
1	pcDNA3.1+*RPGR*	–	pUC19 (11/12)
2	pcDNA3.1+*RPGR*+3G	–	pUC19 (11/12)
3	pcDNA3.1+*RPGR*–2C	–	pUC19 (11/12)
4	pcDNA3.1+*RPGR*+5C	–	pUC19 (11/12)
5	pcDNA3.1+*RPGR*+6C	–	pUC19 (11/12)
6	pcDNA3.1+*RPGR*+7C	–	pUC19 (11/12)
7	pcDNA3.1+*RPGR*+7C	U1-G2	pUC19 (1/12)
8	pcDNA3.1+*RPGR*+7C	U1-G3	pUC19 (1/12)

3.5 Analysis of Splicing Patterns: RNA Extraction and RT-PCR

3.5.1 RNA Extraction and Removal of Contaminating DNA

These next steps will be used to determine the splicing patterns of each *RPGR* minigene.

Total RNA is extracted with the PureLink® RNA Mini Kit.

1. Medium from transfected cells in 12-well plate is removed by aspiration, and then 350 μL of Lysis Buffer is added to each well. In our hands, addition of 2-mercaptoethanol at 10 μL per 1 mL of Lysis Buffer can be omitted.

2. Cell lysate is transferred into clean 1.5 mL tube and one volume of 70 % ethanol is added to each volume of lysate.

3. After vortexing, mixtures are transferred to spin cartridge with attached collection tube and spun in microcentrifuge at $12,000 \times g$ for 15 s at room temperature. Flow-through is discarded and spin cartridge is reinserted into the same collection tube.

4. 700 μL Wash Buffer I is added to spin cartridge and tubes are spun at $12,000 \times g$ for 15 s at room temperature. The collection tube and flow-through are discarded, and the spin cartridge is inserted into a fresh collection tube.

5. A total of 500 μL of Wash Buffer II (with added ethanol) is added to spin cartridge and tubes are spun at $12,000 \times g$ for 15 s at room temperature, discarding flow-through. This step is repeated once.

6. The spin cartridge is spun at $12,000 \times g$ for 2 min to dry the membrane. The collection tube is discarded and the spin cartridge is placed into clean 1.5 mL recovery tube.

7. Between 30 and 100 μL of nuclease-free water is added to the center of the spin cartridge, and tubes are incubated at room

temperature for 1 min, then spun at $12,000 \times g$ for 2 min at room temperature to elute RNA.

8. The concentration of RNA is then measured with Nanodrop. RNA is stored at −20 °C for the short-term or at −80 °C for the long-term.

9. Next, residual DNA is eliminated from samples by RQ1 RNase-Free DNaseI digestion using the following mixture:

   (a) 2 µL of 10× RQ1 RNase-free DNase reaction Buffer.

   (b) 4 µg of RNA.

   (c) 2 µL of RQ1 RNase-Free DNaseI (final ratio of 0.5 U/µg of RNA).

   (d) Nuclease-free water to a final volume of 20 µL.

10. Samples are mixed well, briefly spun down, and then incubated at 37 °C for 1 h.

11. To stop the digestion, 2 µL of RQ1 DNase stop solution are added (stop solution and reaction at 1:10 v/v) and samples are heated at 65 °C for 10 min. At the end of the reaction, the volume of the RNA samples is increased to 100 µL by adding 78 µL of nuclease-free water.

12. RNA is then precipitated, by adding 0.1 volumes (10 µL) of 3 M sodium acetate, pH 5.0, and then 2.5 volumes (250 µL) of 100 % ethanol. Mixtures are kept at −20 °C for at least 1 h, preferably overnight.

13. Tubes are spun at $12,000 g$ for 30 min at 4 °C and supernatant is carefully discarded. A total of 500 µL of 70 % ethanol is added to pellets, and tubes are spun at $12,000 g$ for 15 min at 4 °C. Supernatant is removed by pipetting and pellets are air-dried for ~5 min.

14. Pellets are resuspended in 10 µL water, and RNA concentration is derived using Nanodrop.

3.5.2 Preparing cDNA from RNA

The RNA is used for reverse transcription with Moloney Murine Leukemia Virus Reverse Transcriptase.

1. Reverse transcription reactions consist of the following:

   (a) 1 µg of RNA (maximum volume is 6 µL).

   (b) 0.5 µL of dNTPs (final concentration of 0.5 mM).

   (c) 2 µL of Oligo-dT primer at 40 µM dilution (final concentration of 8 mM).

   (d) Nuclease-free water to a final volume of 8.5 µL.

2. After mixing the reaction mixture, tubes are incubated for 5 min at 65 °C, chilled on ice, and then spun down briefly. Reaction mixtures are completed by adding:

(a) 1 μL of 10× RT Buffer.

(b) 0.5 μL of Reverse Transcriptase (final concentration of 100 U/μL).

3. Reactions are incubated for 1 h at 42 °C and inactivated by heating at 65 °C for 20 min.

3.5.3 PCR Using Radioactive Primers

The resulting cDNAs are then used as templates for semiquantitative or radioactive PCR using radiolabeled primers. To avoid amplification of the endogenous *RPGR* cDNA, the chosen primers anneal to the transcribed portion of the plasmids upstream of the 5′ exon and downstream of the 3′ exon in the minigene.

1. Before PCR, one of the primers is labeled with γ-^{32}P-ATP as follows:

(a) 1 μL of 10× PNK buffer.

(b) 1 μL pcDNA-F (or pcDNA-R) at 10 μM (for a total of 10 pmol).

(c) 1 μL of water.

(d) 6 μL of γ-^{32}P-ATP.

(e) 1 μL of PNK (10 U).

2. Reactions are incubated at 37 °C for 10 min, and heat-inactivated at 80 °C for 2 min.

3. The reactions are then purified using Microspin G-25 columns. In brief, 90 mL of nuclease-free water is added to each reaction and the total 100 μL are pipetted into pre-equilibrated G-25 resin tubes. The tubes are spun at 12,000g for 1 min, and the collected flow-through is then ready for the next step.

4. Then, the primers are mixed as follows: 98.1 μL (~10 pmol) of the radiolabeled primer with 0.9 μL of 100 μM cold pcDNA-F (90 pmol) and 1 μL of 100 μM cold pcDNA-R (100 pmol), to make 100 μL of an equimolar mixture of primers.

5. The PCR is performed using the following recipe:

(a) 2.5 μL of 5× Colourless GoTaq reaction buffer (with 1.5 mM MgCl$_2$ at 1×).

(b) 0.5 μL of 25 mM MgCl$_2$ (final concentration of 1 mM).

(c) 2.5 μL of primer mix (final concentration of 200 nM each).

(d) 0.25 μL of dNTP mix (final concentration of 200 μM each).

(e) 1 μL of cDNA template.

(f) 5.625 μL of nuclease-free water to a total volume of 12.5 μL.

(g) 0.125 μL GoTaq DNA polymerase (0.12 U).

6. The PCR program is as follows:

(a) 5 min at 94 °C.

(b) 30 s at 94 °C.

(c) 40 s at 58 °C.

(d) 50 s at 72 °C.

(e) Repeat **steps 2–4** a total of 22 times (23 cycles).

(f) 5 min at 72 °C.

(g) Forever at 4 °C.

7. With only 23 cycles, the PCR amplification remains within the exponential phase, ensuring that amplicon abundances correspond to the abundances of their templates.

3.5.4 Analyze PCR Products by Gel Electrophoresis

PCR products are separated by electrophoresis in vertical, native (non-denaturing) 6 % polyacrylamide gels.

1. The notched and unnotched glass plates, spacers, and rubber seal are all assembled. The entire setup is clamped together and placed upright.

2. The gel mixture is made in a 50 mL tube by combining 4 mL of 10× TBE buffer and 8 mL of 30 % acrylamide/bis-acrylamide solution 29:1, and then topped up to 40 mL with distilled water.

3. Right before casting, the polymerization is triggered by adding 400 μL of 10 % w/v ammonium persulfate and 40 μL of TEMED, and then mixed well by inverting the tube.

4. The gel solution is immediately poured between the glass plates by pipetting. The comb is inserted in the notched front of the plates, and gel is allowed to polymerize for about 1 h at room temperature in horizontal position.

5. When the gel is polymerized, the clamps and rubber seal are removed. The glass plates are clamped to the running tanks, which are filled with 1× TBE buffer so as to submerge the wells. Unpolymerized acrylamide left in wells is carefully flushed out using syringe and needle.

6. The gel is pre-run at 10 V/cm for 15 min.

7. 2.5 μL of 6× loading dye is added to the 12.5 μL of each radioactive PCR sample and mixed well. 3 μL of the mixture is then loaded into the well. The gel is run at constant voltage of 10 V/cm for 6 h.

8. After electrophoresis is complete, the two plates are carefully separated so that the gel stays on the plate without the notch. Then a Whatman filter paper is laid over the gel avoiding wrinkles. The plate is removed and the gel is covered with plastic wrap.

9. Gels are vacuum-dried with a Model 583 gel-dryer for 2 h with 80 °C heating.

10. After that, the gels are exposed to a storage phosphor screen. The screen is then scanned with a Typhoon Trio variable mode imager, and band intensity is quantified by 1D gel analysis using ImageQuant TL software.

11. The mean percentage of inclusion is then derived from three experimental replicas, which are three RT-PCRs from RNA samples acquired in three independent transfections. In our hands, the test-exon inclusion percentages are highly reproducible as the standard deviations are equal or below 5 %. In general, if the mean percentages of inclusion between two experiments are separate enough so that the standard deviations do not overlap, these values can be deemed "different." Alternatively, statistical tests such as Student's t-test can also be used to compare mean exon inclusion levels across samples.

12. Finally, high-quality figures are generated by exposing Medical X-ray Film General Purpose Green to the radioactive gels at –80 °C and developing them with a Kodak Model 2000 X-Ray Film Processor. Developed films are scanned at the highest possible resolution with a GS-800 Calibrated Densitometer.

3.6 Data Interpretation

The results of the *RPGR* minigene analysis prove that the exon 4 5′ss base pairs to U1 snRNA by bulging out the fourth intronic nucleotide (which is a U), in a register that we named "bulge +4" (Fig. 4, Ref. 24). The wild-type minigene results in almost full exon inclusion, as seen with the predominant band of 444 bp that contains all three exons (lane 1). Consistent with its association with retinitis pigmentosa, the +3G mutation completely abrogates inclusion of this exon as this sample only shows the RT-PCR band of 381 bp corresponding to exon 4 skipping (lane 2). The –2C, +6C, and +7C mutations partially disrupt exon inclusion (lanes 3 and 5), while the +5C does not, perhaps because the +5A-Ψ5 base pair is not essential for recognition of this exon. As the +6C and +7C mutations only disrupt a base pair in the bulge +4 register, the reduced exon inclusion upon these mutations strongly suggests that these nucleotides are base-paired to U1 in this noncanonical register. For +6C the suppressor U1s do not rescue inclusion (data not shown, and *see* **Note** 7 for general discussion about the caveats of U1 suppressor experiments). For +7C, the G3 suppressor restores a base pair in the bulge register, and G2 does so in the canonical register. Thus, the finding that suppressor G3 but not the G2 rescue exon 4 inclusion (lanes 6–8) demonstrates that this 5′ss is recognized by the bulge +4 register.

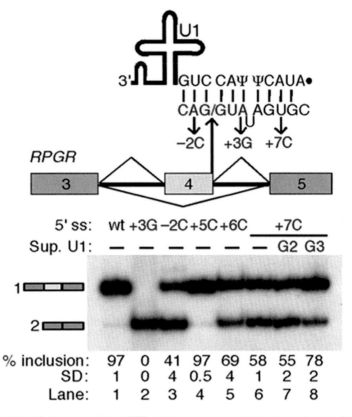

Fig. 4 (*Top*) Schematic of the *RPGR* splicing minigene with the bulge base-pairing register and some of the point mutations. (*Bottom*) Mutational analysis and suppressor U1 experiments. The mutant 5′ss and suppressor U1 are indicated above each lane. RT–PCR products are indicated on the left and correspond to inclusion (1) and skipping (2) of the middle exon. From Roca et al. [24]

---

4   Notes

1. The test 5′ss for the analysis of noncanonical registers are selected from a list of all annotated 5′ss in the human genome [24]. For each 5′ss, the U1 base-pairing register is predicted based on thermodynamic estimates obtained with UNAFold [26]. Nevertheless, it is worth checking that the selected 5′ss is the major 5′ss used for that given exon. To this end, the ENSEMBL [27] database can be searched to confirm that the selected 5′ss is annotated as the reference 5′ss for that particular exon. Otherwise, it is advised to choose another candidate 5′ss, because in this case there is a chance that the selected 5′ss might not be used in the minigene. It is also preferred to exclude 5′ss within first exons because terminal exons cannot be skipped. For terminal exons it might be more difficult to disrupt splicing by

5'ss mutations, as the aberrant splicing patterns upon mutation can only be caused by intron retention or cryptic 5'ss activation. Finally, the identities of all types of RT-PCR products for each splicing minigene should be determined by isolating the bands from agarose gels and sequencing with pcDNA-F.

2. Splicing minigene construction should indeed be easier if the exon that bears the test 5'ss is flanked by one or two short introns (less than 500 bp). In the case of two short introns, no overlap PCR is necessary, as the whole three-exon and two-intron fragment can be amplified with one pair of primers bearing the restriction sites. In the case of one short and one long intron flanking the middle exon, the overlap is only necessary to shorten the long intron, so two PCR reactions are done instead of the three in the original design (Subheading 3.1).

3. An alternative strategy for faster construction of splicing minigenes is to use a "pre-made" minigene template with common "outer" exons and common intronic sequence in which to insert the specific exon with its flanking intron ends (Fig. 5a). For convenience, a Multiple Cloning Site with unique restriction sites can be introduced in the middle of the common intron in the original plasmid. With this design, a single PCR from genomic DNA can generate the full-length fragment containing the test exon and at least 200 nt of intronic sequence on each side of the exon, with primers bearing the restriction sites to clone the fragment into the splicing minigene. In our hands, such hybrid splicing minigenes reliably reflect the splicing patterns of the endogenous substrates (unpublished data).

4. Instead of making new splicing minigenes, the *SMN1* and *SMN2* minigenes can also be used to test noncanonical 5'ss recognition registers in a heterologous context (Fig. 5b) [23, 24]. Human *SMN1* and *SMN2* are nearly identical paralogous genes, and their major difference is that *SMN2* makes less full-length SMN protein because of a translationally silent C to T transition in the sixth nucleotide of *SMN2* exon 7 that results in predominant skipping of this exon [28]. The pCI.*SMN1* and pCI.*SMN2*(C6T) minigenes have been previously described [23, 24, 29]. In this case the pCI mammalian expression vector is used, and the *SMN1/2* minigenes include exons 6-7-8 and flanking introns, with a truncated intron 6 to retain 62 nt of its 5' end and 139 nt of its 3' end. These minigenes only differ at the sixth position of exon 7, with C in *SMN1* and T in *SMN2*. The *SMN1/2* minigenes are used to study 5'ss selection by replacing their native 5'ss in exon 7 (GGA/GUAAGUCU, with '/' as the exon:intron boundary) with test 5'ss sequences via PCR mutagenesis. If the resulting mutant substrates show at least 20 % exon 7 inclusion, then mutational

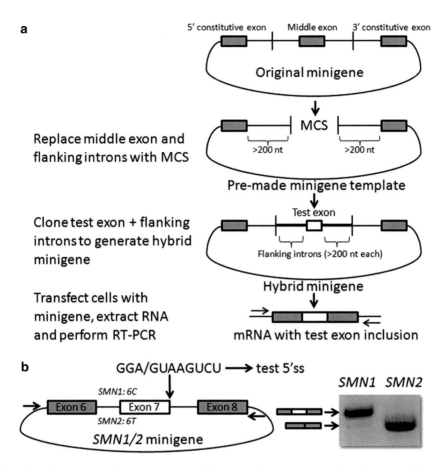

Fig. 5 Alternative designs for splicing minigenes. (**a**) A pre-made minigene template is made, with two flanking constitutive exons and one middle intron with an MCS. Test exons and their flanking intronic segments are cloned into the minigene by a single PCR followed by cloning. The resultant pre-mRNA should include the middle exon by using the correct 5′ss. (**b**) *SMN1/2* minigenes, in which the natural 5′ss is replaced by the test 5′ss. The right figure shows efficient inclusion of exon 7 in *SMN1*, and predominant skipping of this exon in *SMN2*

analysis and suppressor U1 experiments can be used [23, 24]. Last, for RT-PCR the recommended primers are pCI.Fwb (5′GACTCACTATAGGCTAGCCTCG3′) and pCI.Rev (5′GTATCTTATCATGTCTGCTCG3′) [27].

5. The choice of 5′ss mutations needs careful planning. Typically, the selected mutations are substitutions of the natural nucleotide to C at different positions of the 5′ss. Cs are the ideal disruptive nucleotides because such nucleotides are not found in the 5′ss motif, with the exception of the terminal −3C that base pairs to G11 of U1 snRNA in the canonical register. As such, Cs cannot establish canonical (Watson-Crick or wobble) base pairs with any of the nucleotides in the 5′ end of U1, again with the exception of G11. In addition, the three hydrogen bonds that can be established between the mutant C and

the opposed suppressor U1-G mutation maximize the chances that the suppressor would rescue splicing. For instance, for the *RPGR* +7C mutation the suppressor U1-G3 restores one base pair in the bulge register and significantly increases inclusion by 23 % (Fig. 4). Nevertheless, other nucleotide substitutions and corresponding suppressor U1s can be used in some other instances like in the shifted base-pairing tests [23]. If the test 5′ss has documented mutations causing genetic disease, it would be worth reproducing in the splicing minigene the splicing changes associated with the mutation.

6. Typically, suppressor U1 experiments are done by cell transfection because this method is much more convenient than in vitro systems. For cell-free approaches, suppressor U1 snRNAs need to be expressed and assembled into functional U1 snRNP particles, and then added to Nuclear Extract (NE) to influence splicing in vitro, as an adaptation of U1-reconstitution assays [30]. Alternatively, NE can be made from cells expressing the suppressor U1 from a plasmid. As in this case the splicing patterns would be compared across different NE without or with different suppressor U1s, the splicing results could be attributed to differences in NE preparation rather than to the efficacy of a particular suppressor U1. Thus, the best approach for informational suppression in 5′ss:U1 base-pairing is the use of expression plasmids and cell transfection.

7. Our experiments and those from others showed that in some cases the suppressor U1 snRNAs cannot rescue 5′ss recognition at all (Ref. 15 and references therein). First, some 5′ss positions are difficult or refractory to U1 suppression, perhaps because the splicing disruption by the mutation is too strong. Second, suppressor U1s do not rescue inclusion even if they restore a base pair that enhances 5′ss recognition in other contexts. This is consistent with the notion that 5′ss selection—as splicing overall—is entirely context-dependent, so certain mutations have a stronger effect in some exons than in others [31]. Furthermore, endogenous U1 is one of the most abundant RNAs in mammalian cells after ribosomal RNAs, with an estimate of 1 million copies per cell [32]. Thus, plasmid-driven suppressor U1s have to compete with endogenous U1 for snRNP biogenesis and function, and the exogenous U1 snRNP never accumulates as much as the endogenous. Expression of the suppressor U1 can be assessed by primer extension in the absence of certain nucleotide triphosphates [33]. Nevertheless, the lack of splicing rescue by a suppressor U1 alone does not prove or disprove any model. Instead the conclusions are drawn from those suppressor U1s that rescue 5′ss selection, and also from the comparison of the effects between different suppressors (Fig. 4).

Acknowledgments

XR acknowledges funding from Academic Research Fund Tier 1 grant (RG 20/11) from Singapore's Ministry of Education, as well as a Startup Grant from School of Biological Sciences at Nanyang Technological University, Singapore. The authors thank Ms Jia Xin Jessie Ho for details of some protocols.

References

1. Woolford JL Jr, Baserga SJ (2013) Ribosome biogenesis in the yeast *Saccharomyces cerevisiae*. Genetics 195:643–681

2. Fischer U, Englbrecht C, Chari A (2011) Biogenesis of spliceosomal small nuclear ribonucleoproteins. Wiley Interdiscip Rev RNA 2:718–731

3. Brow DA (2002) Allosteric cascade of spliceosome activation. Annu Rev Genet 36:333–360

4. Wahl MC, Will CL, Lührmann R (2009) The spliceosome: design principles of a dynamic RNP machine. Cell 136:701–718

5. Roca X, Krainer AR, Eperon IC (2013) Pick one, but be quick: 5′ splice sites and the problems of too many choices. Genes Dev 27:129–144

6. Sheth N, Roca X, Hastings ML et al (2006) Comprehensive splice-site analysis using comparative genomics. Nucleic Acids Res 34:3955–3967

7. Heinrichs V, Bach M, Winkelmann G et al (1990) U1-specific protein C needed for efficient complex formation of U1 snRNP with a 5′ splice site. Science 247:69–72

8. Kondo Y, Oubridge C, van Roon AM et al (2015) Crystal structure of human U1 snRNP, a small nuclear ribonucleoprotein particle, reveals the mechanism of 5′ splice site recognition. Elife 4

9. Murgola EJ (1985) tRNA, suppression and the code. Annu Rev Genet 19:57–80

10. Prelich G (1999) Suppression mechanisms: themes from variations. Trends Genet 15:261–266

11. Mount SM, Anderson P (2000) Expanding the definition of informational suppression. Trends Genet 16:157

12. Zhuang Y, Weiner AM (1986) A compensatory base change in U1 snRNA suppresses a 5′ splice site mutation. Cell 46:827–835

13. Séraphin B, Kretzner L, Rosbash MH (1988) A U1 snRNA:premRNA base pairing interaction is required early in yeast spliceosome assembly but does not uniquely define the 5′ cleavage site. EMBO J 7:2533–2538

14. Siliciano PG, Guthrie C (1988) 5′ splice site selection in yeast: genetic alterations in base-pairing with U1 reveal additional requirements. Genes Dev 2:1258–1267

15. Carmel I, Tal S, Vig I et al (2004) Comparative analysis detects dependencies among the 5′ splice-site positions. RNA 10:828–840

16. Cohen JB, Snow JE, Spencer SD et al (1994) Suppression of mammalian 5′ splice-site defects by U1 small nuclear RNAs from a distance. Proc Natl Acad Sci U S A 91:10470–10474

17. Lo PC, Roy D, Mount SM (1994) Suppressor U1 snRNAs in Drosophila. Genetics 138:365–378

18. Pinotti M, Bernardi F, Dal Mas A et al (2011) RNA-based therapeutic approaches for coagulation factor deficiencies. J Thromb Haemost 9:2143–2152

19. Kandels-Lewis S, Séraphin B (1993) Involvement of U6 snRNA in 5′ splice site selection. Science 262:2035–2039

20. Lesser CF, Guthrie C (1993) Mutations in U6 snRNA that alter splice site specificity: Implications for the active site. Science 6:1982–1988

21. Wassarman DA, Steitz JA (1992) Interactions of small nuclear RNA's with precursor messenger RNA during in vitro splicing. Science 257:1918–1925

22. Hwang DY, Cohen JB (1996) U1 snRNA promotes the selection of nearby 5′ splice sites by U6 snRNA in mammalian cells. Genes Dev 10:338–350

23. Roca X, Krainer AR (2009) Recognition of atypical 5′ splice sites by shifted base-pairing to U1 snRNA. Nat Struct Mol Biol 16:176–182

24. Roca X, Akerman M, Gaus H et al (2012) Widespread recognition of 5′ splice sites by noncanonical base-pairing to U1 snRNA involving bulged nucleotides. Genes Dev 26:1098–1109

25. Murphy JT, Skuzeski JT, Lund E et al (1987) Functional elements of the human U1 RNA promoter. Identification of five separate regions required for efficient transcription and template competition. J Biol Chem 262:1795–1803

26. Markham NR, Zuker M (2008) UNAFold: software for nucleic acid folding and hybridization. Methods Mol Biol 453:3–31

27. Flicek P, Amode MR, Barrell D et al (2014) Ensembl 2014. Nucleic Acids Res 42:D749–D755

28. Lorson CL, Hahnen E, Androphy EJ et al (1999) A single nucleotide in the SMN gene regulates splicing and is responsible for spinal muscular atrophy. Proc Natl Acad Sci U S A 96:6307–6311

29. Will CL, Rümpler S, Klein Gunnewiek J et al (1996) In vitro reconstitution of mammalian U1 snRNPs active in splicing: the U1-C protein enhances the formation of early (E) spliceosomal complexes. Nucleic Acids Res 24:4614–4623

30. Roca X, Olson AJ, Rao AR et al (2008) Features of 5′-splice-site efficiency derived from disease-causing mutations and comparative genomics. Genome Res 18:77–87

31. Baserga SJ, Steitz JA (1993) In: Gesteland RF, Atkins JF (eds) The RNA world. Cold Spring Harbor Laboratory Press, Cold Spring Harbor, NY, p 359–381

32. Sharma S, Wongpalee SP, Vashisht A et al (2014) Stem-loop 4 of U1 snRNA is essential for splicing and interacts with the U2 snRNP-specific SF3A1 protein during spliceosome assembly. Genes Dev 28:2518–2531

33. Cartegni L, Hastings ML, Calarco JA et al (2006) Determinants of exon 7 splicing in the spinal muscular atrophy genes, SMN1 and SMN2. Am J Hum Genet 78:63–77

Chapter 20

Analysis of Alternative Pre-RNA Splicing in the Mouse Retina Using a Fluorescent Reporter

Daniel Murphy, Saravanan Kolandaivelu, Visvanathan Ramamurthy, and Peter Stoilov

Abstract

In vivo alternative splicing is controlled in a tissue and cell type specific manner. Often individual cellular components of complex tissues will express different splicing programs. Thus, when studying splicing in multicellular organisms it is critical to determine the exon inclusion levels in individual cells positioned in the context of their native tissue or organ. Here we describe how a fluorescent splicing reporter in combination with in vivo electroporation can be used to visualize alternative splicing in individual cells within mature tissues. In a test case we show how the splicing of a photoreceptor specific exon can be visualized within the mouse retina. The retina was chosen as an example of a complex tissue that is fragile and whose cells cannot be studied in culture. With minor modifications to the injection and electroporation procedure, the protocol we outline can be applied to other tissues and organs.

Key words In vivo splicing reporter, Alternative splicing, Retina, Photoreceptor

---

1 Introduction

Alternative splicing is a major mechanism for generating protein diversity in higher eukaryotes. Expression of tissue and cell type specific splicing isoforms is critical for the development and maintenance of differentiated cell types such as neurons, muscle, and epithelial cells. On aggregate each of these differentiated cell types expresses a distinct splicing program and a set of splicing regulators. However, the gross picture painted by tissue-wide analysis of splicing and expression masks the significant variability of specialized cellular subtypes, particularly among neurons and epithelial cells. To understand the organization and functions of complex vertebrate tissues, gene expression and splicing need to be analyzed at the level of the individual cell within the context of the native tissue. Naturally, such studies face significant technical challenges in segregating the signals derived from individual cells. One approach to address these challenges is the use of fluorescent

Ren-Jang Lin (ed.), *RNA-Protein Complexes and Interactions: Methods and Protocols*, Methods in Molecular Biology, vol. 1421, DOI 10.1007/978-1-4939-3591-8_20, © Springer Science+Business Media New York 2016

reporters which provide a convenient way to visualize gene expression and splicing at the individual cell level. The recent development of two-color fluorescent splicing reporters that produce either Green (GFP) or Red Fluorescent Protein (RFP) depending on the splicing of an alternative exon has provided an opportunity to study alternative splicing in vivo [1–6]. Transgenes expressing fluorescent splicing reporters have successfully been used to visualize FGFR2 splicing in mice and *let-2* splicing in C. elegans [3, 6, 7]. However, the relatively high cost and long time frames required for the establishment of mouse transgenes have limited the utility of the splicing reporters in vivo.

Here, we describe a protocol that uses electroporation to introduce splicing reporters into the retina of mice, followed by RT-PCR analysis and fluorescent microscopy imaging to investigate alternative splicing. The protocol is based on an in vivo subretinal injection and electroporation method [8]. The technique involves injection of the plasmid DNA into the space between the retinal pigmented epithelial (RPE) cell layer and retina followed by electroporation using tweezer-type electrodes. Electroporation facilitates the movement of plasmid DNA into the retinal precursor cells. This system is highly customizable, allowing for multiple DNA constructs driven by various promoters to be expressed in rod, bipolar, amacrine, and muller glial cells, with expression lasting upwards of 50 days [8]. It has also been adapted for gain of function and loss of function studies using plasmids and siRNAs, and has been demonstrated in both neonatal and adult mice as well as other species [8, 9]. The in vivo electroporation approach is remarkably tolerant to variations in the size of the DNA. We and others have reproducibly electroporated bacterial artificial chromosome (BAC) clones reaching 100 Kbp in size [10, 11].

In vivo electroporation of DNA constructs provides a rapid and cost-efficient alternative to the use of transgenic animals. The time frame of a typical experiment is dictated mainly by the developmental timeline of the mouse retina and typically lasts 20–30 days. A significant advantage over the use of transgenic animals is the ability to quickly analyze the splicing of a large number of minigenes. This approach has allowed us to carry out extensive mutagenesis experiments involving more than 20 different constructs that were tested in multiple replicates.

A key component of the protocol described here is a two-color fluorescent splicing reporter. The splicing reporter, which was described previously, expresses either GFP or RFP fluorescent protein depending on the splicing of an alternative exon [1]. The reporter is designed to accommodate virtually any alternative exon. The test exon is cloned in an intron that splits the AUG translation initiation codon of the GFP reading frame (Fig. 1). When the exon is skipped from the mature RNA, the AUG codon of the GFP reading frame is joined together by splicing out the intervening sequences and GFP protein is expressed. In this scenario the trans-

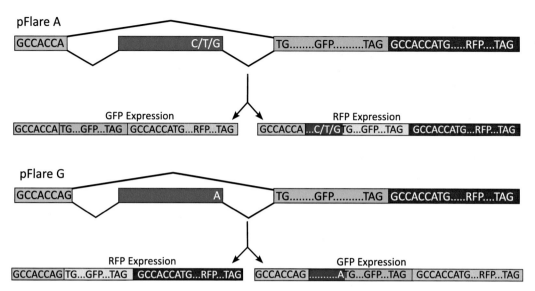

Fig. 1 Two-color fluorescent splicing reporter. Diagram of pFlare splicing reporters. The alternative exon being tested is shown in *blue*, constitutive 5′ and 3′ exons of the backbone are shown in *cyan* and *green/red*, respectively. The *upper panel* shows the pFlare A backbone, which accommodates exons ending in C, T, or G. The *lower panel* depicts the pFlare G backbone for use with exons ending in adenosine. In each panel, transcripts resulting from alternative splicing and the subsequent fluorescent protein expression are shown below. For pFlare A, exon inclusion abolishes the AUG start codon for GFP, resulting in RFP expression, whereas exon skipping reforms the AUG, allowing for GFP expression. Conversely, in pFlare G, exon inclusion establishes the AUG start codon for GFP translation, while exon skipping prevents GFP expression and results in expression of a downstream RFP

lation of the downstream RFP reading frame, which lacks an internal ribosome entry site, is suppressed. In the alternative scenario, the exon is spliced in, leading to disruption of the translation initiation codon of the GFP reading frame, and the downstream RFP reading frame is translated instead. This reporter design (pFlare A in Fig. 1) accommodates exons that do not terminate on adenosine or carry a translation initiation codon with good match to the Kozak consensus. These types of exons are accommodated by a second backbone (pFlare G on Fig. 1) in which the upstream exon no longer carries the adenosine of the GFP reading frame AUG codon. Under this design GFP will be expressed upon inclusion of the alternative exon. This can be achieved by either recreation of the AUG codon of the GFP reading frame by exons terminating on adenosine or initiation of GFP translation from an AUG codon encoded by the exon. In the latter case care must be taken to ensure that GFP is in frame with the AUG codon carried by the exon.

While the focus of this protocol is on studying splicing in the retina, the combination of in vivo electroporation and fluorescent reporters can be applied with some modifications to other tissues and organs. Reliable electroporation protocols have been established for introducing DNA in brain, lung, liver, muscle, heart, skin, and testis [12–19].

2 Materials

2.1 Laboratory Animal and Regulatory Approval

1. Pregnant (untimed pregnancy) mice such as CD-1 IGS (Charles River Laboratories) (*see* **Note 1**).

2. All procedures that are carried out on mice must be approved by the institutional regulatory bodies that oversee the use and human treatment of laboratory animals.

2.2 Tools and Equipment

2.2.1 Surgery and Dissection

1. Microscope for performing surgery and dissections, such as Zeiss Stemi DV4.

2. Hamilton blunt end syringe, 33 or 32 G.

3. Water heated warming blanket such as HTP-1500, Adroit Medical.

4. Square Wave Electroporation System, Nepagene CUY21, or BTX ECM 830.

5. Tweezer-type electrodes: BTX model 520, 7 mm diameter.

6. Two pairs of tweezers, and scissors for dissection of retina prior to both RNA isolation and sectioning and imaging.

7. Chamber for euthanization by asphyxiation from CO_2.

8. CO_2 tank and regulator.

2.2.2 Microscopy

1. Cryostat such as Leica cm 1850.

2. Zeiss LSM 510 laser scanning confocal microscope or equivalent.

2.2.3 RNA Isolation, cDNA Synthesis, PCR Amplification, and Gel Electrophoresis

1. Disposable Pellet Mixers and Cordless Motor (VWR).

2. Tabletop microcentrifuge.

3. Nanodrop spectrophotometer (Thermo Scientific).

4. Thermocycler.

5. Vertical gel electrophoresis apparatus (Labrepco model v16 or equivalent).

6. High voltage power supply capable of producing 500 V DC, such as Bio-Rad Power Pac 3000.

7. 18 G disposable needle.

8. 50 ml syringe.

9. Laser scanner capable of imaging two fluorescence channels, such as Typhoon 9410 Phosphorimager (GE).

2.3 Reagents and Consumables

2.3.1 Subretinal Injection

1. DNA at approximately 2–6 μg/μl diluted in PBS with 0.1 % fluorescein sodium.

2. Ice.

3. Sterile single use needle 22 G1 for dissection (Becton Dickinson).

4. Phosphate-Buffered Saline (PBS).

5. Absolute, Anhydrous ACS/USP grade Ethanol.

6. Cotton swabs.

7. 1:40 dilution (0.05 % final concentration) of Nolvasan (2 % Chlorhexidine (1,1'-Hexamethylenebis [5-(p-chlorophenyl) biguanide]) diacetate, Zoetis) in water.

2.3.2 RNA Isolation, cDNA Synthesis, PCR Amplification, and Gel Electrophoresis

1. Sterile Polystyrene Petri Dish.

2. Filter paper.

3. Ultrapure water.

4. Dry Ice.

5. TRIzol reagent.

6. Chloroform, ACS Spectrophotometric grade.

7. Isopropanol HPLC Grade.

8. Absolute, Anhydrous ACS/USP grade Ethanol.

9. 1.5 ml microcentrifuge tubes.

10. DNAse 1 RNAse-Free enzyme and 10× DNAse 1 Buffer (Roche).

11. 3 M sodium acetate solution, pH 5.2.

12. GlycoBlue™ coprecipitant (Life Technologies).

13. PCR tube strips and caps (8 or 12 tube).

14. 10 mM dNTP mix (dATP, dCTP, dGTP, and dTTP at 10 mM concentration each).

15. Reverse transcription primer mix: 10 µM anchored oligo dT ($dT_{24}VN$) and 50 µM random hexamers.

16. 10× Reverse Transcriptase Buffer: 500 mM Tris–HCL (pH 8.3), 750 mM KCL, 30 mM $MgCl_2$.

17. RNAse H (−) Reverse Transcriptase.

18. Primers combined at 10 µM final concentration each:

pFlare-BGL-F: aaacagatctaccattggtgc

EGFP-N carrying 5′ end 6-FAM label: [6-FAM] cgtcgccgtccagctcgacca

19. 10× Taq buffer: 500 mM KCL, 100 mM Tris–HCl (pH 9.0), 15 mM $MgCl_2$, and 1 % Triton X-100.

20. Taq polymerase at 15 U/µl.

21. Sigmacote (Sigma Aldrich) or equivalent siliconizing reagent.

22. 10 % Ammonium Persulfate (APS) solution in water.

23. Tetramethylenediamine (TEMED).

24. 1× Tris–Borate EDTA Buffer (TBE): 89 mM Tris, 89 mM Boric Acid, 2 mM EDTA. This buffer can be made as a 5× stock solution and diluted before use.

25. Acrylamide gel solution: 4 % Acrylamide/Bis-acrylamide (19:1 crosslink ratio), 1× TBE, 7.5 M Urea. Filter through 0.45 μM filter and store in a dark bottle at 4 °C.

26. Clear formamide loading buffer: Deionized formamide, 2 mM EDTA.

27. Fluorescently labeled size standards: Life technologies/ABI GeneScan 1000 Rox or GeneScan 1200 Liz. Alternatively custom size standards can be prepared by a simple PCR amplification with Rox-labeled primer [20].

2.3.3 Tissue Sectioning and Fluorescent Microscopy

1. Sterile polystyrene petri dish.

2. 24 well polystyrene tissue culture plate.

3. Filter paper.

4. Phosphate-buffered Saline (PBS).

5. 4 % paraformaldehyde (PFA) in PBS.

6. 20 % sucrose (Jt Baker 4097) solution in PBS.

7. Hanks Balanced Salt Solution (HBSS).

8. Sterile single use needle 22 G1 for dissection.

9. Dry ice.

10. Optimal Cutting Temperature Compound (O.C.T.) (Tissue Tek).

11. Cryomolds (Tissue Tek).

12. Frosted microscopy slides.

13. ProLong Gold Antifade reagent with DAPI (Life Technologies).

14. Cover slips.

15. Clear nail polish.

3   Methods

3.1   Generation of Reporter Construct

The procedures for cloning the exon to be tested for inclusion can be found elsewhere [21]. The cassette exon cloned into the reporter construct should contain part of flanking native introns to ensure that all elements necessary for the regulation of its splicing are present. Typically, intronic regulator elements will be located within 100 nt from the exon borders. The sequence conservation of the flanking introns should serve as a guide in the identification of such cis-acting sequences [22]. We recommend incorporating 200–300 nucleotides of each flanking intron. If the sequence conservation extends beyond this range, larger portions of the introns will need to be incorporated into the minigene to ensure that all regulatory sequences are present. It is important to confirm that the regulation of the alternative exon in the minigene construct mirrors the regulation of the exon in the native gene. This can be achieved by transfecting the minigene in cell lines that differentially splice the exon of the native gene and using RT-PCR to confirm that the splicing of the minigene is regulated similarly to the native transcript. It is possible that suitable cell lines may not exist for some cell type specific exons (the *Bbs8* exon 2A example used in this protocol is one such exon). Should this be the case, testing the minigenes will have to be carried out directly in animal tissue and suitable controls will need to be considered to ensure that negative results are not due to failure to incorporate the required regulatory sequences into the minigene (*see* **Note 2**).

3.2 Subretinal Injection and Electroporation

This technique requires a minor surgery to open the eye in a neonatal pup followed by an incision into the sclera. A blunt end syringe is then used to deliver the DNA through the incision created in the sclera. The use of a blunt end needle is necessary to prevent piercing through the back of the eye. The depth at which the DNA needs to be injected is determined by the slight increase in resistance which is felt when the blunt end needle reaches the end of scleral wall. Following injection in the subretinal space, electroporation is used to introduce the DNA into the retinal precursor cells. The steps in this procedure will require some practice, patience, and a steady hand in order to deliver the DNA at the correct site. We typically see the expression of the minigene constructs in 60–70 % of the injections. The entire process (excluding time for anesthetization) should take less than 3–5 min per pup, cause little if any bleeding, and result in a 100 % survival rate.

1. Prepare purified DNA according to plasmid purification kit. For large DNA constructs such as Bacterial Artificial Chromosomes (BACs), we recommend using CsCl gradient to purify the DNA.

2. Resuspend DNA at 2–6 µg/µl in PBS.

3. Monitor untimed pregnant moms daily to check for delivery of pups.

4. To 15–20 µl of DNA in PBS, add 1/10 volume of 1 % fluorescein sodium.

5. Remove newborn pups from cage and anesthetize individually on ice for several minutes. To avoid frostbite from direct contact with ice, place each mouse in a latex sleeve made from the finger of an examination glove (*see* Fig. 2a). Mice are fully anesthetized when they no longer respond to a toe pinch.

6. Clean the eyelid with ethanol using a cotton swab.

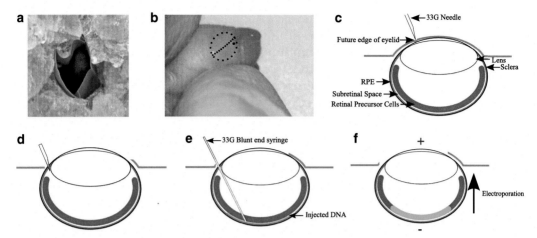

Fig. 2 Subretinal injection and electroporation at postnatal day 0. (**a**) Image of neonatal mouse pup in examination glove sleeve. (**b**) Image of neonatal mouse pup, using thumb and forefinger to gently spread the skin around the eye. The outline of the eye and the future edge of the eyelid are shown by *dotted lines*. The approximate location of the scleral incision is denoted by a *round dot*. (**c**) Cartoon of an aerial view of neonatal mouse eye showing position of future edge of the eyelid and needle used for incision. Skin is shown in *beige*, future edge of the eyelid, lens, sclera (*black*), RPE (*grey*), subretinal space, and retinal precursor cells (*orange*) are denoted by *arrows*. (**d**) Cartoon showing the location of pinpoint incision through the sclera. (**e**) Cartoon showing the DNA injection using blunt end needle. The injected DNA is shown in *green*. (**f**) Cartoon showing direction of current in electroporation. Cells that have taken up injected DNA are shown in *green*

7. Repeat the cleaning with a 1:40 solution of Nolvasan.

8. Under the microscope using a 32-G needle, carefully make an incision along the future edge of the eyelid (Fig. 2b, c).

9. Expose the eye by spreading the eyelids.

10. Make a pinpoint incision through the sclera using the tip of the needle (Fig. 2c, d).

11. Insert a blunt end syringe containing 0.5 μl of the DNA and dye solution into the incision, carefully maneuver the syringe around the lens toward the back of the eye until the retinal pigmented epithelium (RPE) is reached, you should feel some resistance (Fig. 2e).

12. Slowly inject the DNA into the subretinal space. The dye should be visible as it spreads across a small area of the retina (Fig. 2e).

13. Dip tweezer-type electrodes in PBS and apply them to the head of the pup with light pressure. The positive electrode is indicated by the screw, and should be oriented on the same side as the injection (Fig. 2f).

14. Apply five pulses of 80 V at 50 millisecond durations with 950 millisecond duration between pulses.

15. Place pups on a warming blanket for recovery before returning them to their mother.

3.3 Dissection of the Retina for RNA Isolation

At postnatal day 16 (P16), the photoreceptors should be fully developed (although the outer segments may not be fully elongated) and retinae can be harvested for RNA isolation. The following protocol outlines the dissection of the eye in order to isolate the retina. All procedures should be carried out in an RNAse-free environment (*see* **Note 3**).

1. Euthanize animals according to IACUC approved protocol.

2. To remove injected eye spread eyelids with thumb and forefinger. With the other hand, apply tweezers over and around the eyeball to pop it out of the eye socket. Grasp the optic nerve and enucleate the eye (*see* **Note 4**) (Fig. 3a).

3. Place eye on a small square of filter paper dampened with HBSS, under microscope (Fig. 3a).

4. Using a needle, puncture the eye at the border of the cornea and sclera. This will release pressure and allow for further dissection (Fig. 3b).

5. Using the needle incision as a starting point, cut out the cornea with scissors (Fig. 3c).

6. Carefully pull the lens out with tweezers. Before and after images shown on Fig. 3d, e.

7. Place eye in 1 drop of RNAse-free water or HBSS on a plastic petri dish.

8. Using two tweezers, separate the retina from the sclera as shown on Fig. 3f–h.

9. With one tweezers, grip the sclera near the optic nerve. With the second, lightly clamp down on the back of the eye near optic nerve (Fig. 3g).

10. Gently push the second pair of tweezers away from the optic nerve into the drop of water or HBSS. The retina (clear) will separate from the sclera (Fig. 3h).

11. Place retina into a 1.5 ml centrifuge tube containing 200 µl TRIzol. Freeze on dry ice while collecting other samples.

12. To serve as a negative control, collect the retina of the other, uninjected eye as described above and add to a separate TRIzol containing tube.

3.4 RNA Isolation and DNAse Treatment

1. Homogenize each retina by pulsing 10–15 times for 1–2 s with handheld homogenizer.

2. Add additional 200 µl of TRIzol to each tube containing homogenized retina and mix by vortexing.

3. Add 1/10 volume of Chloroform to each sample.

4. Vortex well.

5. Spin samples at $15,000 \times g$ for 10 min.

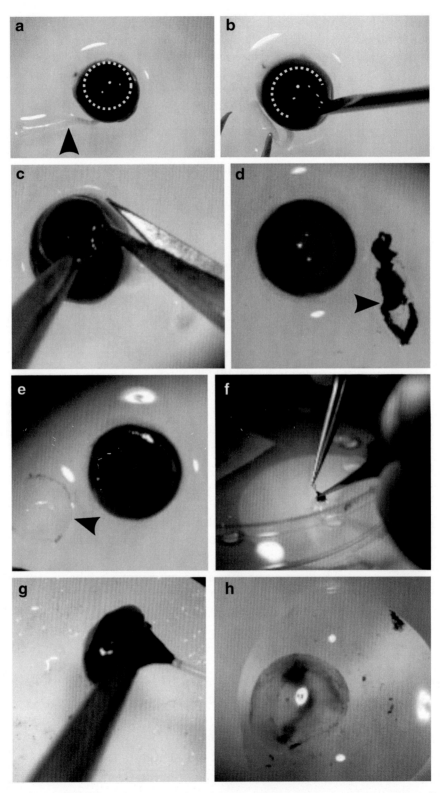

Fig. 3 Retinal Dissection. (**a**) Image of adult pigmented mouse eye. The border of the cornea and sclera is indicated by a *dotted line. Black arrow* indicates optic nerve. (**b**) Image of needle puncture at border of cornea

6. Transfer the aqueous (top) phase, which should be approximately 150–200 µl, to a new tube (*see* **Note 5**).

7. Add 1.5 vol of isopropanol to each sample and mix by vortexing.

8. Incubate at room temperature for 15 min.

9. Spin each sample at 15,000×*g* for 10–15 min to precipitate RNA.

10. Carefully remove the supernatant.

11. Add 0.5 ml of 75 % ethanol.

12. Centrifuge at 15,000×*g* for 5 min.

13. Remove ethanol and briefly air-dry pellets.

14. Resuspend each pellet in 100 µl of ultrapure water.

15. Make a 2× concentrated master mix containing 20 µl of 10× DNAse1 Buffer, 79 µl of water, and 1 µl DNAse 1 per tube.

16. Add 100 µl of the master mix to each sample.

17. Mix samples and incubate at 37 °C for 20 min.

18. Add 200 µl of chloroform and vortex well.

19. Centrifuge at 15,000×*g* for 10 min.

20. Transfer the aqueous (top) phase to a new tube. Take care not to carry over the interphase, containing the denatured proteins.

21. Add 1/10 volume, or 20 µl of 3 M sodium acetate pH 5.2.

22. Add 3 vol or 600 µl ice-cold ethanol and mix well.

23. Incubate each sample at −20 °C for at least 30 min.

24. Resuspend RNA in 25 µl of ultrapure water.

25. Determine the RNA concentration on a Nanodrop spectrophotometer (Thermo Fisher). The typical yield is approximately 1 µg of RNA per retina.

3.5 cDNA Synthesis

1. Prepare a 2× Reverse transcription master mix (2 µl of 10× RT buffer; 1 µl of 10 mM dNTPs; 0.5 µl Oligo dT/random hexamer primer mix; 6 µl water, and 0.5 µl reverse transcriptase per reaction).

2. Dispense 10 µl of master mix into each tube of a PCR tube strip.

3. Add 10 µl of RNA solution containing up to 1 µg of RNA to each reaction and mix by pipetting. 100 ng of RNA is typically sufficient to reliably quantify the splicing of the minigene.

Fig. 3 (continued) and sclera. (**c**) Image of removing the cornea. (**d**) Image of eye once cornea (*black arrow*) is removed. (**e**) Image of eye once lens (*black arrow*) is removed. (**f**) Expansive view of using tweezers to separate the retina from the sclera. (**g**) magnified view of the same image as in panel (**f**). (**h**) Image of retina in a drop of HBSS

4. Cap the strip and incubate in a thermal cycler under the following conditions: 25 °C for 5 min; 43 °C for 40 min; 75 °C for 15 min; hold at 10 °C until the strip is removed.

3.6   PCR

1. Prepare a master mix (2 μl of 10× Taq PCR buffer; 0.4 μl of 10 mM dNTPs, 2 μl of 10 μM primer mix, 0.2 μl of Taq DNA polymerase per reaction).

2. Dispense 19.5 μl to each tube of a PCR tube strip.

3. Transfer 0.5 μl of cDNA to each reaction.

4. Mix by pipette and cap the strip. Make sure there are no air bubbles at the bottom of each tube.

5. Amplify the templates using the following conditions. Initial denaturation step at 94 °C for 4 min, followed by 20–30 cycles of 94 °C for 30 s, 62 °C for 30 s, and 72 °C for 30–60 s, followed by a final extension at 72 for 5 min. Then hold the reactions at 10 °C until ready to remove.

3.7   Electrophoresis and Imaging

The electrophoresis protocol described here uses denaturing urea/polyacrylamide gels to resolve the PCR products. The gels are then imaged on a fluorescent imager without being disassembled.

1. Clean slides of each glass plate with ethanol and dry with paper towels.

2. If the plates have not been siliconized before, apply Sigmacote to the cleaned plates and spread it with a paper towel until dry. Clean the plates with ethanol again as in **step 1** described above.

3. Place spacers on the clean side of the larger of the two glass plates—place the two side spacers with the foam dam toward the top, and the bottom spacer across the bottom edge. Place the smaller of the two glass plates, clean side facing down, on top of the larger glass plate to create a plate-spacer-plate sandwich.

4. Clip the sandwich together with binder clips and set aside.

5. Repeat for as many gels as needed.

6. In a clean flask mix acrylamide solution (25 ml per gel) with 1/100 vol of 10 % APS (250 μl per gel) and 1/1,000 vol TEMED (25 μl per gel) and pour the gels.

7. Holding the plate sandwich at approximately 15–20° from horizontal, pour the gel solution in a steady stream along one of the side spacers allowing it to flow smoothly between the two glass plates while ensuring that no air bubbles are formed. Once filled to the top, place the gel horizontally to insert the comb. Leave the gel in this position until the gel polymerizes (approximately 20–30 min).

8. After the gel has polymerized, remove the clips and the bottom spacer, as well as the well comb.

9. Place the sandwich onto the running apparatus with the larger glass plate facing out. Use two clips on each side of the gel to clip the sandwich to the gel.

10. Fill the upper and lower reservoirs with 1× TBE buffer ensuring that the gel is covered. Flush with TBE any air that was trapped at the bottom of the gel.

11. Immediately rinse the wells using an 18 G needle on a 50 ml syringe filled with 1× TBE.

12. Apply a piece of clear adhesive tape such as Scotch tape, to the outside glass plate directly over the wells. Use a sharpie pen to label on the tape each well with a number that corresponds to the sample that will be loaded in the well. It is critical to label the wells as the loading buffer contains no running dye and it is impossible to determine which lanes have been loaded.

13. Attach the cover of the gel apparatus and pre-run the gel for 30–50 min at 400 V. While the gel is running, prepare the PCR amplicons for loading as described in **steps 14–16** below.

14. Prepare a loading buffer mix containing 10 μl of clear formamide loading buffer and 0.3 μl fluorescent size standard for each sample loaded onto the gel.

15. Dispense 10 μl of loading buffer mix to each tube on a PCR tube strip, according to the number of samples being analyzed.

16. Transfer 1 μl of the PCR amplicon to each of the tubes containing the loading buffer mix.

17. Seal the tubes and incubate in a thermal cycler at 95 °C for 5 min to denature the samples. Place the tubes on ice.

18. Turn off the gel power supply. Remove the urea that has accumulated in the wells by rinsing them again as described in **step 11**. Failing to remove the urea will result in distorted bands.

19. Load 10 μl of the denatured samples in each well.

20. Run the gel at 400 V for 55 min then turn off the power supply.

21. Remove the gel sandwich from the apparatus.

22. Remove the side spacers, but do not disassemble the gel sandwich.

23. Clean the plates with deionized water to remove any acrylamide or urea attached to the outside of the plates.

24. Dry the plates with a paper towel.

25. Clean the surface of the phosphorimager with a moist paper towel and dry it.

26. Place the gel sandwiches on the phosphorimager glass plate.

27. Select the scan area in the typhoon software and close the lid.

28. Select the appropriate combination of excitation lasers and bandpass emission filters depending on the labels present in the samples and the size standards. For FAM-labeled primers, we recommend blue laser (480 nm) excitation and 520 nm bandpass emission filter.

29. Set the focal plane to +3 mm to adjust the focal point above the surface of the phosphorimager, to account for the width of the glass plate and gel.

30. Choose the appropriate orientation for your output image.

31. Scan the gel. While scanning, make sure that there are no saturated pixels, indicated by red dots on the preview window. If there are saturated pixels, rescan the gel after lowering the photomultiplier (PMT) voltage for the appropriate channel (*see* Fig. 4 for example results).

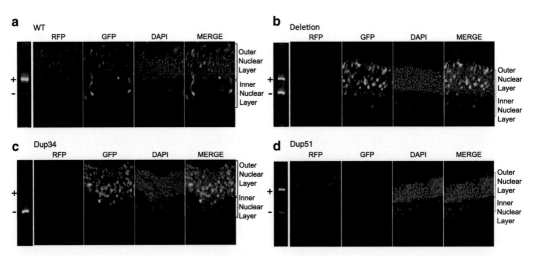

Fig. 4 RT-PCR and Fluorescence microscopy. The *left panel* in each set shows RT-PCR results for injected eyes after gel electrophoresis and imaging. "+" denotes exon inclusion and "−" denotes exon skipping. The *green* and *red bands* are PCR amplicons and ROX size standard, respectively. The *right panel* in each set shows fluorescence microscopy images of sections taken from injected eyes. The layers of the retina are indicated with *brackets* on the side of the image. (**a**) Example of photoreceptor specific *Bbs8* Exon 2a wild-type (WT) minigene. RT-PCR indicates a very high level of exon inclusion. Fluorescence microscopy shows high RFP fluorescence in the Outer Nuclear Layer (ONL) containing the photoreceptor cell nuclei. Few neuronal cells expressing the minigene in the Inner Nuclear Layer (INL) produce GFP which indicates exon skipping. (**b**) Example of *Bbs8* Exon 2a minigene once the critical intronic splicing enhancers have been deleted. RT-PCR indicates majority of transcripts skip exon 2a. Fluorescence microscopy shows high levels of GFP in all cell layers, and low levels of RFP in the ONL. (**c**) Control minigene containing the Dup34 synthetic exon. RT-PCR and fluorescence microscopy show the exon being skipped in all cell layers. (**d**) Control minigene containing the Dup51 synthetic exon. RT-PCR and fluorescence microscopy show high levels of exon inclusion regardless of cell type

3.8 Dissection of the Retina for Tissue Sectioning

To ensure that photoreceptor maturation is complete, we allow 20 days post-injection before the injected eyes are taken for tissue sectioning. The procedure outlined here is very similar to the retinal dissection for RNA isolation described above and shown in Fig. 3a–e. Here, however, maintaining retinal morphology throughout the process is vital for obtaining unblemished tissue sections. Excessive pressure or shearing forces on the eye can cause the retina to detach or tear, making interpretation of results difficult. While some retinal detachment and malformation can be expected from the injection process, most of the tissue expressing the reporter should still display normal morphology.

1. In a petri dish, soak a piece of filter paper in HBSS.

2. Carefully enucleate the injected eye, and place it on the filter paper. This may require removal of fascia surrounding eye. Retinal detachment may occur if excess pressure is exerted on the eye.

3. Puncture with a 33 G needle at the border of the cornea and sclera (Fig. 3b).

4. Place the eye in a well of a 24 well plate containing 500 μl of 4 % PFA in PBS. Incubate 5–10 min at room temperature (RT).

5. Place the eye back under the microscope and remove the cornea and lens to prepare eye cup (Fig. 3c). Alternatively, the lens can be removed at the end of **step 9**.

6. Place the eye cup into the well containing 4 % PFA. Incubate for 30–60 min at RT for fixation.

7. Wash the eye cup three times for 5 min each with PBS at RT.

8. Remove PBS and immerse the eye cup in 20 % sucrose in PBS solution and incubate with shaking overnight at 4 °C.

9. Replace the 20 % Sucrose solution with a 1:1 mixture of O.C.T. compound and 20 % sucrose in PBS. Incubate with shaking for 1 h.

10. Prepare an ethanol/dry ice bath in a container with wide opening.

11. Fill a labeled cryomold container with O.C.T. compound. Rather than writing on the plastic itself, label a piece of tape and attach it to the lip of the cryomold to prevent the ethanol from washing off marker.

12. Remove the eye and blot off excess O.C.T./Sucrose in PBS. Place the eye in the cryomold containing O.C.T.

13. Orient the eye such that the opening (where cornea and lens were removed) is facing to the right (*see* **Note 6**).

14. Snap freeze the eye in O.C.T. by submerging the well of the cryomold in ethanol/dry ice bath.

15. Store the eye at –80 °C. Eyes can be stored at –80 °C for more than a year.

3.9 Sectioning and Staining

Once the dissected eye is properly fixed and frozen in a block of O.C.T. compound, it can be cut into 16 μm thick sections using a cryostat. Cryostat operation will require training which is beyond the scope of this protocol. Properly oriented, the sectioned retina should resemble a "C" with the opening formed from the removal of the cornea and lens. As many "C" sections as is practical for mounting can be arranged on each slide. For mounting, we use a reagent containing DAPI to stain the nuclei. The RFP and GFP expressions from the minigene are visible without any further staining (Fig. 4). If necessary, immunofluorescence staining following standard protocols can be used to visualize cell type specific markers or other proteins of interest. We recommend using secondary antibodies labeled with Cy5 or equivalent infrared fluorescent dye, to prevent overlap with the fluorescence of the GFP and RFP proteins.

1. Using a cryostat, cut sections at 16 μm.

2. Arrange 5–10 tissue sections on the stage of the cryostat.

3. Quickly but carefully apply the slide (treated side facing down) to the sections. The tissue sections will adhere to the slide. The slide should be at room temperature. This process should be done quickly to ensure that the slide will not cool, which can inhibit proper attachment of sections. Do not press the slide firmly onto the stage.

4. Air-dry sections on each slide for 30 min and mount or store at −20 °C. Slides can be stored for several months.

5. To room temperature slides add a drop of ProLong Gold with DAPI. To avoid bubbles, use a cut pipette tip to apply the Pro-Gold to the retinal sections. Very little is needed as the solution will spread once cover slip is applied.

6. Carefully apply the cover slip and ensure each retina is coated with Pro-Gold.

7. Air-dry for 30 min.

8. Seal the cover and slide edge with clear nail polish and store at 4 °C.

9. Image the slides on a confocal microscope.

4 Notes

1. Many mouse strains carry mutations causing retinal degeneration, such as *rd1* and *rd8*. In some cases the same mouse strain may or may not carry retinal degeneration mutations depending on the commercial source. It is critical to ensure that the experimental animals do not carry *rd* alleles.

2. The nature of these controls is beyond the scope of this protocol as it strongly depends on the exon and tissue being studied.

3. Preventing RNase contamination:

- Maintain clean bench surfaces.
- If necessary RNase contamination on equipment and bench surfaces can be inactivated by treating with RNaseZap reagent (Life Technologies).
- Always wear gloves.
- Use ultrapure or DEPC treated water to prepare reagents.
- Where applicable autoclave or filter sterilize the reagents.
- Use RNase- and DNase-free plasticware. If possible use aerosol resistant tips.

4. Keeping the optic nerve intact provides a convenient method for handling the eye throughout the procedure with minimal risk for causing damage.

5. Adding 1 μl GlycoBlue™ Coprecipitant to the supernatant (optional) helps to visualize the RNA pellet after the precipitation steps.

6. Ensure that there are no air bubbles inside or surrounding the eye cup. Trapped air can make it impossible to obtain intact sections.

Acknowledgements

We thank Zachary Wright and Abigail Hayes for advice and assistance in taking pictures. This work was supported by grants from the National Institutes of Health (EY017035), West Virginia Lions, Lions Club International Fund, and an internal grant from West Virginia University. Imaging experiments were performed in the West Virginia University Microscope Imaging Facility, which has been supported by the Mary Babb Randolph Cancer Center and NIH grants P20 RR016440, P30 GM103488, and P20 GM103434.

References

1. Stoilov P, Lin C-H, Damoiseaux R et al (2008) A high-throughput screening strategy identifies cardiotonic steroids as alternative splicing modulators. Proc Natl Acad Sci 105:11218–11223

2. Orengo JP, Bundman D, Cooper TA (2006) A bichromatic fluorescent reporter for cell-based screens of alternative splicing. Nucleic Acids Res 34:e148

3. Kuroyanagi H, Ohno G, Sakane H et al (2010) Visualization and genetic analysis of alternative splicing regulation in vivo using fluorescence

reporters in transgenic Caenorhabditis elegans. Nat Protoc 5:1495–1517

4. Somarelli JA, Schaeffer D, Bosma R et al (2013) Fluorescence-based alternative splicing reporters for the study of epithelial plasticity in vivo. RNA 19:116–127

5. Newman EA, Muh SJ, Hovhannisyan RH et al (2006) Identification of RNA-binding proteins that regulate FGFR2 splicing through the use of sensitive and specific dual color fluorescence minigene assays. RNA 12:1129–1141

6. Takeuchi A, Hosokawa M, Nojima T et al (2010) Splicing reporter mice revealed the evolutionarily conserved switching mechanism of tissue-specific alternative exon selection. PLoS One 5:e10946

7. Ohno G, Hagiwara M, Kuroyanagi H (2008) STAR family RNA-binding protein ASD-2 regulates developmental switching of mutually exclusive alternative splicing in vivo. Genes Dev 22:360–374

8. Matsuda T, Cepko CL (2004) Electroporation and RNA interference in the rodent retina in vivo and in vitro. Proc Natl Acad Sci 101:16–22

9. Nickerson JM, Goodman P, Chrenek MA et al (2012) Subretinal delivery and electroporation in pigmented and nonpigmented adult mouse eyes. Methods Mol Biol 884:53–69

10. Magin-Lachmann C, Kotzamanis G, D'Aiuto L et al (2004) In vitro and in vivo delivery of intact BAC DNA—comparison of different methods. J Gene Med 6:195–209

11. Barnabé-Heider F, Meletis K, Eriksson M et al (2008) Genetic manipulation of adult mouse neurogenic niches by in vivo electroporation. Nat Methods 5:189–196

12. Young JL, Dean DA (2015) Electroporation-mediated gene delivery (chapter three). Adv Genet 89:49–88

13. Heller R, Cruz Y, Heller LC et al (2010) Electrically mediated delivery of plasmid DNA to the skin, using a multielectrode array. Hum Gene Ther 21:357–362

14. Young JL, Barravecchia MS, Dean DA (2014) Electroporation-mediated gene delivery to the lungs. Methods Mol Biol 1121:189–204

15. Aihara H, Miyazaki J (1998) Gene transfer into muscle by electroporation in vivo. Nat Biotechnol 16:867–870

16. Tevaearai HT, Gazdhar A, Giraud M-N et al (2014) In vivo electroporation-mediated gene delivery to the beating heart. Methods Mol Biol 1121:223–229

17. De Vry J, Martínez-Martínez P, Losen M et al (2010) In vivo electroporation of the central nervous system: a non-viral approach for targeted gene delivery. Prog Neurobiol 92:227–244

18. Heller R, Jaroszeski M, Atkin A et al (1996) In vivo gene electroinjection and expression in rat liver. FEBS Lett 389:225–228

19. Michaelis M, Sobczak A, Weitzel JM (2014) In vivo microinjection and electroporation of mouse testis. J Vis Exp 23(90)

20. DeWoody JA, Schupp J, Kenefic L et al. (2004) Universal method for producing ROX-labeled size standards suitable for automated genotyping, BioTechniques 37: 348, 350, 352

21. Green MR, Sambrook J (2012) Molecular cloning: a laboratory manual, 4th edn. Cold Spring Harbor Laboratory Press, Avon, MA, Three-volume set

22. Cooper TA (2005) Use of minigene systems to dissect alternative splicing elements. Methods 37:331–340

INDEX

Ren-Jang Lin (ed.), *RNA-Protein Complexes and Interactions: Methods and Protocols*, Methods in Molecular Biology, vol. 1421,
DOI 10.1007/978-1-4939-3591-8, © Springer Science+Business Media New York 2016